The Telecommunications Fact Book and Illustrated Dictionary

Second Edition

Ahmed S. Khan

THOMSON
DELMAR LEARNING™

Australia Canada Mexico Singapore Spain United Kingdom United States

DELMAR LEARNING

The Telecommunications Fact Book and Illustrated Dictionary

2nd Edition
by Ahmed S. Khan

Vice President, Technology and Trades SBU:
Alar Elken

Editorial Director:
Sandy Clark

Acquisitions Editor:
Steve Helba

Development Editor:
Dawn Daugherty

Marketing Director:
David Gerza

Channel Manager:
Dennis Williams

Marketing Coordinator:
Stacey Wiktorek

Production Director:
Mary Ellen Black

Production Manager:
Andrew Crouth

Production Editor:
Dawn Jacobson

Technology Project Manager:
Kevin Smith

Editorial Assistant:
Dawn Daugherty

Printed in Canada
1 2 3 4 5 XX 07 06 05

For more information contact Thomson Delmar Learning
Executive Woods
5 Maxwell Drive, PO Box 8007,
Clifton Park, NY 12065-8007
Or find us on the World Wide Web at
www.delmarlearning.com

Library of Congress Cataloging-in-Publication Data:

Khan, Ahmed S.
The telecommunications fact book and illustrated dictionary / Ahmed S. Khan. —2nd ed. p. cm
Includes bibliographical references and index.
ISBN 1-4180-1173-8 (pbk.)
1. Telecommunication—Dictionaries. I. Title.
TK5102.K48 2005
621.382'03 – dc22
2005015635

NOTICE TO THE READER

Publisher does not warrant or guarantee any of the products described herein or perform any independent analysis in connection with any of the product information contained herein. Publisher does not assume, and expressly disclaims, any obligation to obtain and include information other than that provided to it by the manufacturer.

The reader is expressly warned to consider and adopt all safety precautions that might be indicated by the activities herein and to avoid all potential hazards. By following the instructions contained herein, the reader willingly assumes all risks in connection with such instructions.

The publisher makes no representation or warranties of any kind, including but not limited to, the warranties of fitness for particular purpose or merchantability, nor are any such representations implied with respect to the material set forth herein, and the publisher takes no responsibility with respect to such material. The publisher shall not be liable for any special, consequential, or exemplary damages resulting, in whole or part, from the readers' use of, or reliance upon, this material.

Contents

Preface v

Acknowledgments vi

PART I
DEFINITIONS

MODULE 0: Numerics 1
MODULE 1: Definitions A 5
MODULE 2: Definitions B 31
MODULE 3: Definitions C 42
MODULE 4: Definitions D 54
MODULE 5: Definitions E 73
MODULE 6: Definitions F 79
MODULE 7: Definitions G 91
MODULE 8: Definitions H 95
MODULE 9: Definitions I 100
MODULE 10: Definitions J 109
MODULE 11: Definitions K 111
MODULE 12: Definitions L 113
MODULE 13: Definitions M 118
MODULE 14: Definitions N 129
MODULE 15: Definitions O 136
MODULE 16: Definitions P 140
MODULE 17: Definitions Q 150
MODULE 18: Definitions R 152
MODULE 19: Definitions S 166
MODULE 20: Definitions T 181
MODULE 21: Definitions U 191
MODULE 22: Definitions V 193
MODULE 23: Definitions W 196

MODULE 24: Definitions X 200
MODULE 25: Definitions Y 201
MODULE 26: Definitions Z 203

PART II

OVERVIEW OF TELECOMMUNICATIONS FACTS, FIGURES, FORMULAS, STANDARDS, AND TECHNOLOGIES

MODULE 27: Fiber Optics Communications 205
MODULE 28: Optical Amplifiers 227
MODULE 29: Fundamentals of Electromagnetic Signals 238
MODULE 30: Digital Logic: A Summary 243
MODULE 31: Units and Physical Constants 246
MODULE 32: Symbols for Electronic Devices 250
MODULE 33: Telecommunications Formulas 253
MODULE 34: Integrated Services Digital Network (ISDN) 269
MODULE 35: Network Architecture 273
MODULE 36: Digital Subscriber Line (DSL): An Overview 276
MODULE 37: IP Addresses: An Introduction 278
MODULE 38: ITU-TS Recommendations 281
MODULE 39: TIA/EIA Standards: 232, 449, 485, 644, and ITU-TS Equivalents 287
MODULE 40: Codes for Information/Data Transmission 294
MODULE 41: International Television Systems and HDTV 300
MODULE 42: International Telephone Dialing Codes 305
MODULE 43: Telecommunication Agencies/Organizations 314
MODULE 44: Telecommunications Journals and Magazines 318
MODULE 45: FCC Frequency Allocations 322
MODULE 46: Satellite Communications: An Overview 331
MODULE 47: Wireless Technologies and Standards: A Summary 343
MODULE 48: Computer Networks: An Overview 346
MODULE 49: Telecommunications Technology Trends and Statistics . 353
MODULE 50: Telecommunications Acronyms 357
Bibliography 369
Index 372

Preface

The rapid growth of the Internet and emerging telecommunications technologies have transformed the way people access information. An enormous amount of information is available to a user as the result of a simple click of a mouse button. But due to the commercial nature of the Internet, many times a user has to sift through a plethora of data before finding the desired item. In other words the content level of the information is low, or in telecommunications jargon, the signal-to-noise ratio (SNR) of the retrieved information is low. The authenticity and accuracy of the information may also be problematic.

A number of telecommunications books are available, but many of these books require some prior knowledge of the subject. Considering the diverse nature of telecommunications technologies, coupled with the commercial nature of the Internet, readers find it difficult to quickly locate, access, and understand a concept, a fact, or a technology.

The second edition of *The Telecommunications Fact Book and Illustrated Dictionary* has been revised to provide a self-contained quick reference to new terms, facts, and technologies related to various facets of telecommunications, including computer networks, wireless networks, fiber optics, satellites, protocols, and analog, digital, and data communications.

The most unique feature of *The Telecommunications Fact Book and Illustrated Dictionary* is its illustrated approach. Wherever possible an illustration is employed in conjunction with text to convey the concept in a clear, concise, and graphical manner, helping readers with different learning styles to grasp the presented materials quickly. The book contains 51 modules. Modules 0–26 (Part I) contain definitions of telecommunications terms in alphabetical order, and Modules 27–50 (Part II) provide a database of facts and figures, and an overview of various telecommunications concepts and technologies.

The Telecommunications Fact Book and Illustrated Dictionary is intended primarily for two-year and four-year college students taking communications courses related to data, analog, digital, wireless, satellite, fiber optics, and computer networks. The book can also be used to supplement a telecommunications textbook. All other eager readers—novices, technicians, engineers, and managers associated with various aspects of telecommunications—can also benefit by using it as a quick reference.

The author welcomes your comments and suggestions for improving *The Telecommunications Fact Book and Illustrated Dictionary.* He can be reached via e-mail: khan@dpg.devry.edu or telecommunications_fact_book@yahoo.com.

Ahmed S. Khan

Acknowledgments

The author and Thomson Delmar Learning wish to express thanks and appreciation to the reviewer panel for its suggestions and comments during development of this second edition. Thanks go to:

William Lin
Indiana University-Purdue University Indianapolis
Indianapolis, Indiana

Justice Anderson
DeVry University
Orlando, Florida

Gordon Snyder
Springfield Technical Community College
Springfield, Massachusetts

I would like to thank my students for their input for improving *The Telecommunications Fact Book and Illustrated Dictionary*. Thanks are also due to Dawn Daugherty, Dawn Jacobson, and Steve Helba at Delmar Learning for their help and support, and to Melissa Mattson at Pre-Press Company for her professionalism in the organization of the book.

I am indebted to my wife Tasneem for her help and encouragement, which made this project possible. And special thanks go to Professor Robert Lawrence of DeVry University, Addison, Illinois, for his suggestions for improving the book and his dedicated work in editing the manuscript.

بسم الله الرحمن الرحيم

TO MY PARENTS, MY WIFE, AND ALL SEEKERS OF KNOWLEDGE, TRUTH, AND WISDOM

活到老。
學到老

MODULE 0 Numerics

0 One of the two digits (bits 0, 1) of the binary number system. In data communications, it represents a low signaling level (also called space level) or state of a binary signal.

1 One of the two digits (bits 0, 1) of the binary number system. In data communications, it represents a high signaling level (also called mark level) or state of a binary signal.

0TLP In a telephone system, a point used as a reference to measure signal and noise levels at other points in the system. *Also see* **zero transmission level point.**

1 Gigabit Ethernet A high-speed Ethernet specified by IEEE 802.3z protocol. Table 0-1 lists the standards for 1 Gigabit Ethernet.

TABLE 0-1 Gigabit Ethernet standards

STANDARD	SPEED	LINK DISTANCE	MEDIUM	IEEE 802.__	OSI LAYER
1000Base-CX	1000 Mbps	25 m	Balanced shielded cable	802.3z	Physical
1000Base-T PMA (physical medium attachment)	1000 Mbps	100 m	4 Pair CAT-5 UTP	802.3z	Physical
1000Base-T PCS (physical coding sublayer)	1000 Mbps	100 m	4 Pair CAT-5 UTP	802.3z	Data link
1000Base-SX	1000 Mbps	220 m	62.5-micron multimode optical fiber	802.3z	Physical
		500 m	50-micron multimode optical fiber	802.3z	Physical

(continues)

TABLE 0-1 (continued)

1000Base-LX	1000 Mbps	550 m	62.5-micron multimode optical fiber/ 50-micron multimode optical fiber	802.3z	Physical
		5 km	10-micron single mode optical fiber	802.3z	Physical

1 Base5 An Ethernet configuration that according to IEEE 802.3 specification uses star topology and transmits data at 1 Mbps over an unshielded twisted pair (UTP) wire. See Table 0-2.

TABLE 0-2 IEEE 802.3 specifications

STANDARD*	SPEED	LINK DISTANCE	MEDIUM	TOPOLOGY	OSI LAYER
1Base5+	1 Mbps	250 m	Unshielded twisted pair (UTP)	Star	Physical
10Base2++	10 Mbps	185 m	50-Ω thin coax	Bus	Physical
10Base5	10 Mbps	500 m	50-Ω thick coax	Bus	Physical
10BaseT++++	10 Mbps	100 m	Unshielded twisted pair (UTP)	Star Bus	Physical
100BaseT+++	100 Mbps	100 m	Unshielded twisted pair (UTP)	Star Bus Hierarchical Hybrid	Physical
10Broad36++++	10 Mbps	1,800 m	75-Ω Coax	Star	Physical

* In the nomenclature of standards, typically the first number represents the speed, the middle word indicates the medium, and the last number stands for link or segment length in 100-m units.

+ Maximum segment length is 500, thus 5.

++ Maximum segment length is 185, rounded up to 200, thus 2.

+++ T represents twisted pair.

++++ Maximum segment length is 3,600 m using two 1.800-m segments, thus 36.

For additional information about IEEE 802. Standards visit http://www.ieee802.org
http://grouper.ieee.org/groups/

2B1Q (2-Binary, 1-Quaternary) A line coding technique used in the ISDN that employs two signaling levels (±2.5 V and ±0.8 V) to achieve encoding of 2 bits per baud.

4B/5B encoding An encoding technique that takes a 4-bit code and maps it to equivalent 5-bit code.

4D-PAMS5 Four-dimensional, five-level pulse amplitude modulation, an encoding scheme used to achieve a data transmission speed of 1 GB over copper wires.

5-4-3 rule In 10Base2 Ethernet configuration, a rule that limits the network to five 200-m segments using four repeaters, with three segments having nodes attached to them.

8B/10B encoding An encoding technique that takes an 8-bit code and converts it to equivalent 10-bit code by splitting the 8 bits into three most-significant bits and five least-significant bits.

8.3 alias In Microsoft Windows, a short file name, an "alias" (with eight characters and a three-character extension) assigned to each file with a long name so that these files can be used in other programs that do not support long file names.

10BaseF A 10 Mbps Ethernet specification that refers to l0BaseFB, l0BaseFL, and l0BaseFP standards for connecting nodes in a local area network (LAN) with the help of fiber optic cable.

10Base2 An Ethernet configuration that according to IEEE 802.3 specification uses bus topology and transmits data at 10 Mbps over a 50-Ω thin coax cable. Also known as *Thinnet*. See Table 0-2.

10Base5 The original standard for Ethernet configuration that according to IEEE 802.3 specification uses bus topology and transmits data at 10 Mbps over a 50-Ω thick coax cable. Also known as *Thicknet*. See Table 0-2.

10BaseT A baseband Ethernet configuration that according to IEEE 802.3 specification transmits data at 10 Mbps using unshielded twisted pair (UTP). See Table 0-2.

100BaseT An Ethernet configuration that according to IEEE 802.3 specification transmits data at 100 Mbps using unshielded twisted pair (UTP) wire. See Table 0-2.

10Broad36 A broadband Ethernet configuration that according to IEEE 802.3 specification transmits data at 10 Mbps using coax cable. See Table 0-2.

100VG-AnyLAN A 100 Mbps Ethernet and token ring architecture, based on IEEE 802.12 and IEEE 802.12 standards, using four pairs of category 3, 4, or 5 unshielded twisted pair (UTP) cabling.

1000Base-CX See Table 0-1. *Also see* **IEEE 802**

1000Base-LX See Table 0-1. *Also see* **IEEE 802**

1000Base-SX See Table 0-1. *Also see* **IEEE 802**

1000Base-T See Table 0-1. *Also see* **IEEE 802**

802 A set of standards developed by the Institute of Electrical and Electronics Engineers (IEEE) for local area networks (LANs). *Also see* **IEEE 802**. For additional information, visit http://www.ieee802.org and http://grouper.ieee.org/groups/

802.1 An IEEE standard for local area networks (LANs) covering OSI layers 3 to 7. *Also see* **IEEE 802**.

802.1x An IEEE standard for wired and wireless local area networks (LANs) that specifies Internet working and link security, and authentication mechanisms.

802.11a An IEEE standard for wireless local area networks (LANs) with a transmission speed of 108 Mbps. Also called *Wi-Fi5*. See Table 48-6.

802.11b An IEEE standard for wireless local area networks (LANs) with a transmission speed of 11 Mbps. Also called *Wi-Fi*. See Table 48-6.

802.11g An IEEE standard for wireless local area networks (LANs) that combines the features of IEEE 802.11a and IEEE 802.11b and allows a transmission speed of 54 Mbps. See Table 48-6.

802.2 An IEEE standard for local area networks (LANs) that describes the implementation of the logical link control (LLC) sub-layer of the data link layer. *Also see* **IEEE 802**.

802.3 An IEEE standard for local area networks (LANs) that describes the implementation of the physical layer and the media access control (MAC) sub-layer of the data link layer using carrier sense multiple access/collision detection (CSMA/CD) method. *Also see* **IEEE 802**.

802.4 An IEEE standard for local area networks (LANs) that describes the implementation of the physical layer and the media access control (MAC) sub-layer of the data link layer using the token passing access method in a bus topology. *Also see* **IEEE 802**.

802.5 An IEEE standard for local area networks (LANs) that describes the implementation of the physical layer and the media access control (MAC) sub-layer of the data link layer using the token passing access method in a ring topology. *Also see* **IEEE 802**.

802.6 An IEEE standard for metropolitan area networks (MANs) based on the distributed queue dual bus (DQDB), which allows multiple systems to interconnect with the help of two unidirectional logical buses. *Also see* **IEEE 802**.

802.7 *See* **IEEE 802**.

802.8 *See* **IEEE 802**.

802.9 *See* **IEEE 802**.

MODULE 1 Definitions A

a Abbreviation for atto; a = 10^{-18}.
A Abbreviation for ampere.
A0 An FCC designation for radio emission of an unmodulated carrier.
A1 An FCC designation for radio emission of a continuous wave carrier.
A2 An FCC designation for radio emission of a tone modulated carrier wave.
A3 An FCC designation for radio emission of amplitude modulated (AM) speech signals.
A4 An FCC designation for radio emission of amplitude modulated (AM) television facsimile.
A5 An FCC designation for radio emission of amplitude modulated television.
A channel A band of microwave frequencies ranging from 0.1 to 0.25 GHz; designated for military use.
AB test A method of comparing two audio systems by switching inputs so the same recording is heard in rapid succession over one system and then the other.
A-law A European version of a companding (*com*pressing/ex*panding*) algorithm used in the telephone system. It allows anolog channels to maintain acceptable signal-to-distortion levels over a wide dynamic range of input signals. *Also see* **μ-law, companding**, *and* Module 33 *for formulas.*
abbreviated dialing A dialing scheme that allows a caller to use a short code in place of a 7-digit number to call an address on a telephone network.
ABC Abbreviation for (1) automatic brightness control in TV circuits, (2) automatic bass compensation, or (3) American Broadcasting Company.
abnormal propagation Electromagnetic wave propagation in which unstable atmospheric/ionospheric conditions interfere with waves, preventing them to follow their normal paths.
abnormal reflection A sharp, intense reflection from the sporadic E layer of the ionosphere occurring at frequencies higher than the critical frequency of the layer. Also known as *sporadic reflection.*
abort The unanticipated termination (due to software or hardware problems or user request) of an application or a program before it is fully executed or completed.
absolute address A pattern of characters that is used to identify a unique storage location in a memory device. Also called *specific address.*
absolute cutoff frequency The minimum value of the frequency at which a waveguide will propagate electromagnetic energy without any attenuation.
absolute delay The time interval between the transmission of two synchronized radar/loran radio signals from the same or different stations.

absorption Dissipation of energy of an electromagnetic wave or acoustic wave into another form due to its interaction with matter. Acoustic energy is lost when acoustic waves travel through a medium. Radio waves also lose energy as they travel through the atmosphere/ionosphere.

absorption circuit A tuned circuit that dissipates energy taken from another circuit.

absorption coefficient The fraction of the intensity of radiation absorbed by a unit thickness of a particular material.

absorption current The component of current in a dielectric that is proportional to the rate of accumulation of electric charge within the dielectric.

absorption loss The part of transmission loss due to dissipation of electrical energy into other forms, such as heat, when radiated energy is transmitted or reflected by a material.

absorption modulation A system of amplitude modulation (AM) in which a variable impedance device is coupled to the output circuit of the radio transmitter to absorb carrier power in accordance with the information signal to be transmitted. Also called *loss modulation.*

absorption wavemeter A device for measuring the wavelength/frequency of a radio wave, by tuning a resonant circuit so that the maximum energy is absorbed from a source to be measured.

AC, ac Abbreviation for alternating current (also written as A.C. or a.c.) (Figure 1-1).

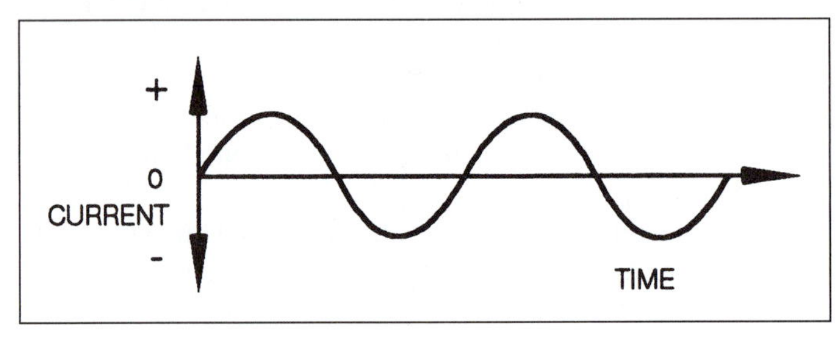

FIGURE 1-1 Alternating current

AC coupling A coupling arrangement between two stages or circuits via a capacitor that passes only time-varying (AC) signals and no DC component (Figure 1-2).

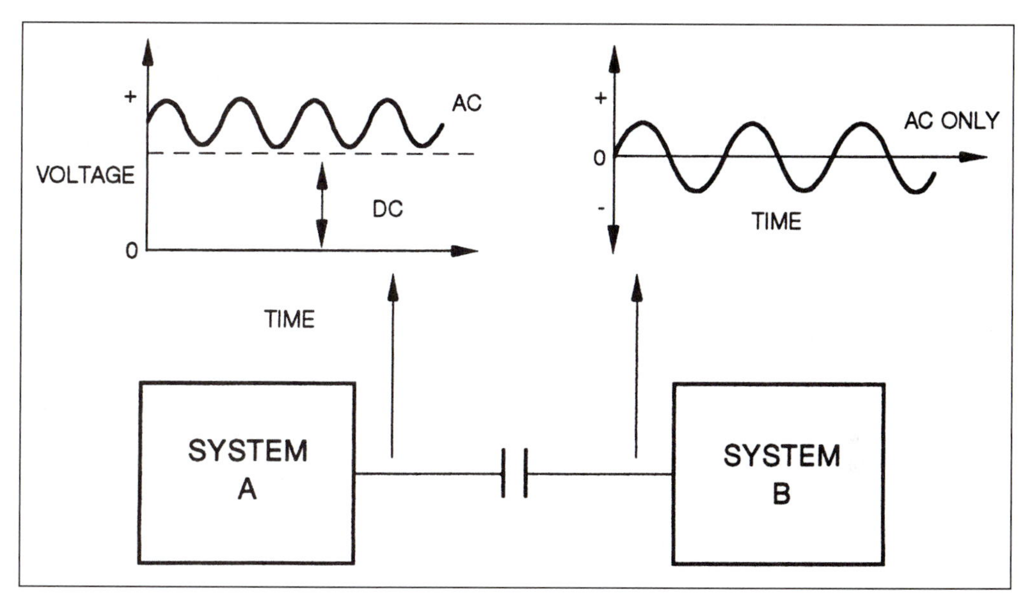

FIGURE 1-2 AC coupling

AC signaling Frequencies that carry the address information or system status information.

accelerated graphics port (AGP) A bus or slot on the computer motherboard that provides an interface to a single video card.

acceptance angle The angle measured from the longitudinal centerline up to the maximum solid angle, within which (1) all incident light entering the fiber will be totally internally reflected and propagated through the fiber without refraction and (2) all received light reaches the sensitive area of the photosensitive device (Figure 1-3). The maximum value of the acceptance angle is given by:

$$\Theta_0 = \sin^{-1}\left(\frac{\sqrt{n_1^2 - n_2^2}}{n_0}\right)$$

Where: Θ_0 = maximum acceptance angle
n_1 = refractive index of fiber core
n_2 = refractive index of fiber cladding
n_0 = refractive index of launching medium (usually air)

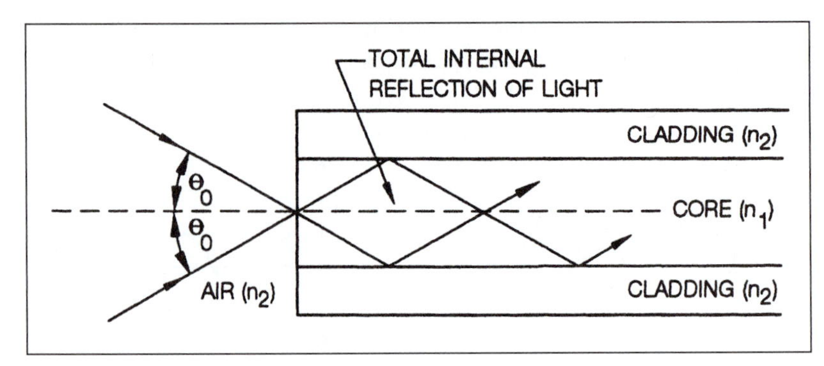

FIGURE 1-3 Acceptance angle

acceptance cone The maximum acceptance angle rotated three dimensionally around the fiber's central axis from the acceptance cone. All the light incident on fiber within the cone will propagate through without refraction of light into the cladding (Figure 1-4).

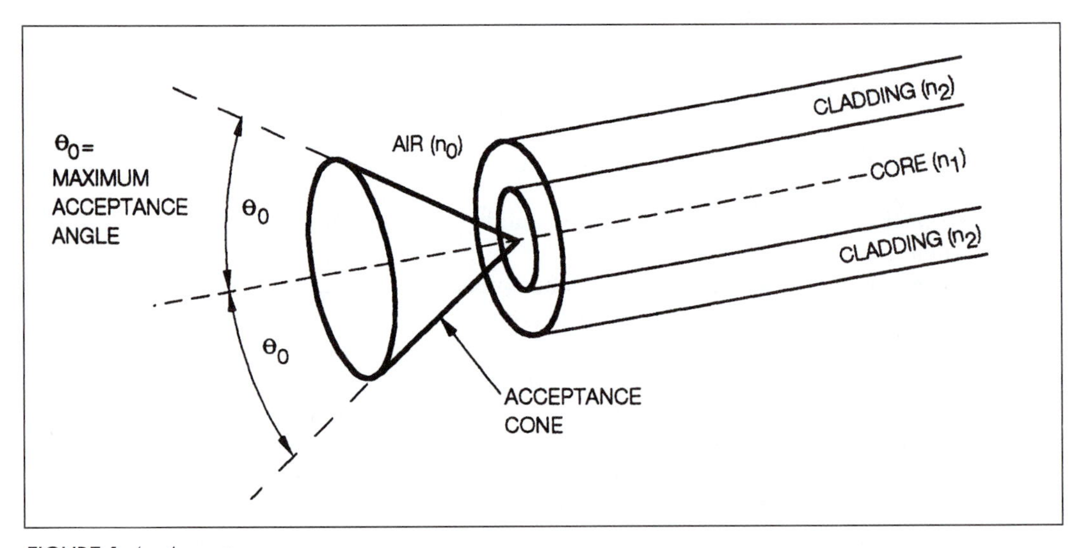

FIGURE 1-4 Acceptance cone

acceptor circuit A series resonant circuit that has a low impedance (Z) at resonant frequency and high impedance at all other frequencies (Figure 1-5).

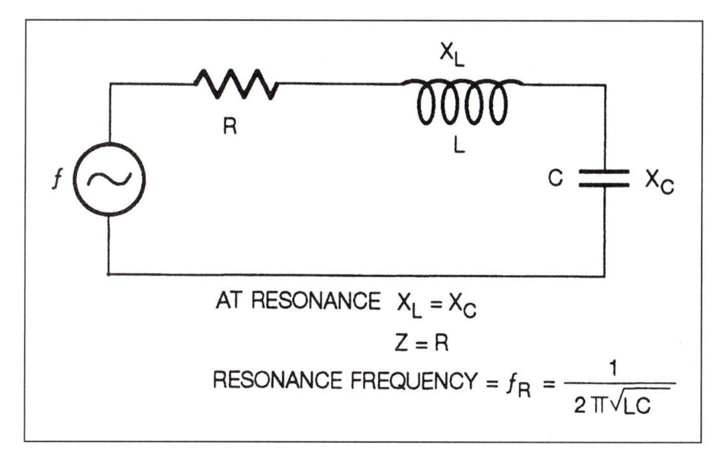

FIGURE 1-5 Acceptance cone

access code Special character-sequence that is dialed-in to get connected to a circuit, feature, or special equipment in a system.

access method *For LAN see* **carrier sense multiple access** with **collision avoidance (CSMA/CA)** *and* **carrier sense multiple access** with **collision detection (CSMA/CD)**, *and* Module 48. *For ISDN see* **BRI** *and* **PRI** in Module 34.

ACD See **automatic call distributor**.

ACF Advanced communications function; products by IBM that support sophisticated computer networking functions for IBM systems and terminals.

ACK Affirmative acknowledgment; characters for acknowledging the successful receipt of a message block.

ACK0 An affirmative reply to acknowledge even-numbered messages in a BSC protocol.

ACK1 An affirmative reply to acknowledge odd-numbered messages in a BSC protocol.

acoustic coupler A device that converts electrical signals into an audio signal and vice versa, enabling data to be transmitted over telephone lines using modems (Figure 1-6).

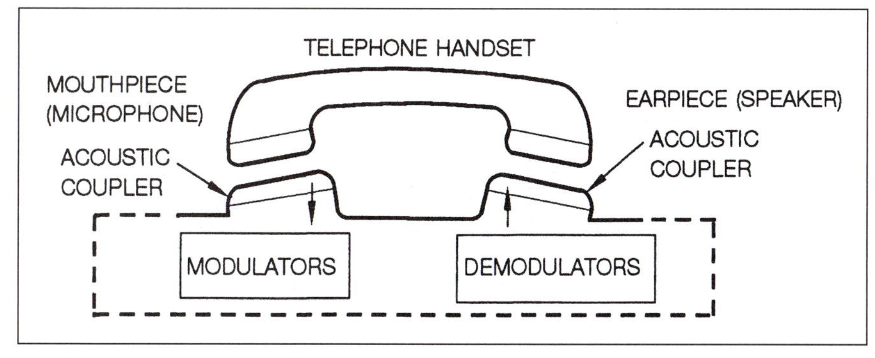

FIGURE 1-6 Acoustic coupler

active attack An attempt by an unauthorized user, such as a hacker, to break into a computer network in order to alter data, to launch a malicious or self-replicating program, or to clog the network. An active attack can be detected, in contrast to a passive attack, which cannot be detected.

active filter A filter consisting of a passive network element and amplifier used for transmitting or rejecting a desired frequency or range of frequencies (Figure 1-7). The gain of frequency depends on the frequency of the input signal.

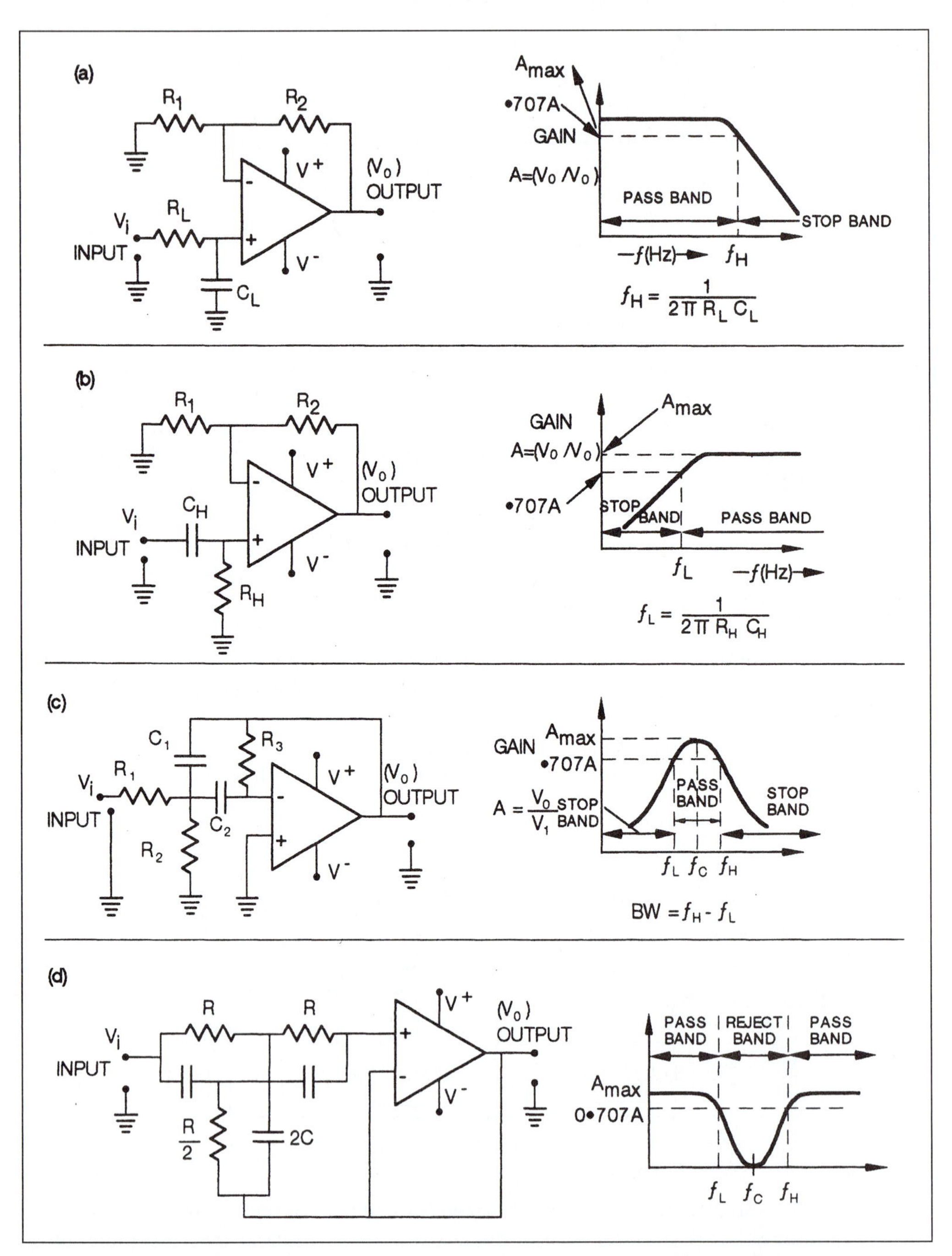

FIGURE 1-7 Active filters: (a) low pass active filter; (b) high pass active filter; (c) band pass; (d) band reject

active hub A hub that regenerates signals, thus performing the function of a regenerator or repeater.

active jamming Intentional radiation or reradiation of a specific band of frequency in order to impair the use of that band of frequency.

active lines The number of horizontal scan lines to form a television picture on a cathode ray cube (CRT) (Figure 1-8). The U.S. standard is 525 lines; in Europe, it is 405 or 625 lines (*also see* **HDTV**).

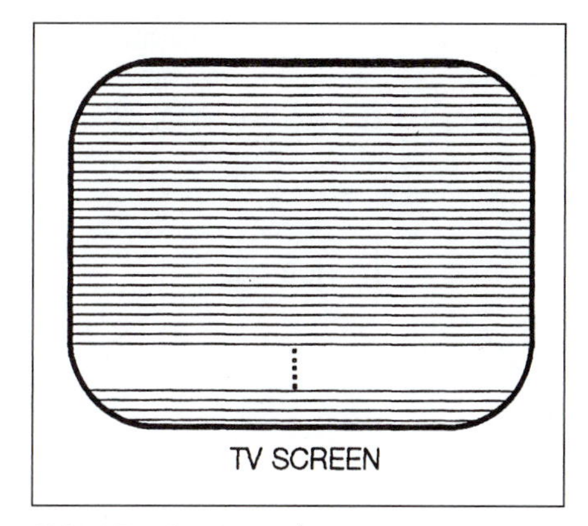

FIGURE 1-8 Active lines

active satellite A type of commercial satellite in which the radio signal received by the satellite is amplified and then retransmitted to earth (Figure 1-9).

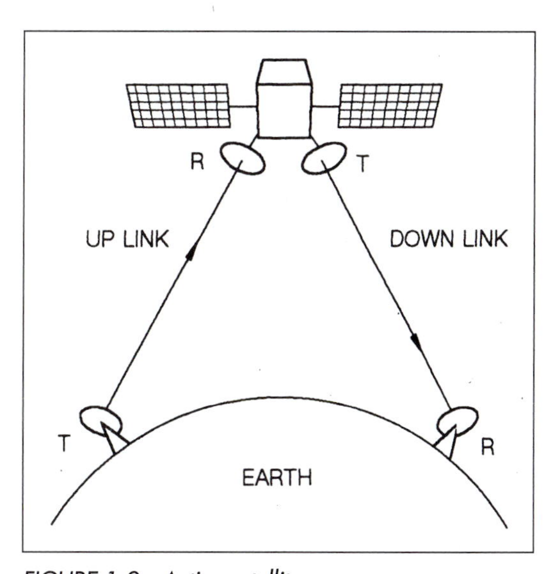

FIGURE 1-9 Active satellite

adaptive delta modulation A method of digitizing analog signals. The analog signal is converted into a weighted digital pulse train that is decoded at the receiver to retrieve the original analog information signal. This technique is similar to delta modulation except a variable gain amplifier is included at the input to increase the step height for faster rising signals. This amplifier enables the step waveform to follow faster waveforms than the ordinary delta modulation system.

ADCCP A standard bit-oriented protocol set by the American National Standards Institute (ANSI).

add/drop multiplexer (ADM) A device used in a synchronous optical network (SONET) that adds (multiplexes) signals to or removes (demultiplexes) signals from the SONET ring (Figure 1-10). *See also* **SONET** *and* Module 33.

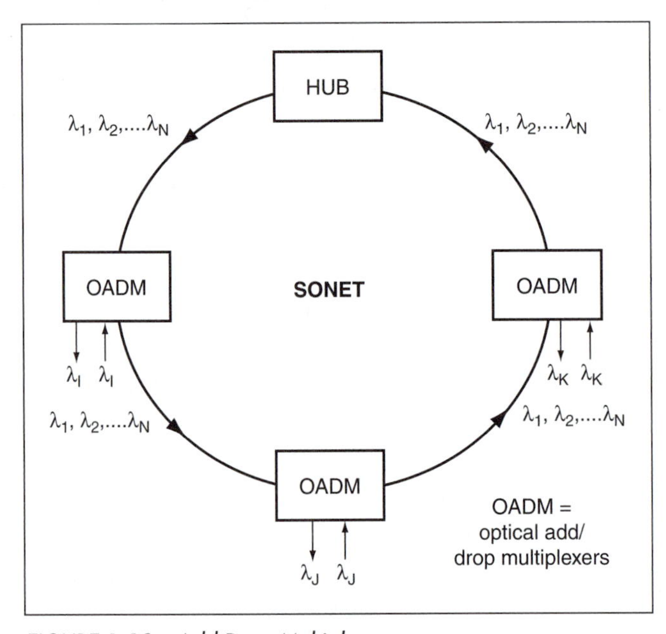

FIGURE 1-10 Add Drop Multiplexers

address A combination of bits/digits that uniquely: (1) represents the destination of data, (2) identifies a specific location of electronic/memory system, and (3) represents a telephone number (Figure 1-11).

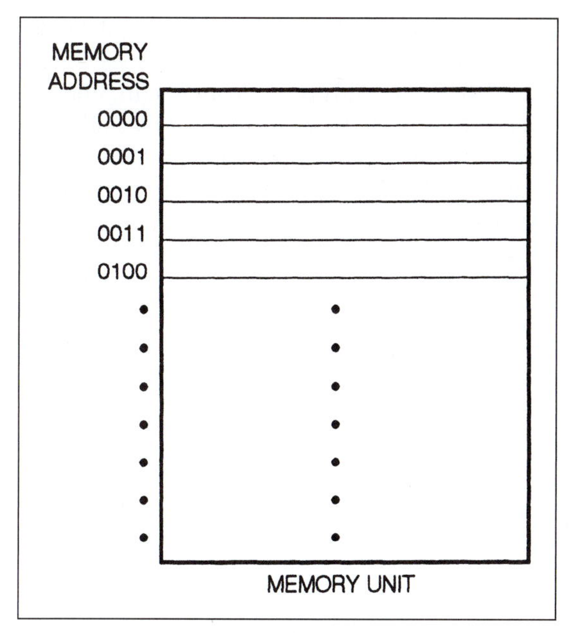

FIGURE 1-11 Memory address

address resolution protocol (ARP) A TCP/IP protocol used to convert an IP address of a node to a physical address.

address spoofing An act falsifying the source IP (Internet Protocol) address in an IP packet and sending it with an expected acknowledgement (ACK), which a hacker can guess or acquire through snooping.

adjacent channel A band of frequencies either immediately after or immediately before the current channel (Figure 1-12).

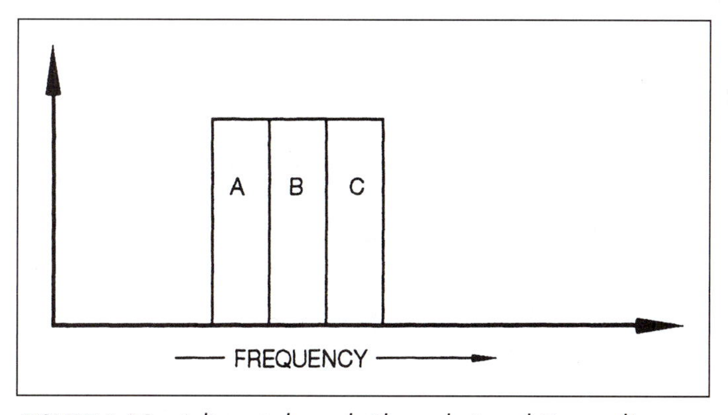

FIGURE 1-12 Adjacent channel: Channels A and C are adjacent to channel B.

ADSL (Asymmetrical Digital Subscriber Line) A digital subscriber line (DSL) that offers high-speed data transfer over telephone lines. *See* Module 36.

Advanced Mobile Phone System (AMPS) An analog mobile phone system standard used in North America and Asia. It was developed by Bell Labs in the 1970s, and used commercially for the first time in the United States in 1983. It offers good voice quality, but low spectrum efficiency. It uses a frequency division multiple access (FDMA) scheme. *Also see* Module 47.

agent A denial of service (DoS) attack program, covertly installed on computers connected to the Internet, which can generate malicious traffic when instructed by a hacker.

algorithm A set of rules or a procedure that defines a solution to a problem.

alias A name that is easy to remember, used for a name that is usually difficult to remember.

aliasing The occurrence of undesirable frequency components in the spectrum of a sampled waveform when the Nyquist sampling theorem is violated, i.e., the sampling rate is not high enough. (Figures 1-13a and 1-13b illustrate aliasing in time and frequency domains.)

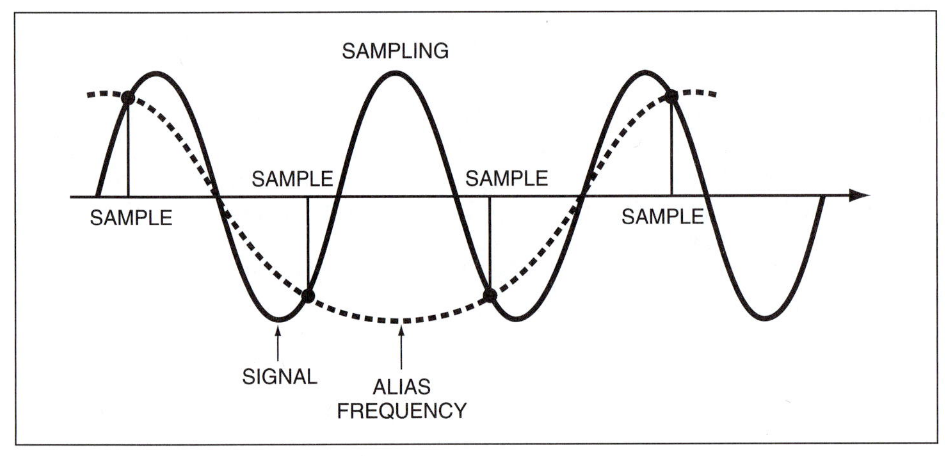

FIGURE 1-13 (a) Aliasing in time domain

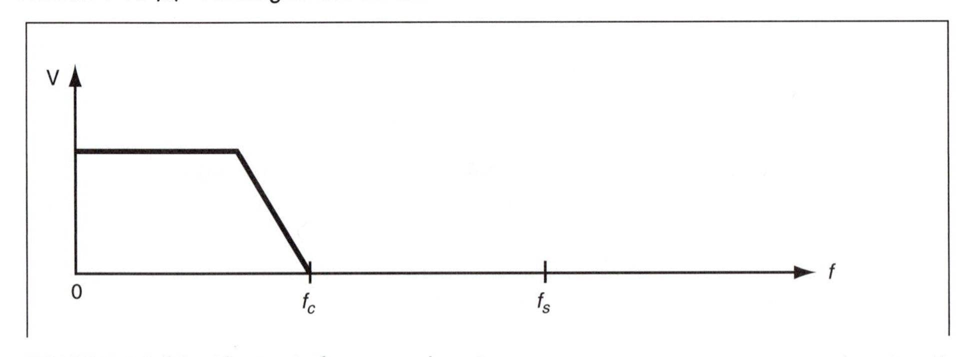

FIGURE 1-13 (b) Aliasing in frequency domain

(continues)

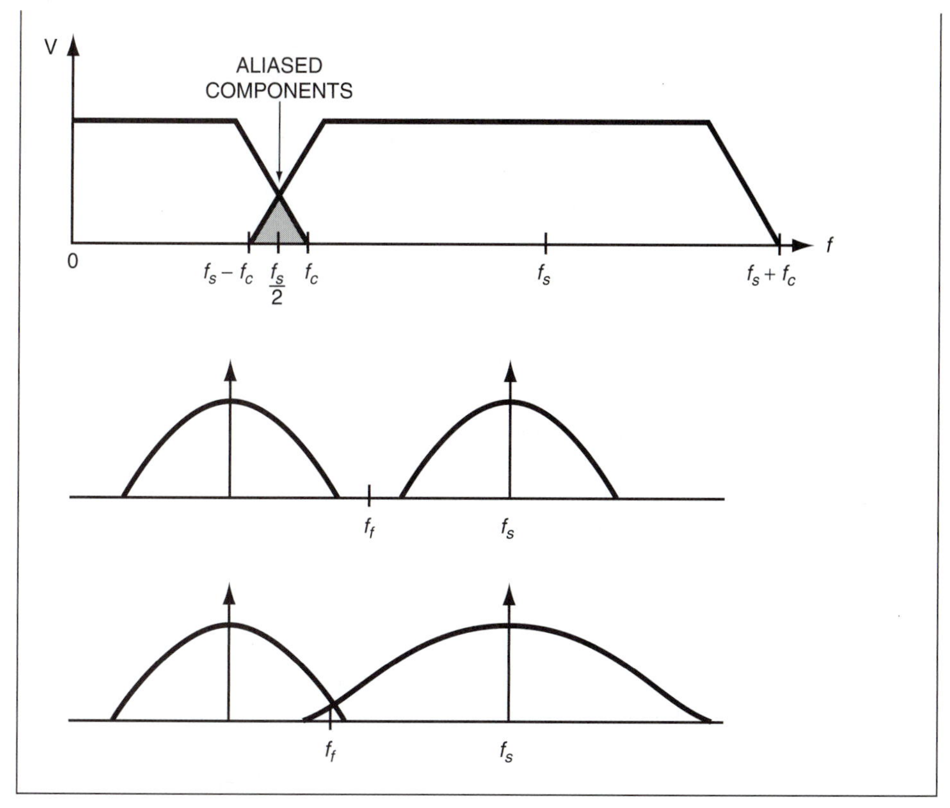

FIGURE 1-13 (b) (continued)

allocated channel A band of frequencies allocated to a specific user.

alphanumeric A symbol that represents a letter or number, either individually or in combination. Table 1-1 illustrates the alphanumerics.

TABLE 1-1 Alphanumerics

ALPHANUMERIC		
CHARACTERS		NUMBERS
A	BLANK	0
B	.	1
C	(	2
D	+	3
E	$	4
F	*	5
G	)	6
H	-	7
I	/	8
J	'	9
K	=	

(continues)

TABLE 1-1 (continued)

CHARACTERS		NUMBERS
L	S	
M	T	
N	U	
O	V	
P	X	
Q	Y	
R	Z	

alteration In computer networks, the failure of security mechanisms to ensure the integrity of data due to an accident or malicious activity.

alternate routing An alternate communication path used when the normal path is not available for transmission.

AM signature A part of the AM signal that distinguishes the signal from other signals. It is used in satellite communication.

American National Standards Institute (ANSI) A regulatory body that coordinates standards setting and approves standards in the United States. It also represents the United States to the International Order for Standardization (ISO). [http://www.ansi.org]

American Standard Code for Information Interchange (ASCII) A 7-bit code for data transfer established by the American Standards Association to achieve compatibility between data services. ASCII is equivalent to the ISO 7-bit code. (*See* Module 40 *for ASCII codes for alphanumerics and special characters.*)

Examples:

		b_6	b_5	b_4	b_3	b_2	b_1	b_0
A	=	1	0	0	0	0	0	1
9	=	0	1	1	1	0	0	1
#	=	0	1	0	0	0	1	1

American Wire Gauge (AWG) A standard for measuring and classifying the thickness of wire conductors. The AWG number and the size of a wire are inversely related. That is, the larger the AWG number, the thinner the wire.

TABLE 1-2 Examples of AWG

AWG (STRANDING) MATERIAL	NUMER OF SHIELDS/ MATERIAL	NOMINAL OUTER DIAMETER (INCH)
11 (7 × 19) bare copper	1/bare copper	0.405
20 (19 × 32) tinned copper	1/tinned copper	0.195
22 (27 × 36) tinned copper	1/tinned copper	0.160
26 (7 × 34) bare copperweld	1/tinned copper	0.100

amplifier A device used to boost the power of an analog signal (Figure 1-14). Analogous to a *repeater* used in digital transmission.

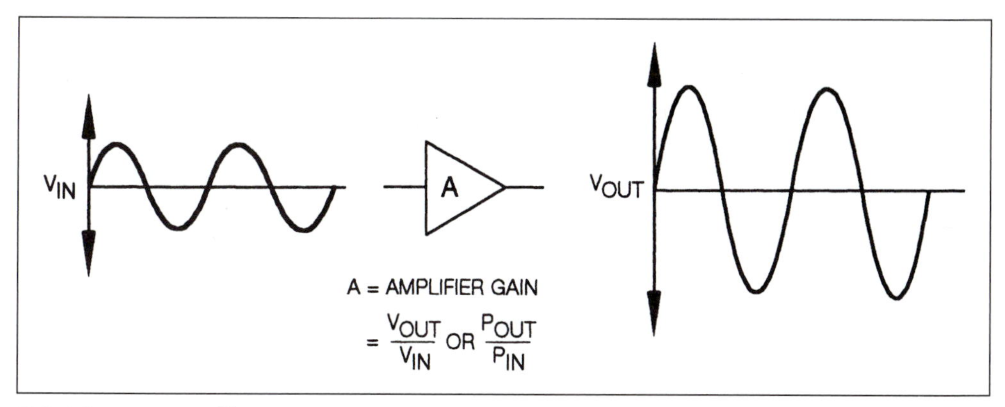

FIGURE 1-14 Amplifier

amplitude The size or magnitude of a signal (Figure 1-15). It is usually expressed in terms of volts or amperes.

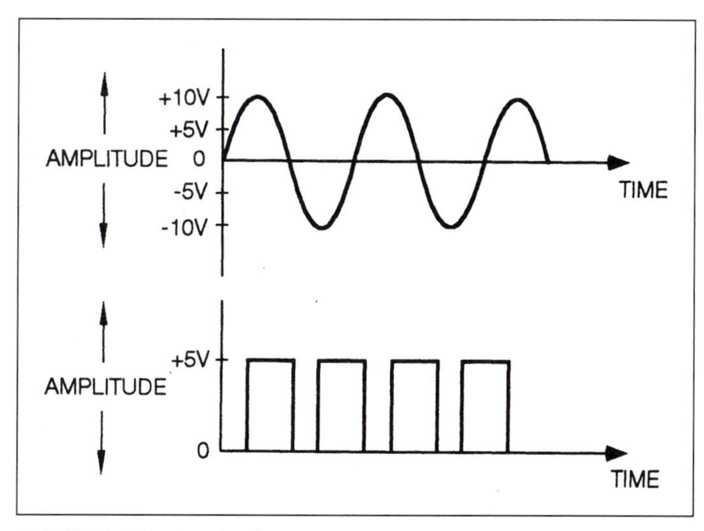

FIGURE 1-15 Amplitude

amplitude distortion The type of distortion occurring in an amplifier or other device in which output is not a linear function of the input amplitude (Figure 1-16).

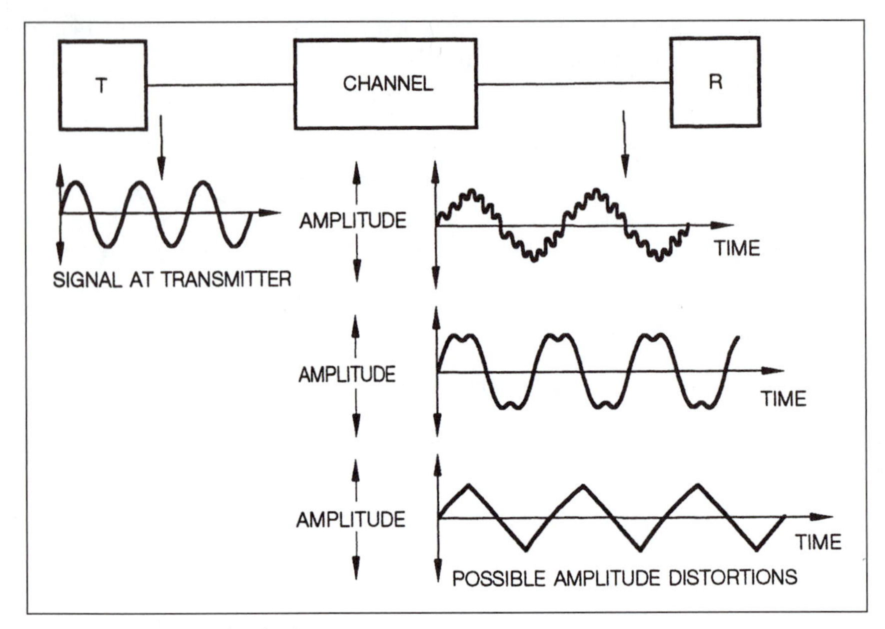

FIGURE 1-16 Amplitude distortion

amplitude modulation A type of modulation in which the amplitude of a high-frequency signal (carrier) is varied in accordance with the variations in the amplitude of a low-frequency signal (modulating signal) (Figure 1-17). *Also see* **modulation**.

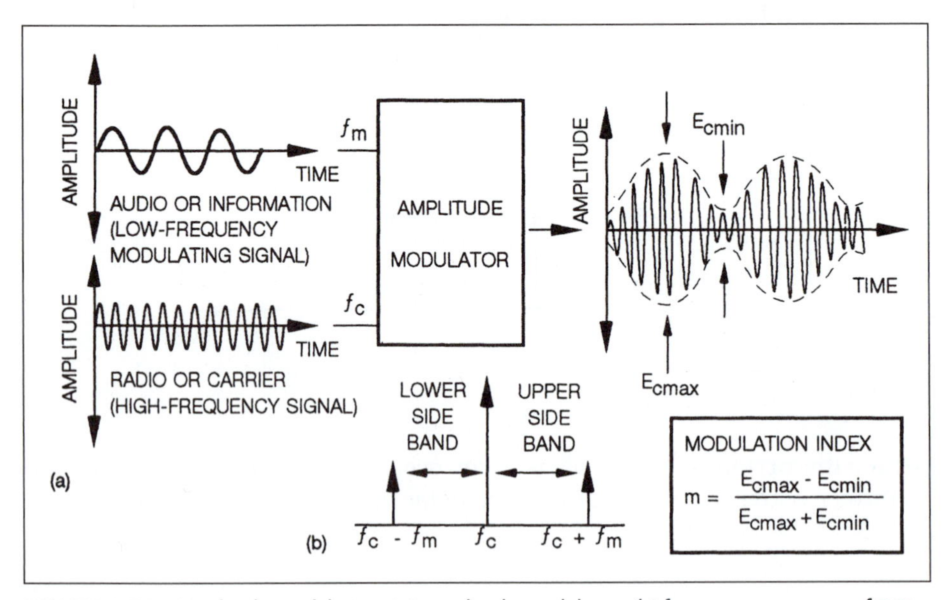

FIGURE 1-17 Amplitude modulation: (a) amplitude modulator; (b) frequency spectrum of DSB AM signal

amplitude shift keying (ASK) A type of RF modulation in which the amplitude of an analog signal is varied in accordance with the changes in a digital information signal. Also known as *on-off keying (OOK)* (Figure 1-18).

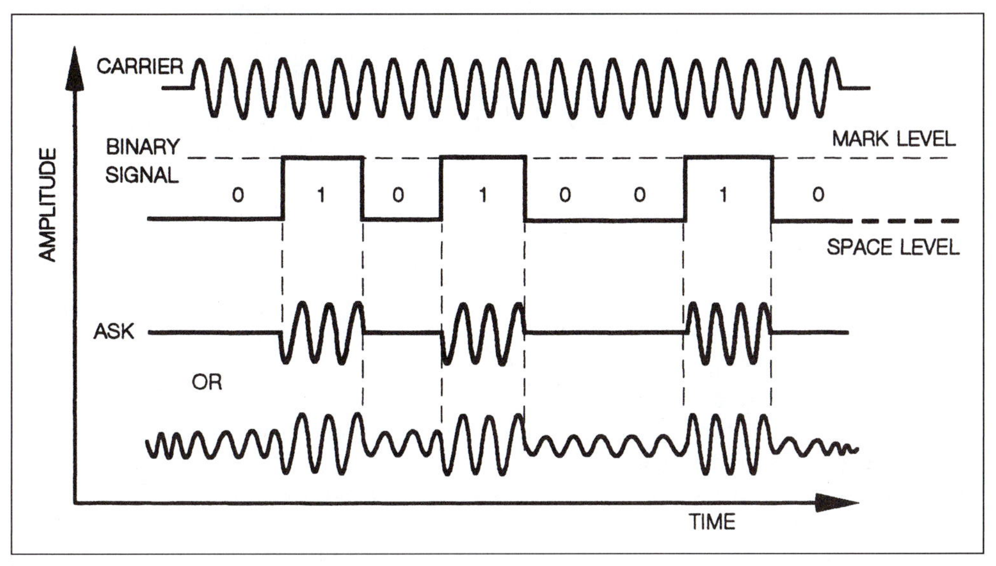

FIGURE 1-18 Amplitude shift keying

analog communications A system of telecommunications that uses continuous electrical signals for the transmission and reception of information.

analog intensity modulation A modulation technique used in optical transmission in which the intensity of the optical source is varied in accordance with the changes in the information signal.

analog signal A signal that is continuous in nature with respect to time and can take values between a maximum and minimum (Figure 1-19).

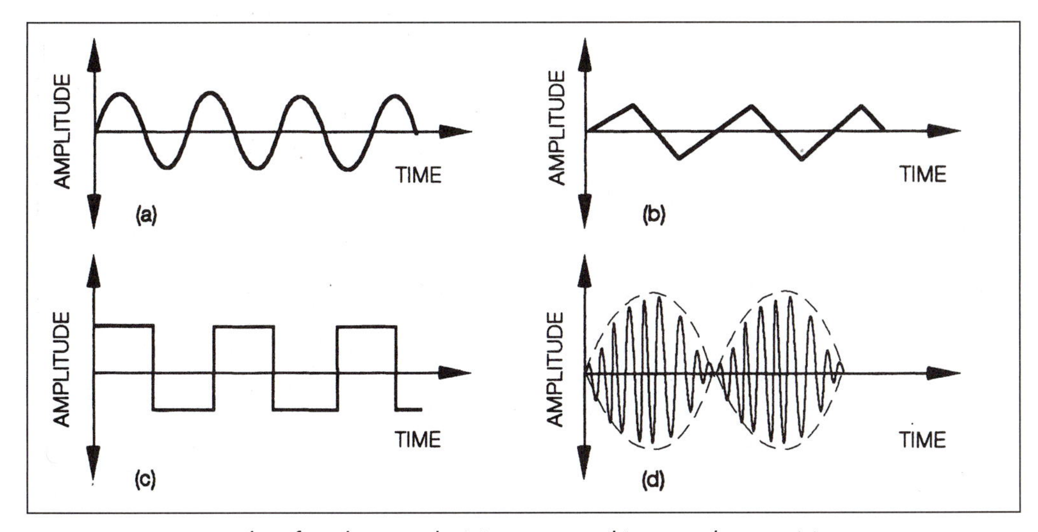

FIGURE 1-19 Examples of analog signals: (a) sine wave; (b) sawtooth wave; (c) square wave; (d) amplitude modulated wave

TABLE 1-3 Analog signals versus digital signals

CHARACTERISTIC	ANALOG SIGNAL	DIGITAL SIGNAL
Nature:	Continuous	Discrete
Noise control:	Poor	Better
Information carrying capacity expressed in:	Bandwidth (Hz)	Channel capacity or speed (bps)
Circuits required to generate signal:	Complex	Simple

analog switch A switch that allows the connection between two circuits for the transmission of analog signals.

analog-to-digital conversion The process of converting an analog signal to digital form by sampling the amplitude of the analog signal and converting the values of amplitude into a digital code.

analog-to-digital converter (ADC) A device that converts an analog signal into a digital signal. Figure 1-20 illustrates an ADC.

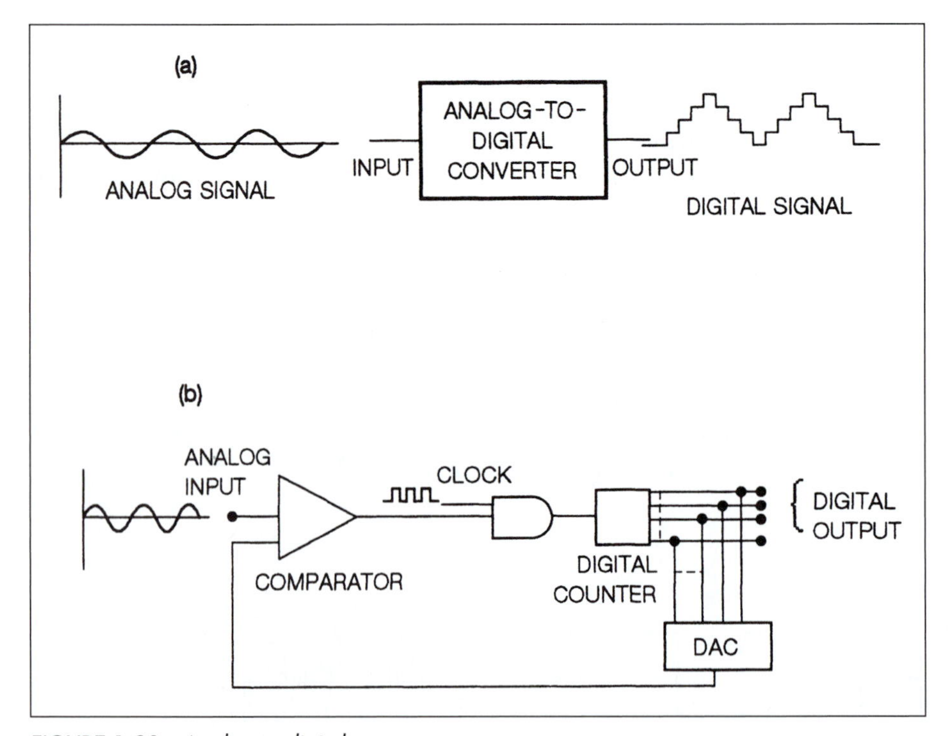

FIGURE 1-20 Analog-to-digital converter

anonymous FTP The process of file transfer protocol (FTP) login for public access with the help of a common username, without a user account on the FTP server.

antenna A metallic structure or wire used for the radiation and reception of radio signals such as ELF, AM, FM, VHF, UHF, microwaves, etc. Table 1-4 gives a summary of the characteristics of various types of antennas.

TABLE 1-4 Antennas

TYPE	STRUCTURE	RADIATION PATTERN	GAIN
DIPOLE	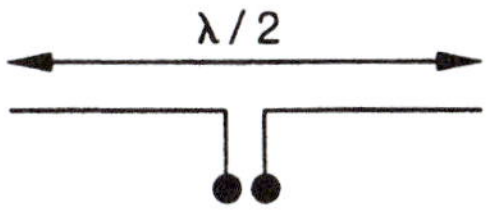 $\lambda/2$		2.14 db
BICONICAL	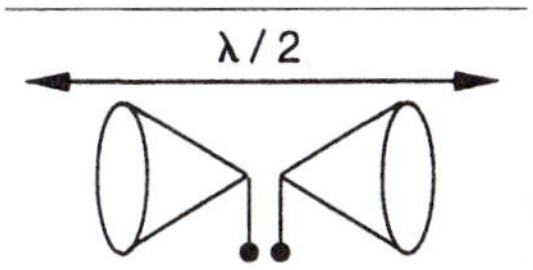 $\lambda/2$		2.14 db
HELICAL	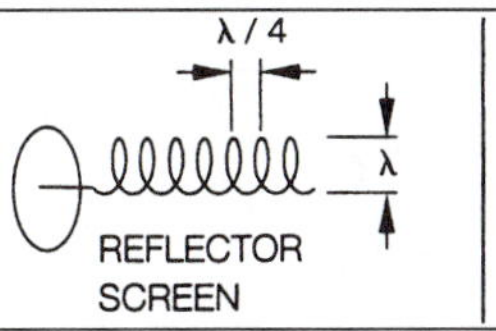 $\lambda/4$, λ, REFLECTOR SCREEN	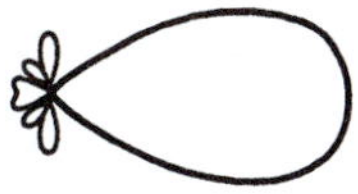	10.14 db
YAGI	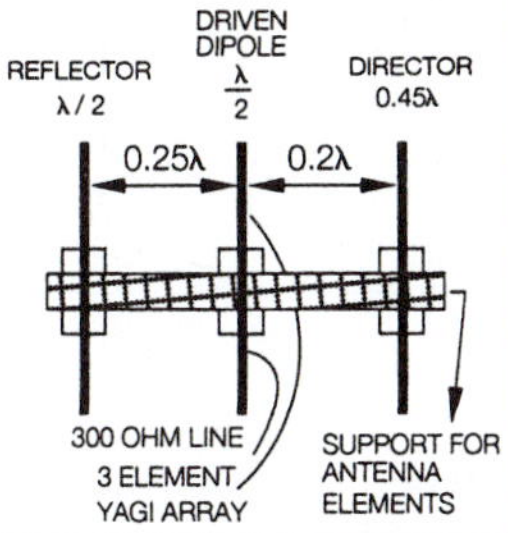 DRIVEN DIPOLE $\frac{\lambda}{2}$, REFLECTOR $\lambda/2$, DIRECTOR 0.45λ, 0.25λ, 0.2λ, 300 OHM LINE, 3 ELEMENT YAGI ARRAY, SUPPORT FOR ANTENNA ELEMENTS	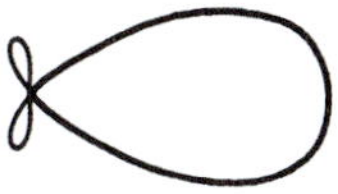	APPROX. 7db, CAN BE INCREASED UP TO 16 db BY INCREASING ARRAY SIZE

antenna efficiency The ability of an antenna to convert RF energy from the transmitter into electromagnetic radiation (Figure 1-21).

$$\eta\,\% = \left[1 - \left(\frac{1}{Q}\right)\right] \times 100$$

where: Q = The quality factor of the resonant network

The ideal is 100 percent, if Q is infinitely high. Practically, however, 50–96 percent is possible. Antenna efficiency in terms of radiation resistance is given by:

$$\eta = \frac{R_r}{R_r + R_L} = \frac{R_r}{Z}$$

where: Z = terminal impedance into antenna
R_r = radiation resistance of antenna
R_L = Loss resistance

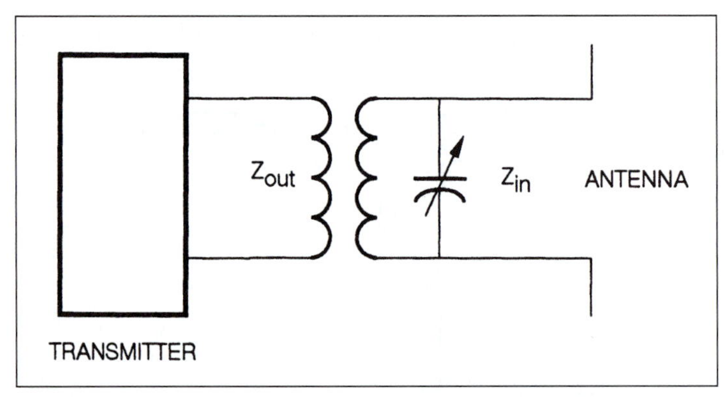

FIGURE 1-21 Antenna efficiency

antivirus software A software program that shields a computer system from a malicious code (virus) by screening and handling files that contain a virus.

apogee The point farthest from the center of the earth in an elliptical orbit of a satellite (Figure 1-22). *Also see* **perigee**.

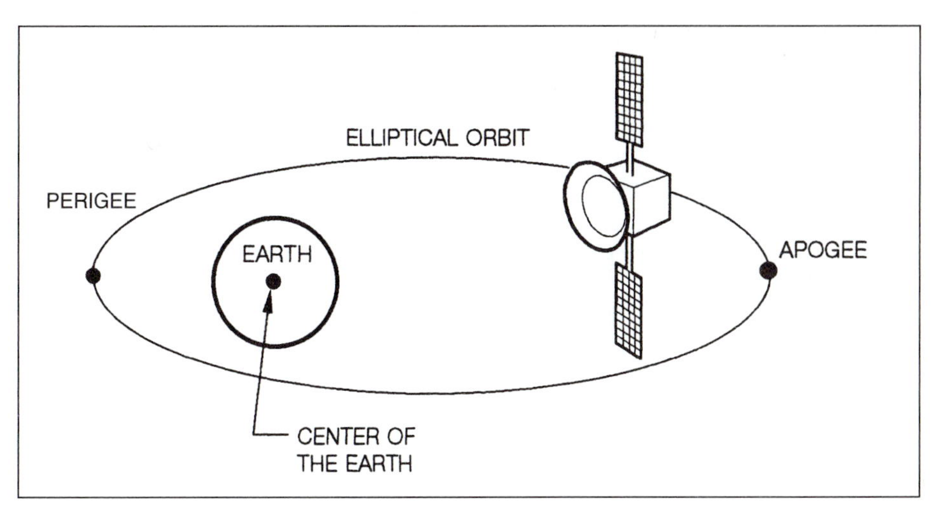

FIGURE 1-22 Apogee

apogee kick motor (AKM) A navigational motor on a communication satellite that is used to correct the satellite's direction.

AppleTalk Computer networking protocol used by Apple Macintosh computers.

application layer The top layer of the OSI model. *See* **OSI model** *and* Module 35.

application program A program that performs a user function, is usually unique to one type of application, and contains no input-output coding.

archive The process of transferring or saving data to a storage device or folder in order to save space or organize the data.

ARPANET A computer network developed by ARPA (Advanced Research Projects Agency of the Department of Defense).

array A group of devices, such as laser diodes, light-emitting diodes, antennas, or computer hard disks, connected together to enhance system performance or capability. In computer software, an array refers to a collection of data elements of the same type that can hold any number of variables of the same type. It can also be described as a container object that contains a list of references to the data stored in specific memory locations.

assembly language A low-level computer language that provides maximum access to all of a computer's internal and external devices. An assembly language requires a significant amount of coding.

asymmetric communication systems/asymmetrical network A communication system or a network in which there is no symmetry between the information exchange capabilities of two points. The amount of information exchanged at any time between two points is not the same (see Figure 1-23). *Also see* **symmetric systems.**

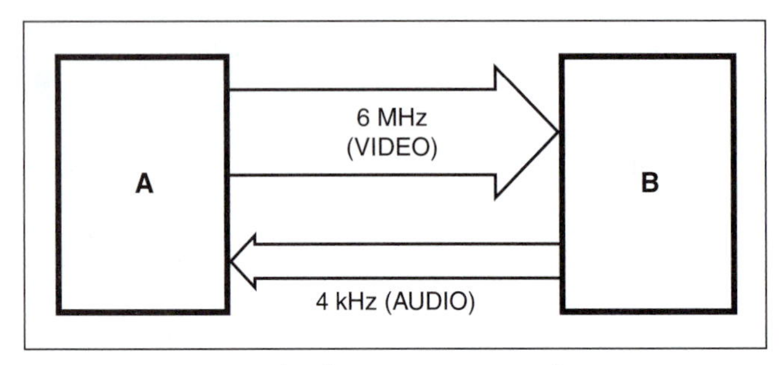

FIGURE 1-23 Example of asymmetric network

asymmetric cryptography Also known as public key cryptography, a cryptosystem in which each user has his or her own pair of public and private keys.

asynchronous A transmission technique or a circuit that does not require clock signals for synchronization purposes.

asynchronous transmission A mode of data transmission that does not require a clock signal for synchronization. It allows one character at a time to be sent and received over a transmission link. Each character contains its own synchronization information in the form of start and stop bits (Figure 1-24). It is also called *start-stop transmission. Also see* **synchronous communication** *and* **second transmission.**

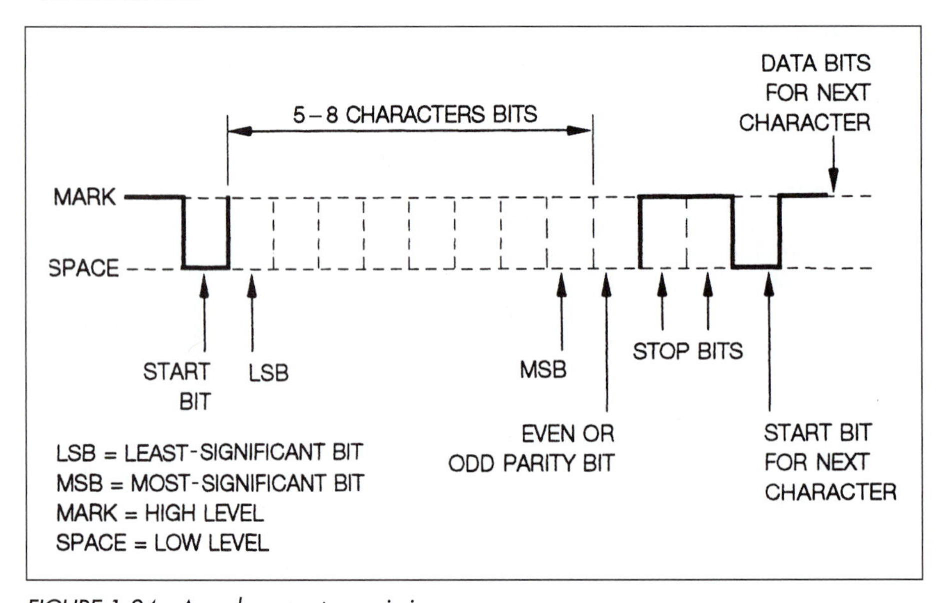

FIGURE 1-24 Asynchronous transmission

TABLE 1-5 Asynchronous versus synchronous transmission

ASYNCHRONOUS	SYNCHRONOUS
Character by character transmission	Data block by block transmission
Each character has its own synchronization information in the form of a start bit and a stop bit.	Each character does not require synchronization in the fom of a start bit and a stop bit. Each data block has its own synchronization elements.
Less efficient	More efficient
Requires simple circuitry	Requires complex circuitry

asynchronous transfer mode (ATM) A high-speed data transmission technology, used for the Internet backbone; operates at 155 Mbps. It is connection-oriented, and uses asynchronous bandwidth allocation and transport rather than time division multiplexing allocation and transport. It can support both connection and connectionless services at both constant bit rate (CBR) and variable bit rate (VBR). It can support transmission speeds of 155 Mbps or higher. It uses 53-byte cells, formatted according to a specific protocol (see Figure 1-25).

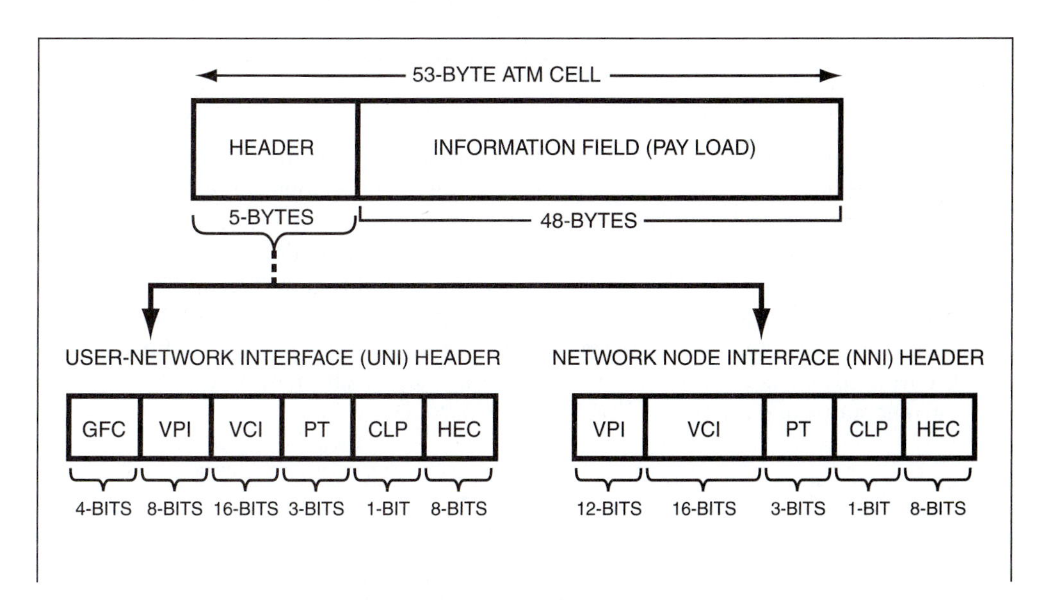

FIGURE 1-25 Asynchronous transfer mode (ATM) *(continues)*

ATM HEADER FIELDS		
GFC	Generic flow control	Provides local functions, e.g., identification of multiple stations connected to a single ATM interface
VPI	Virtual path identifier	In conjunction with VCI, identifies the next destination of a cell as it passes through a series of switches to reach its final destination
VCI	Virtual channel identifier	In conjunction with VCP, identifies the next destination of a cell as it passes through a series of switches to reach its final destination
PT	Payload type	Indicates whether cell contains user or control data
CLP	Cell loss priority	Indicates whether a cell should be discarded if it experiences extreme congestion in the network
HEC	Header error control	Calculates the checksum for the header field

FIGURE 1-25 (continued)

AT&T American Telephone & Telegraph company was formed in 1985 to build and operate long-distance lines and provide nationwide telephone service. Historically, AT&T controlled the Bell Telephone Laboratories, 23 Bell operating companies, and the Western Electric Company. In 1982, the government's antitrust suit of 1974 against AT&T was settled by a modified judgment that required AT&T to divest itself of all its operating companies. The operating companies were divided into seven regional companies to provide local telephone service and switching. Figure 1-26 shows the pre- and post-divestiture organization of AT&T. (For the current status and history of AT&T, visit http://www.att.com.)

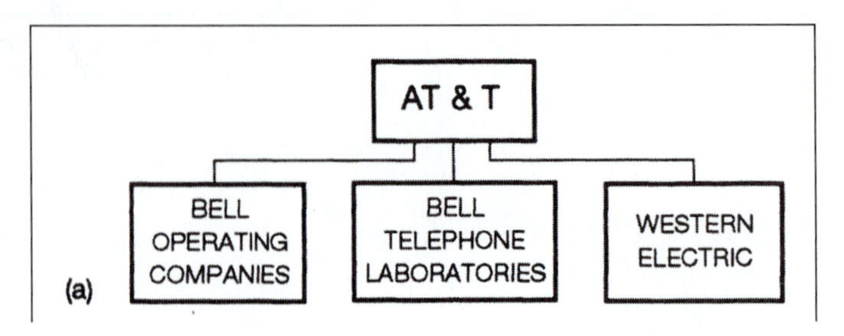

FIGURE 1-26 (a) Pre- and (b) post-divestiture organization of AT&T (1982) *(continues)*

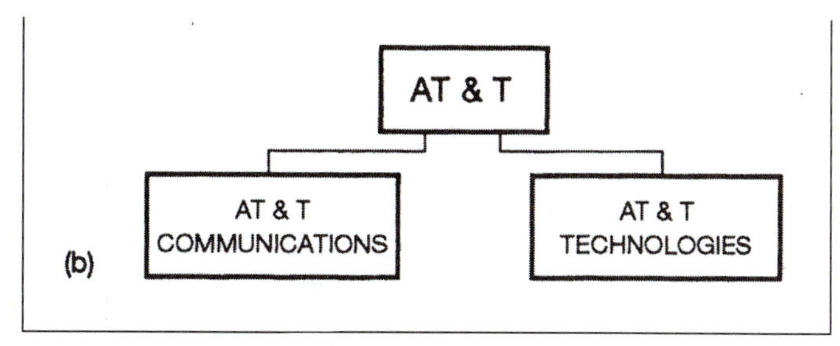

FIGURE 1-26 (continued)

atmospheric noise *See* **noise**.

attack An attempt by an unauthorized user to gain unauthorized access to a computer network or to deny authorized users access. An attack could be active or passive. *See* **active attack**.

attacker A person or a group who attempts to violate the security and integrity of a computer network.

attenuation A decrease in current, voltage, or power of a signal in its transmission between two points due to signal loss in the transmission medium (Figure 1-27). Usually expressed in decibels (db). *Also see* **decibel (db)**.

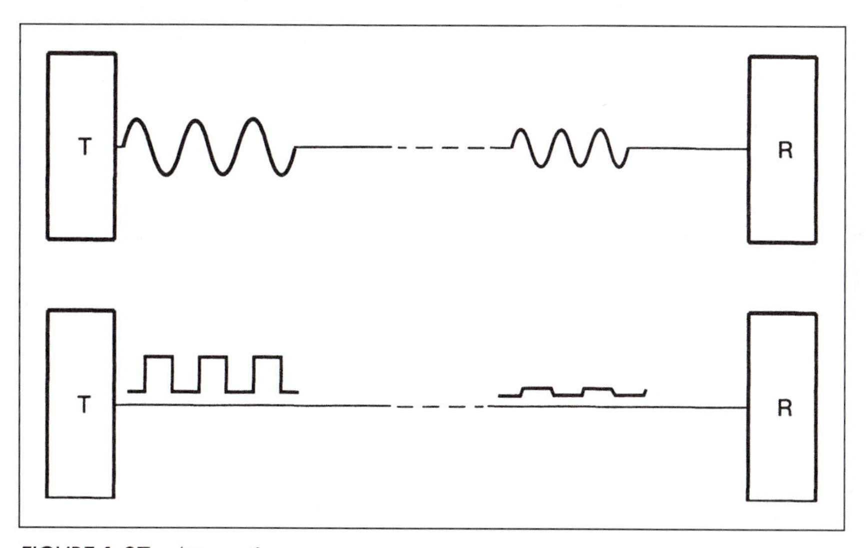

FIGURE 1-27 Attenuation

attenuation factor A factor that expresses the power loss per unit length of a transmission medium, such as two-wire line, coaxial cable, optical fiber, etc. Usually expressed in db per kilometer or mile.

audio frequencies Acoustic (sound) frequencies that can be heard by the human ear. Usually audio frequencies cover the range from 20 Hz to 20 kHz (Figure 1-28). Also called *sonic frequencies*.

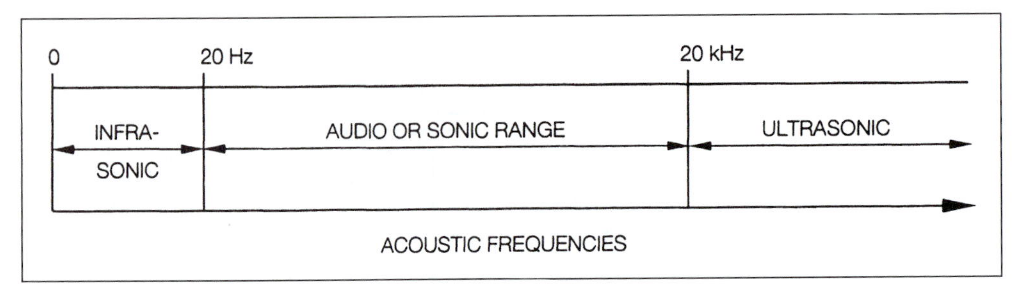

FIGURE 1-28 Audio frequencies

authenticate A security measure that establishes and verifies a user's identity with the help of unique information that distinguishes a user from others.

automatic call distributor (ACD) A device that switches incoming calls to answering positions automatically.

automatic calling unit A device that enables a business machine to dial calls automatically.

automatic dialer A device that automatically generates dialing digits.

automatic direction finder A device that automatically and continuously monitors the direction of arrival of electromagnetic signals.

automatic identification of outward dialing (AIOD) A method of automatic message accounting in which toll calls placed at different calling stations in a telephone system are automatically recorded.

automatic retransmission request A technique used for error detection and correction that involves the retransmission of data when error is detected.

automatic route selection A telephone used to select the best trunk for a call going out of PBX.

automatic send and receive A teletype terminal with paper tape or magnetic storage capabilities that allows a message to be originated off-line for later transmission.

availability Continuous operation of a device/system/network. Also expressed as a probability—how consistently/reliably a device/system/network can be accessed for operation. (*Also see* **reliability**, *and* Module 33.) Availability is represented by:

$$A = \frac{MTBF}{MTTR + MTBF}$$

Where: A = availability of a device
MTBF = mean time before failure (hours)
MTTR = mean time to repair (hours)

avalanche photodiode (APD) A photodiode that takes advantage of avalanche multiplication of photocurrent. As the amplitude of the reverse bias across a photodiode is increased to the breakdown voltage level, avalanche multiplication occurs (Figure 1-29). Electron-hole pairs generated by the absorption of photons acquire enough energy to create additional electron-hole pairs by collision with the atoms, thus resulting in a much larger current gain than other photodiodes. APDs have a low noise figure over a wide bandwidth and their typical gain-bandwidth product is 100 GHz. The multiplication factor is given by:

$$M = \frac{I_M}{I_{UM}} = \frac{I_{ph} - I_d}{I_{uph} - I_{ud}} = \frac{1}{1 - \left(\frac{V_R}{V_b}\right)^n}$$

Where: I_M = total multiplied current
I_{UM} = total unmultiplied current
I_{ph} = multiplied photocurrent
I_d = multiplied dark current
I_{uph} = unmultiplied photocurrent
V_R = reverse bias voltage
V_b = breakdown voltage
n = a constant (1, 2, . . .)

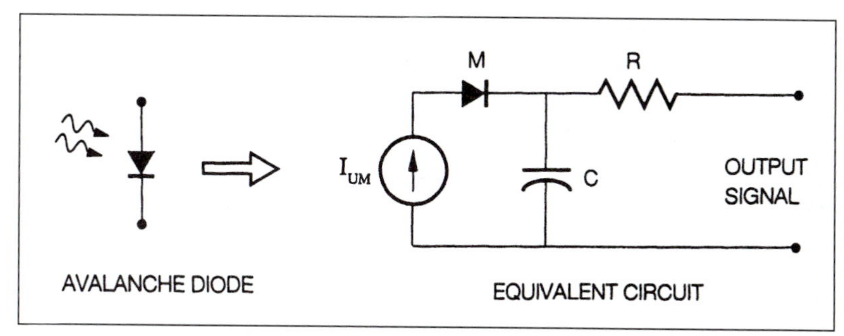

FIGURE 1-29 Avalanche photodiode

average busy-hour traffic count An average of traffic counts taken during peak hours on business days when the largest volume of traffic is handled. *Also see traffic formulas in* Module 33.

average call distribution A statistic used in network traffic, equal to the total number of minutes of conversation divided by the number of effective calls. *Also see traffic formulas in* Module 33.

average noise factor The ratio of available SNR at the input to the SNR at the output of a device or circuit, over the frequency range of interest. *Also see* **noise** *and noise formulas in* Module 33.

$$F_{avg} = \frac{\text{input SNR}}{\text{output SNR}} = \frac{P_{si}P_{so}}{P_{ni}P_{no}}$$

Where: SNR = signal-to-noise ratio $= \frac{\text{signal power}}{\text{noise power}}$

P_{si} = signal power at the input
P_{so} = signal power at output
P_{ni} = noise power at input
P_{no} = noise power at output

average noise figure A figure of merit used to compare the noise in a network with a noiseless network. It is the measure of the degree of degradation in SNR between input and output of a network over a frequency range. *Also see* **signal-to-noise ratio (SNR)** *and noise formulas in* Module 33.

$$\text{Noise figure} = \text{NF} = 10 \log \text{F}$$

$$\text{Where: F} = \frac{\text{input SNR}}{\text{output SNR}}$$

axial ray A light ray that travels along an optical fiber axis (Figure 1-30). *Also see* **optical fiber** *and* Module 27.

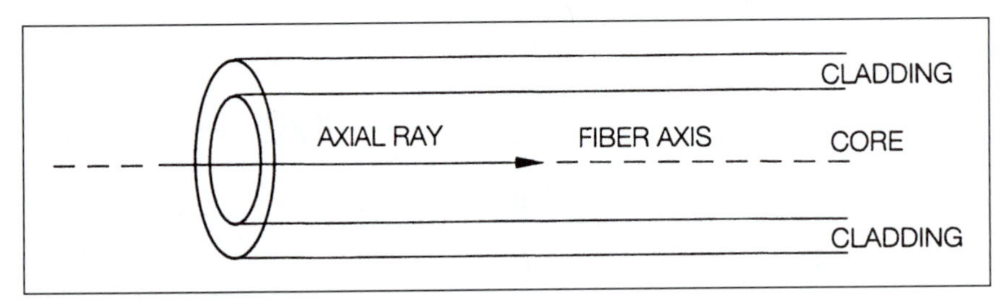

FIGURE 1-30 Axial ray

MODULE 2 Definitions B

B-band The frequencies in the range 0.25–0.50 GHz; used by the military as a microwave band.

B-channel In Integrated Digital Services Network (ISDN), the bearer channel, which handles circuit-switched voice and circuit- or packet-switched data at 64 kbps. *Also see* Module 34.

B-ISDN Broadband integrated services digital network, a digital service that uses broadband technology (asynchronous transfer mode [ATM]) to offer high-speed data transfer to support interactive services (e.g., video conferencing, video mail) and distribution services (e.g., television broadcast). *Also see* **integrated services digital network (ISDN), asynchronous transfer mode (ATM)**, *and* Module 34.

back scattering Radiation of unwanted electromagnetic signals to the back side of an antenna. The radio signal gets scattered in the opposite direction of the incident wave.

backward wave oscillator (BWO) A type of oscillator used to generate microwave frequencies. It consists of a traveling wave tube in which oscillating currents are produced by using an oscillating electron tube to bunch the electrons as they travel from cathode to anode.

balanced line A transmission line consisting of two conductors that (1) have equal resistance per unit length, (2) have equal impedance to ground, and (3) carry equal but 180-degrees-out-of-phase signals with respect to ground. In contrast to *unbalanced line*, which is preferred for data communications (Figure 2-1). Also called *balanced-to-ground.*

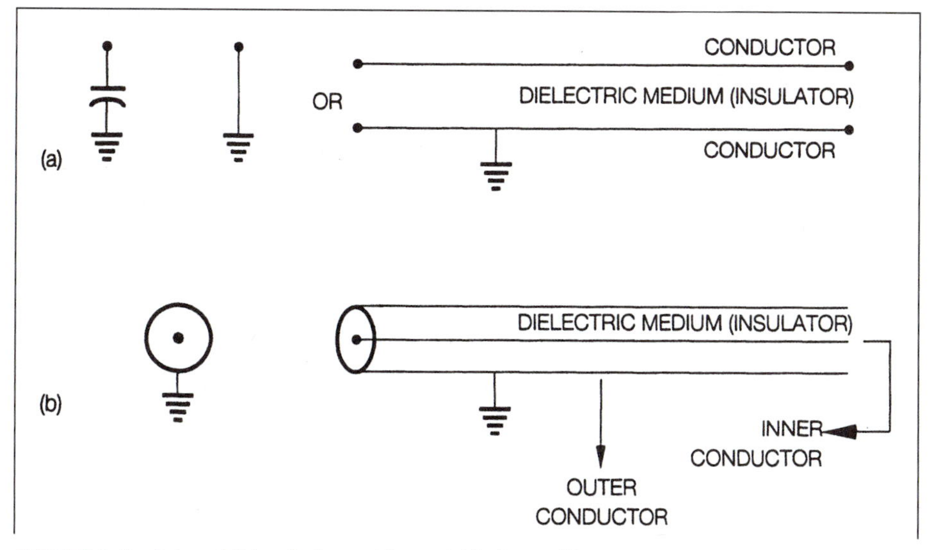

FIGURE 2-1 (a) and (b) unbalanced lines; (c) balanced lines *(continues)*

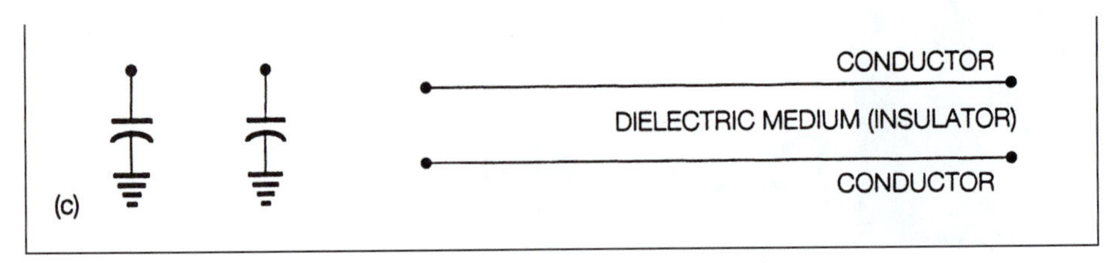

FIGURE 2-1 (continued)

balun *Bal*anced *Un*balanced: A device that connects a balanced line with an unbalanced line or vice versa.

band A range of frequencies between two specific limits.

bandgap energy The difference in energy between the two energy bands in solids. It represents the range of energies forbidden for electrons to have (Figure 2-2).

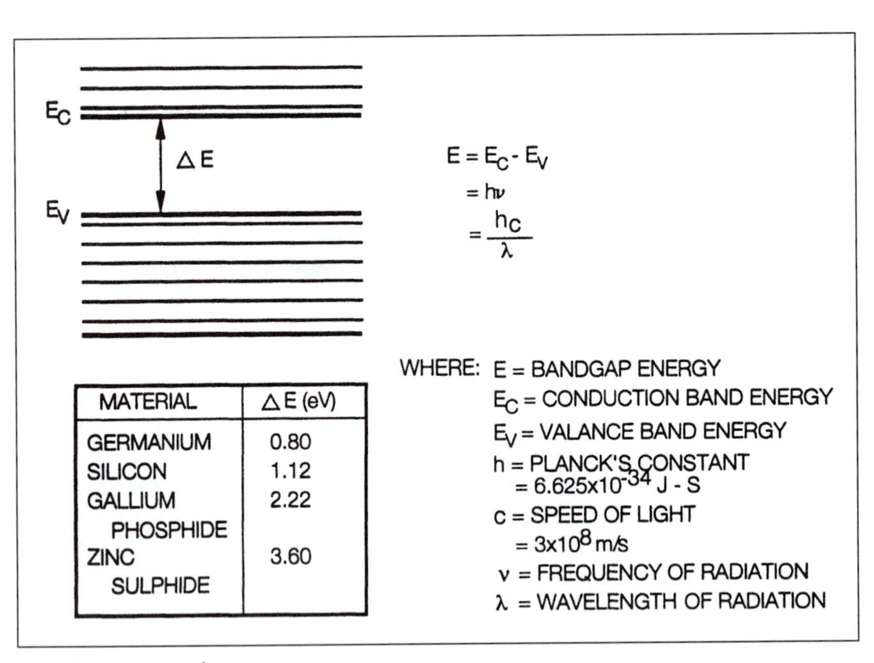

MATERIAL	ΔE (eV)
GERMANIUM	0.80
SILICON	1.12
GALLIUM PHOSPHIDE	2.22
ZINC SULPHIDE	3.60

FIGURE 2-2 Bandgap energy

bandpass filter A circuit that allows a band of frequencies to pass. The band is the difference between the upper and lower cutoff frequencies (Figure 2-3). *Also see* **Active filter** *and* Figure 1-7.

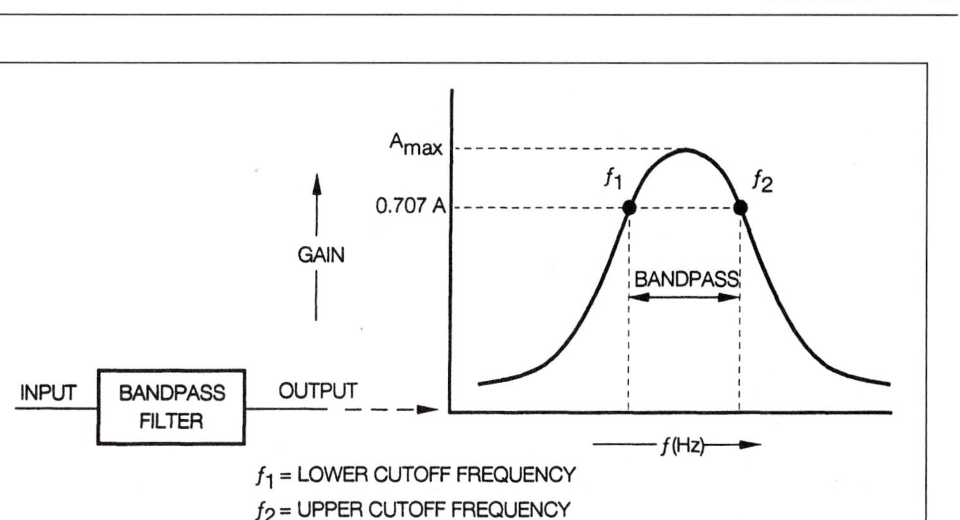

FIGURE 2-3 Bandpass filter

bandwidth (BW) A range of frequencies that can pass through a device, system, or channel without significant attenuation. It is expressed as the difference of upper and lower cutoff frequencies. Bandwidth is a measure of the data transmission capacity of a system or a channel (Figure 2-4).

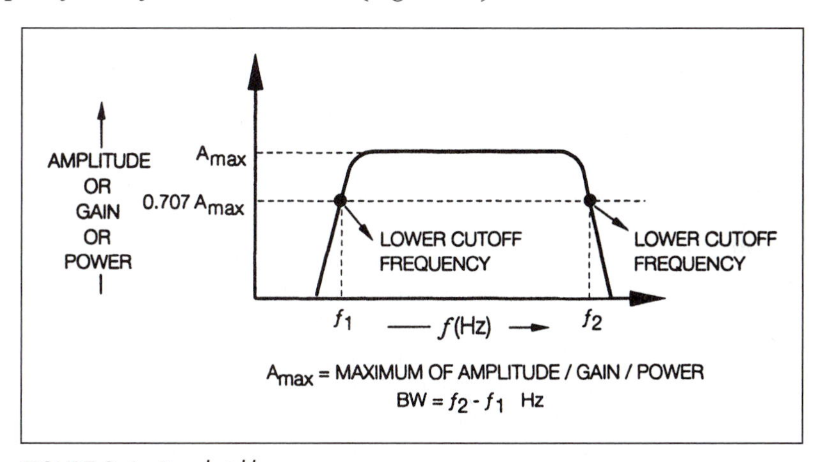

FIGURE 2-4 Bandwidth

BARITT *BAR*rier *I*njected *T*ransit *T*ime diode. The newest addition to the family of microwave diodes used for the generation of microwave frequencies.

baseband The frequency band occupied by signals before they are modulated with a carrier (Table 2-1).

TABLE 2-1 Baseband versus broadband transmission

BASEBAND	BROADBAND
Single-channel transmission Modulation is not required	Multiple-channel transmission Modulation required

baseband signaling The transmission of analog or digital information at original frequencies, i.e., without modulation with a high-frequency carrier.

basic telecommunications access method (BTAM) An IBM lowest level access-method used to control the transfer of data between main storage and a terminal (local or remote). It supports binary synchronous communications and start/stop communications. It supplies the application program with micro instructions for using the capabilities of the supported device.

batch environment A type of data processing in which a computer receives instructions and program from a terminal or a peripheral device and executes the requested operation at its own convenience. It is in contrast to *real-time* or *on-line processing.*

baud A unit of signaling speed. It represents the number of discrete conditions or elements per second.

baud rate The number of discrete conditions or signal elements per second. It can be determined by taking the reciprocal of the shortest signaling element (Figure 2-5). It is different from the *information rate. Also see information rate or data rate in* Module 33.

$$\text{baud rate} = \frac{1}{T}\,\text{baud}$$

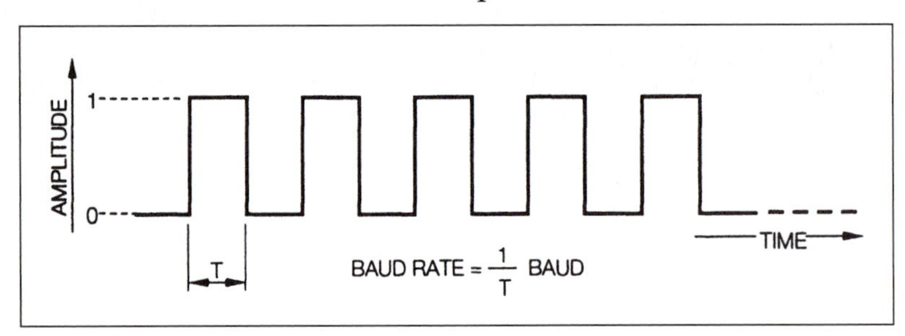

FIGURE 2-5 Baud rate

Baudot code A 5-bit, 32 character alphanumeric code used in transmission of character information in telex. It uses 1 start element and 1.43 stop elements. (*See* Module 40.)

BCD Binary Coded Decimal. A code that represents decimal digits in binary. A group of four binary bits is used to represent a decimal digit. (*See* Module 40.)

BCDIC Binary Decimal Coded Interchange Code. An IBM 8-bit character code. It is an extension of the BCD code to represent a more complete set of characters.

BCH An error detection and correction technique named after BOSE-CHAUDHURI-HOCQUENGHEN.

BDLC An IBM procedure for data transmission. A complete message or block is transmitted with control characters for synchronization, error detection, etc.

beam Concentrated electromagnetic radiation, such as the flow of electrons within a cathode ray tube (CRT) or vacuum tube, the flow of photons in a laser, etc.

beam divergence A process in which the concentrated flow of radiation splits.

beam splitter A device in optoelectronics to divide an optical beam into two or more separate beams.

beam transmission The transmission of a radio signal in which the signals are sent in a narrow beam rather than broadcast. Microwave and satellite systems use beam transmission (Figure 2-6).

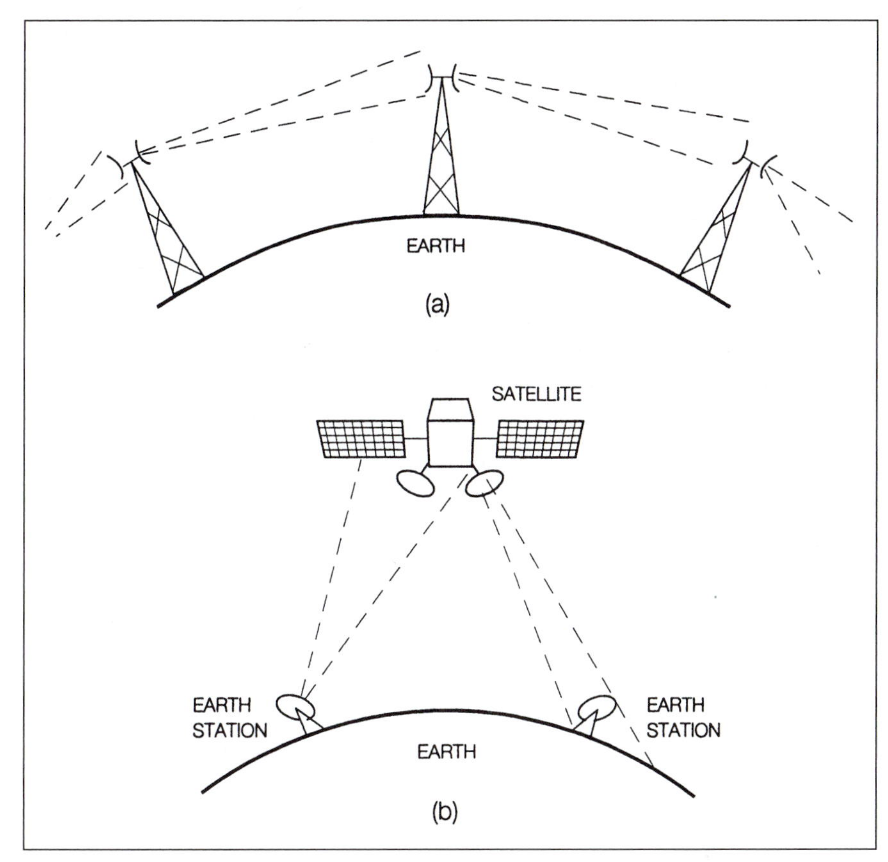

FIGURE 2-6 Examples of beam transmission: (a) microwave transmission; (b) satellite transmission

beeper A small transceiver that produces a beep for the purpose of alert for two-way contact.

beat frequency A frequency produced resulting from the mixing or interference of two signals whose frequencies are close to each other. Used in heterodyne receivers. *See also* **heterodyning**.

bias

1. In electronics, the voltage or current applied to an active device to set a steady state operating condition of the circuit where the operation is most linear.
2. In data communications, signal distortion with respect to bit timing.

bidirectional The property of a network that allows the user to transmit and receive information at the same time, i.e., full duplex operation.

binary synchronous communications (BSC) A data link protocol developed by IBM that uses character synchronization. *See also* **synchronous communication**.

biometric Detection and classification of physical attributes such as fingerprints, voice pattern, or iris or retina of the eye. Biometrics are used for authentication.

BIOS (Basic Input/Output System) Firmware that controls a computer's input/output functions.

bipolar Having two poles, such as north and south, positive and negative. Figure 2-7 illustrates a bipolar signal.

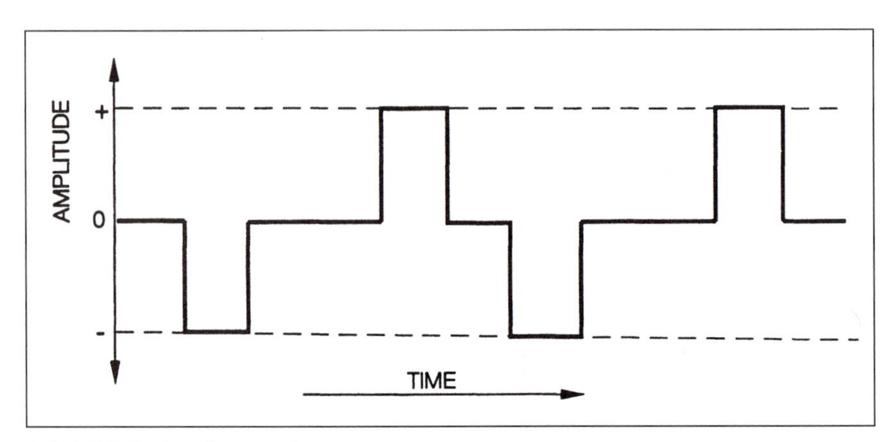

FIGURE 2-7 Bipolar signal

bit A contraction of "binary digit." One of two coefficients of the binary numbering system, i.e., "1" and "0." It represents a value between the one (mark) and zero (space) levels.

bit error rate A ratio used to express the erroneous bits with respect to a certain number of bits received. It is expressed as a power of ten. For example:

$$\text{BER} = \frac{1}{10^9} = 10^{-9}$$

represents that one bit out of one billion is erroneous.

bit error rate test (BERT) A test performed to determine the number of erroneous bits with respect to a certain number of bits received.

bit-oriented protocol A protocol of data communication in which a fixed bit pattern is added to each character for synchronization. A commonly used bit-oriented protocol is SDLC. *See also* **synchronous communication**.

bit rate The rate of data transmission, expressed as bps (bits per second). Usually expressed as: kbps, Mbps, Gbps.

bit stream A string of binary information in which each bit position is considered as an independent unit (Figure 2-8).

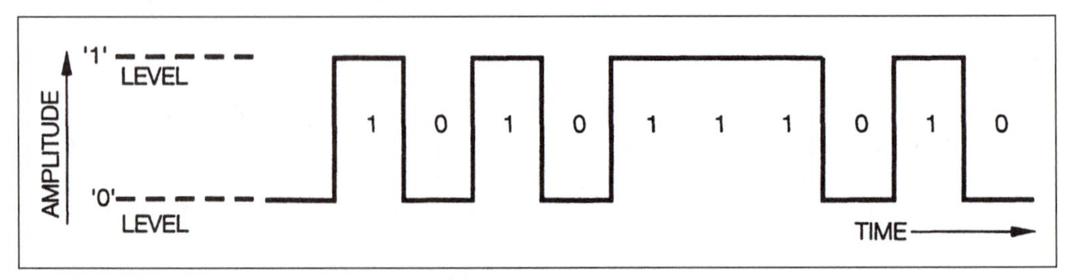

FIGURE 2-8 Bit stream

bit time The time occupied by a bit in a stream of digital data (Figure 2-9).

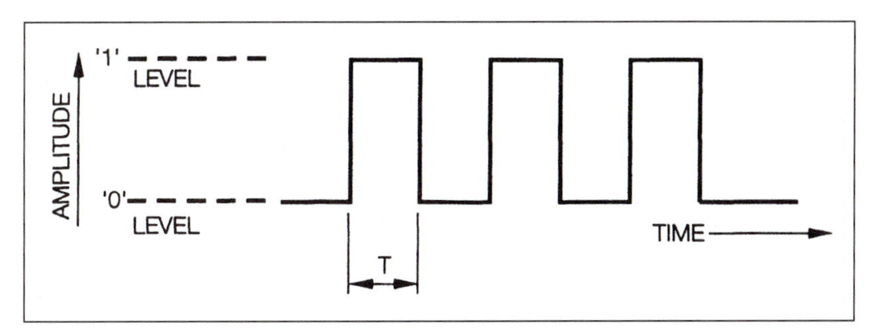

FIGURE 2-9 Bit time

block In data communications, a collection of characters that form a message or part of a message. Each block is appended with control information regarding source, destination, error detection and correction, address, etc. (Figure 2-10).

SYN	STX	DATA BLOCK 1	BCC	STX	DATA BLOCK 2	STX	BCC

FIGURE 2-10 Block

block check character (BCC) A number of characters added to a message or data block that facilitates error control in date communications. The value of the BCC is determined in accordance with an algorithm to help in detecting errors.

block error rate A figure of merit used to express the performance of a blocked data transmission by errors in blocked transmission.

block length The length of a block expressed as the number of characters, bits, bytes, or words contained in a group.

block transmission A method of transmitting data in blocks. Block transmission allows high data rates in contrast to *character-by-character transmission.*

blocked call The inability of a calling party to get connected to the party being called over a telephone network.

blocking Division of a long information message into two or more message blocks. Block size and the number of blocks affect the transmission rate of a communication system.

Bluetooth A low-power wireless technology that provides a connectivity standard for network devices. It operates at 2.4 GHz (the ISM band of the radio spectrum) and provides point-to-point as well as point-to-multipoint connectivity between devices (laptops, printers, cell phones, PDAs, modems, etc.) simultaneously. *Also see* **industrial, scientific,** *and* **medical (ISM) bands**.

BNC connector A connector used on the ends of a thin coaxial cable (Figure 2-11).

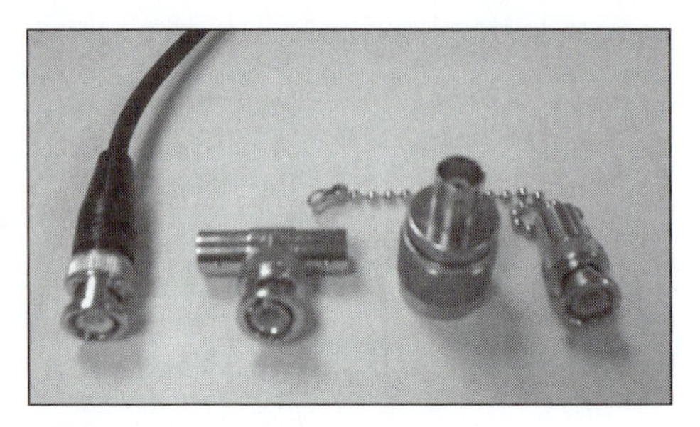

FIGURE 2-11 BNC connector

BNC T connector The connectors used on a thin coaxial cable that allow a device to connect to the medium (Figure 2-12).

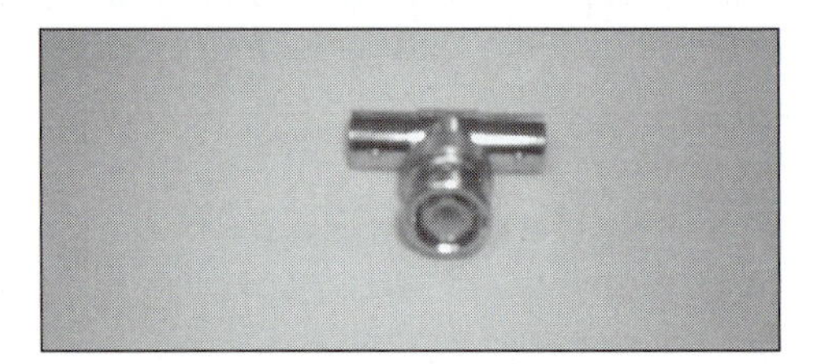

FIGURE 2-12 BNC T connector

BOC Bell Operating Company, one of the divested operating telephone companies in the United States.

bolometer A device used for measuring the intensity of thermal radiation.

boot (boot-up) The process of turning on or resetting a computer; it involves the loading of an operating system into the computer memory.

bps Bits per second. Represents the unit for measuring data transmission capacity. For example:

kbps = kilo or thousand bits per second
Mbps = Mega or million bits per second
Gbps = Giga or billion bits per second
Tbps = Tera or trillion bits per second

break condition A feature in communication that permits the receiving station to interrupt the transmitting station and take control of the circuit.

breakout box A device that allows access to various points of a physical interface connection (such as TIA/EIA-232, TIA/EIA-449, etc.) for testing and monitoring purposes.

breakout panel A box from which one or more conductors are brought out.

bridge

1. To connect a load across a circuit.
2. A type of circuit used to measure quantities like resistance, capacitance, inductance, and admittance.
3. In local area networks (LANs), an interface that connects two similar LANs (Figure 2-13).

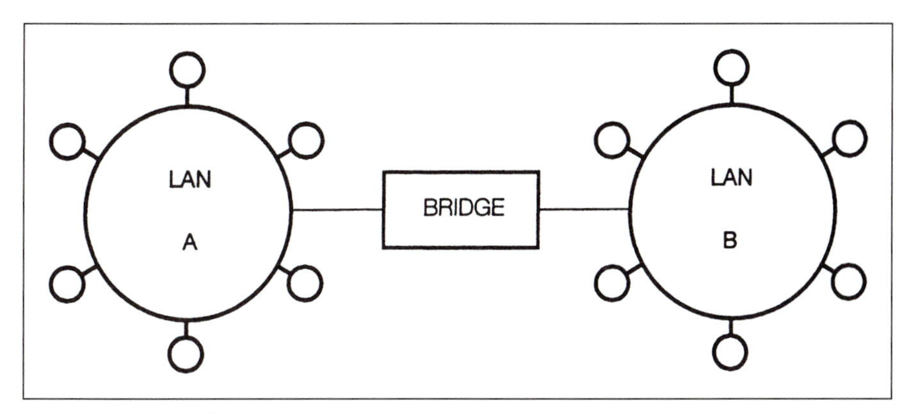

FIGURE 2-13 Bridge

broadband (wideband) Transmission equipment and media that support a wide range of electromagnetic frequencies, usually higher than the voice band frequencies. Broadband facilities can carry many voice and data characters simultaneously, using different carrier frequency signals. *See also* **baseband**.

broadband Transmission technologies (service and applications) that offer high data rates, generally exceeding 1.544 Mbps. *Also see* **baseband**.

TABLE 2-2 Broadband technologies

BROADBAND TECHNOLOGY	DATA RATE
HDSL	1.544 Mbps
T-l	1.544 Mbps
FDDI	100 Mbps
ATM	155 Mbps
SONET	51.84 Mbps (OC-1)–2.488 Gbps (OC-48)

broadband signaling A type of transmission technique in which a large number of channels can be transmitted by a wide range of electromagnetic frequencies. It consists of multiple frequency carriers; each carrier signal is modulated by an information channel (Figure 2-14).

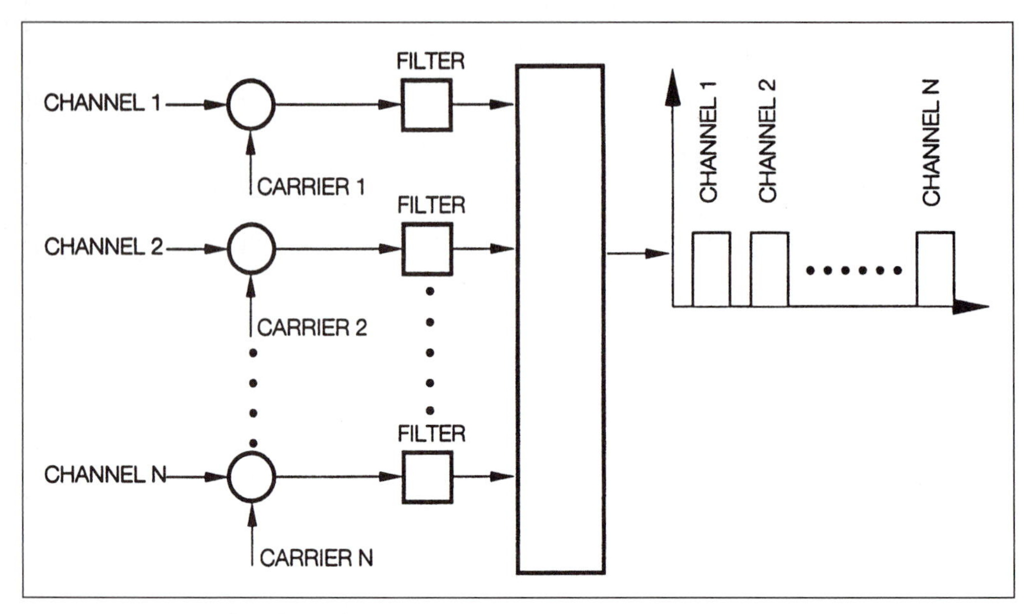

FIGURE 2-14 Broadband signaling

broadcast A transmission of a signal to many receivers/nodes/users in a computer network.

brouter A device that combines the functionality of a bridge and a router. *Also see* **bridge** *and* **router**.

browser An application software program that allows a user to retrieve documents containing text and/or graphics from the World Wide Web (WWW) and displays them using Hypertext Markup Language (HTML). Microsoft's Internet Explorer and Netscape's Navigator are the most widely used Internet browsers.

BT Busy tone. A tone generated by combining 480 Hz and 620 Hz, used in a telephone system to signal that the call cannot be completed.

buffer A storage device/circuit used (1) in data transmission to compensate for a difference in rate of data flow or time of occurrence of events, or (2) for temporary storage of data.

bug A mistake or malfunction in the software or hardware of a system.

bundle A number of optical fibers combined together.

burst

1. A sequence of signals in data communications regarded as one unit in accordance with some specific criterion or measurement.
2. A color signal from the composite video signal in color television.

burst error A sequence of consecutive errors in data communication.

bus

1. A conductor or group of conductors used for the transmission of one or more signal.
2. In data communications, a network topology in which all stations are connected to a common path (Figure 2-15).

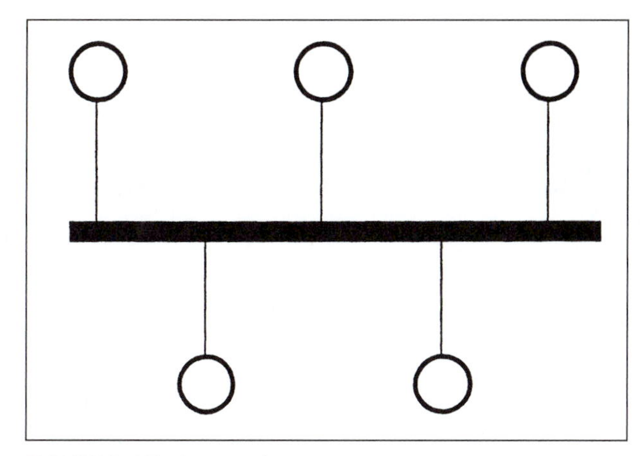

FIGURE 2-15 Bus topology

3. In computers, a path over which data transmission takes place between various components of system.

Also see **Topology** *and* Module 48.

busy hour A continuous period of one hour during which the average traffic in a telecommunication system is maximum. *Also see traffic formulas in* Module 33.

busy signal An audible tone or flashing signal that indicates the unavailability of a number.

by Busy. Represents a link that is in use.

byte A group of eight bits considered as a logical unit. For example:

$$\underbrace{11010100}_{\text{byte}},\ \underbrace{1010001}_{\text{byte}}$$

MODULE

3 Definitions C

C band The portion of the electromagnetic spectrum in the range of 4–6 GHz, used for microwave and satellite communications.

C conditioning A type of line conditioning that controls distortion and delay in a transmission line to keep it within specific limits.

cable One or more conductors assembled within an insulated jacket. It allows the conductor to be used separately or in a group. The most commonly used cables are coaxial cable and two-wire twisted pair (Figure 3-1). Cables can be aerial or underground.

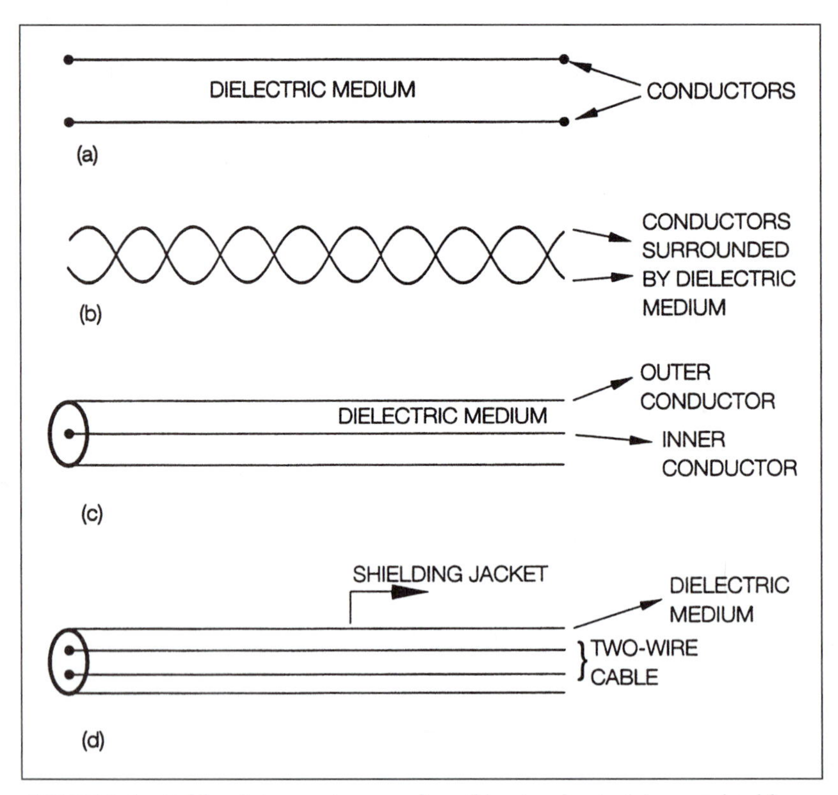

FIGURE 3-1 Cables: (a) two-wire open line; (b) twisted pair; (c) coaxial cable; (d) shielded two-wire cable

cable drop A cable (coax or fiber) that connects a neighborhood cable node to the customer's premises.

cable modem A device that allows a high-speed connection to the Internet via a cable television (CATV) connection where the cable television company acts as the Internet service provider (ISP).

call A demand to set up a connection over the telephone network. Also used as a unit in telephone traffic.

call capacity The capacity of a switching system in routing calls at the same or one time.

call duration The time interval between the establishment and the termination of a call by the calling station.

call forwarding A feature that permits automatic forwarding of an incoming call to another telephone number by dialing a special code.

call waiting A feature that allows a customer engaged in a telephone call to receive another call. The customer hears a beep or a series of beeps as the indication of an incoming call. The customer can place the first call on hold and answer the second call.

calling rate The number of calls attempted per unit time in a telephone network.

camera phone A cellular phone with built-in camera to take digital photos and transmit them over the wireless connection.

CAP Carrierless amplitude. A type of quadrature amplitude modulation (QAM) in which the modulated signal is transmitted and the carrier is suppressed.

carrier

1. A high-frequency signal that is modulated with a low-frequency information signal.
2. A commercial company that offers circuits to carry voice/data. Also called *common carrier*.

carrier frequency Frequency of carrier signal that is used in modulation. One of the characteristics of the carrier (such as amplitude, frequency, or phase) is varied in accordance with the changes in the amplitude of the information signal (Figure 3-2). *Also see* **Modulation**.

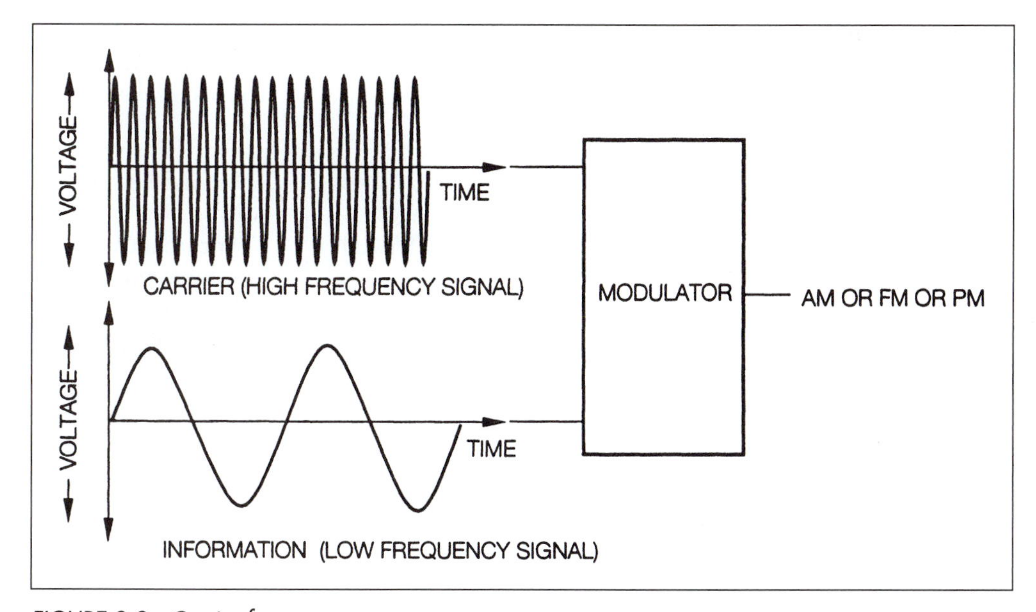

FIGURE 3-2 Carrier frequency

carrier sense multiple access with collision avoidance (CSMA/CA) An access method used in local area networks (LANs) that requires devices to sense traffic on the network before transmitting the data in order to avoid the collision of data packets.

carrier sense multiple access with collision detection (CSMA/CD) An access method used in local area networks (LANs) that requires devices to wait (to re-transmit) for a random duration of time after a collision of data packets.

carrier signaling A technique used in a multichannel carrier transmission.

carrier system A system in which a number of channels are transmitted over a link by modulating each channel with a different carrier frequency. The information is retrieved at the receiver by a demodulation process. *See also* **broadband signaling**.

CAT Abbreviation for category, describes the type of twisted pair cable. Table 3-1 lists the characteristics of CAT 1– CAT 7 cables.

TABLE 3-1 Characteristics of CAT cables

CATEGORY (CAT)	CABLE TYPE	DATA RATE	TYPICAL APPLICATION
Category 1 (CAT 1)	Unshielded Twisted Pair (UTP) contains two wire pairs	20 kbps	Voice communications
Category 2 (CAT 2)	Unshielded Twisted Pair (UTP) contains four wire pairs	4 Mbps	Data communications
Category 3 (CAT 3)	Unshielded Twisted Pair (UTP) contains four wire pairs	10 Mbps (BW of 16 MHz)	• 10 Mbps Ethernet • 4 Mbps Token Ring
Category 4 (CAT 4)	Unshielded Twisted Pair (UTP) contains four wire pairs	16 Mbps	• 10 Mbps Ethernet • 16 Mbps Token Ring
Category 5 (CAT 5)	Unshielded Twisted Pair (UTP) contains four wire pairs	100 Mbps (BW of 100 MHz)	100 Mbps Ethernet • ATM (Asynchronous Transfer Mode) • FDDI (Fiber Data Distribution Interface)
Enhanced Category 5 (CAT 5e)	Enhanced version of CAT 5 wire contains high quality copper, and has high twist ratio.	BW of 200 MHz	

(continues)

TABLE 3-1 (continued)

Category 6 (CAT 6)	Shielded Twisted Pair (STP) contains four wire pairs	250 Mbps	Gigabit Ethernet
Category 7 (CAT 7)	Shielded Twisted Pair (STP) contains multiple wire pairs	1 GHz	IP Video applications

CATV

1. Community antenna television
2. Cable television

CBX

1. Centralized branch exchange
2. Computerized branch exchange

CCIR Comite Consultatif International de Radio Communication (International Radio Consultative Committee). One of the committees set up by the International Telecommunication Union (ITU) for radio communications. Former name for ITU-R. *Also see* Module 38.

CCITT Comite Consultatif International Telephonique et Telegraphique (The Consultative Committee on International Telephone and Telegraph). An advisory committee established under the United Nations within the ITU to make recommendations for international communications systems. Former name for the ITU-TS (International Telecommunication Union), a specialized agency of the United Nations. (*See* Module 38).

CCS Acronym for (1) continuous commercial service, which categorizes the safe operating parameters of an electronic component and communication equipment operation over a long, uninterrupted period, or (2) hundred calls per second, a unit of telephone traffic load. *Also see traffic formulas in* Module 33.

CCTV Closed-circuit television.

CD-ROM Compact Disk-Read Only Memory. A computer storage medium that can store 650+ MB of data. (*See* Table 3-2).

TABLE 3-2 Types of CD-ROM

CD-R	Compact Disk-Recordable
CD-ROM	Compact Disk-Read Only Memory
CD-RW	Compact Disk-Rewritable

cell A data unit consisting of a header and a fixed-length information field. Typical cell size is smaller than a frame.

cells

1. A memory device/circuit that holds one byte, one character, one word, or one record.
2. A cellular telephone service area with low-power equipment to switch, transmit, and receive calls to/from any mobile unit within the service area (Figure 3-3).

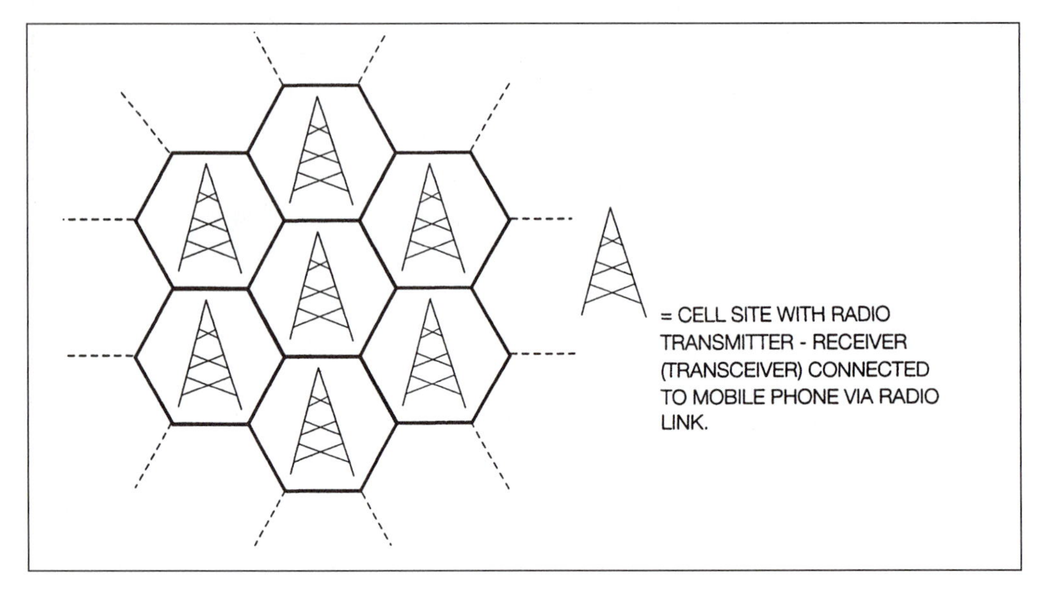

FIGURE 3-3 Cells

cellular radio/telephone Mobile telephone service with many service areas (called cells) that reuse transmission frequencies to increase the number of possible subscribers. The service area is divided into hexagonal cell sites that fit together to define a honeycomb. Each cell contains a radio transceiver and a controller that controls the transfer of calls from one call to another under the direction of a central switch (Figure 3-4). *Also see* Module 47.

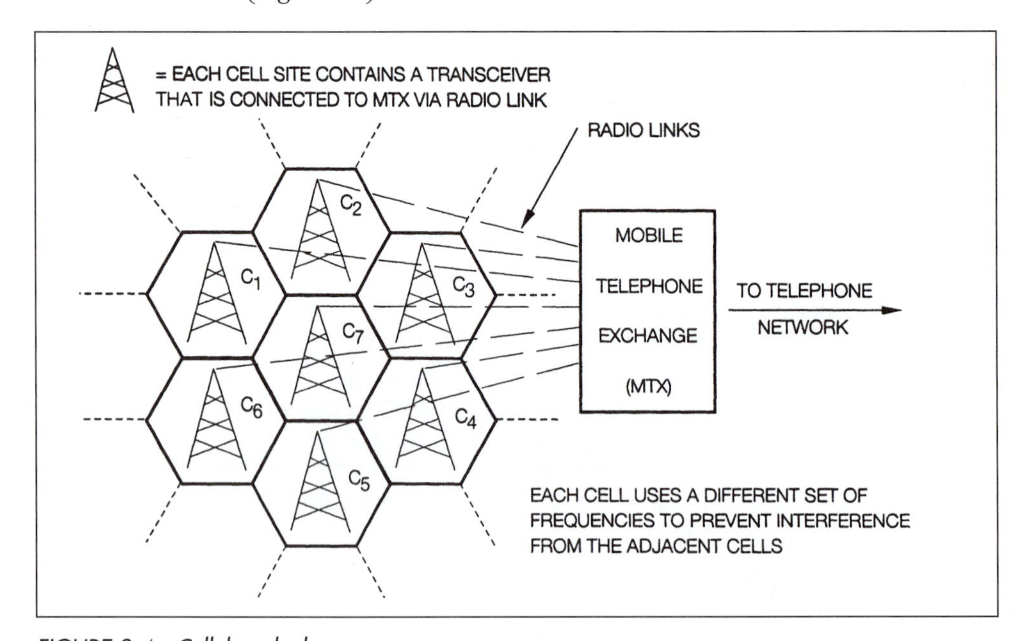

FIGURE 3-4 Cellular telephone

central office (CO) A local telephone company (communications common carrier) building that houses switching equipment that provides the local exchange telephone service for a geographical area. All customer lines are terminated at the central office (Figure 3-5). A central office provides service for up to ten thousand customers, and is designated by the first three digits of a telephone number. Also called exchange or end office (EO). *Also see* **Voice-over-IP**.

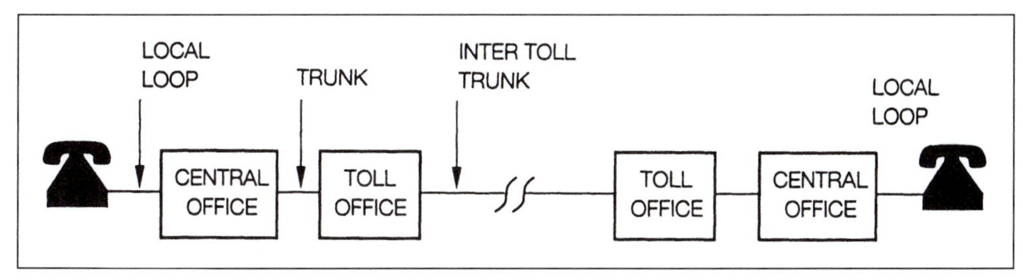

FIGURE 3-5 Central office

centrex A type of private branch exchange service in which an incoming call may be dialed direct to an extension without operator assistance. (Switching facilities at the central exchange allow interconnection of various lines or circuits.)

centum call seconds (CCS) A unit of traffic intensity measurement. One CCS refers to a 100-second call or to an aggregate of 100 call-seconds (100 [C] call seconds). Thirty-six CCS equal one erlang or 60 MOU (minutes of use). *Also see* **erlang**, **minutes of use (MOU),** *and traffic formulas in* Module 33.

CF Call forwarding.

channel A medium or path used for the transmission of signals between the transmitter and the receiver (Figure 3-6). Also called *circuit, link, line, path,* or *facility.*

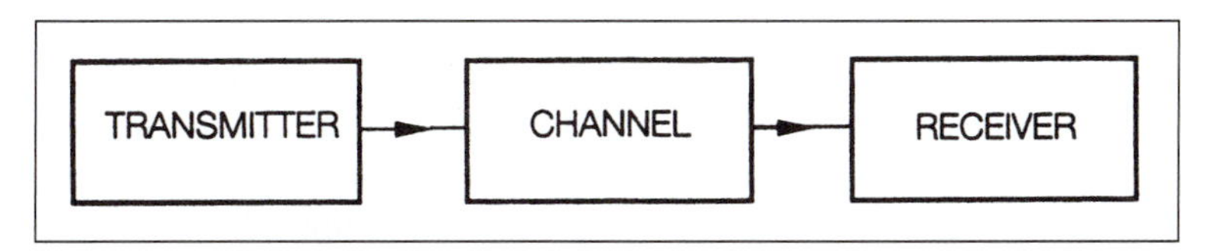

FIGURE 3-6 Channel

channel bank Communication equipment used in a telephone central office that performs the multiplexing of low-speed voice-grade channels into high-speed composite channels. It also transmits information for each channel for demultiplexing of the channel at the receiver.

channel capacity A measure of the maximum possible information rate at which data can be transmitted over a channel. Also called channel speed. Table 3-3 lists the formulas for predicting channel speed.

TABLE 3-3 Channel capacity or speed

Nyquist Theorem	In 1928, Nyquist proposed that a noiseless channel with bandwidth (W) can transmit 2W binary voltage values per second. $channel\ speed = 2W\ bps$ Nyquist's theorem assumes the ideal case (noiseless channel); it gives a theoretical maximum for the capacity of a channel. The presence of noise on a channel limits the transmission capacity below 2W bps.
Shannon's Law	In 1948, Shannon proposed that the maximum capacity of a line: $channel\ speed = W \log_2 [\ 1 + SNR\ ]$ where $W = Bandwidth\ of\ channel$ $SNR = Signal\text{-}to\text{-}Noise\ Ratio = \frac{signal\ power}{noise\ power}$ Shannon's law is one of the most important laws of telecommunications. It sets the maximum limit for the transmission capacity of a channel.

channel, voice-grade A channel that can transmit speech, analog data, or facsimile generally in the range of 300–3400 Hz.

character A letter, number, figure, control function, punctuation, or any special sign contained in a message.

character-oriented protocol A protocol or transmission technique in which control information is encoded in fields of one or more bytes.

circuit A communication path between two or more components in a communication system. Also called *channel.*

circuit switching A public switch telephone network capability that allows a signal path to be established and maintained until released. In case of the unavailability of a path, the circuit is said to be busy. It is contrasted with *message* and *packet switching.*

Class A A type of Internet Protocol (IP) address. *See* Module 37.

Class B A type of Internet Protocol (IP) address. *See* Module 37.

Class C A type of Internet Protocol (IP) address. *See* Module 37.

Class D A type of Internet Protocol (IP) address. *See* Module 37.

Class E A type of Internet Protocol (IP) address. *See* Module 37.

clicking The action of pressing and releasing the left button of a mouse in order to initiate a task/process represented on the computer screen.

client A device/workstation that requests services from a server on a network.

closed-circuit television (CCTV) A point-to-point television transmission, as opposed to a point-to-multiple-point television transmission.

CMOS (Complementary Metal-Oxide Semiconductor) A technology used to manufacture integrated circuits (ICs). CMOS chips require less electrical power compared to TTL (transistor-transistor logic) chips.

TABLE 3-4 Comparison of TTL and CMOS technologies

	ACCEPTABLE INPUT SIGNAL LEVELS	POWER CONSUMPTION	SPEED
TTL	High: 2–5 V Low: 0–0.8 V	High	High
CMOS	High: 11–15 V Low: 0–4 V	Low	Low

CO Central office.

coaxial cable A type of transmission line consisting of an outer conductor surrounding an inner conductor separated by a dielectric (Figure 3-7). It has a higher bandwidth than a two-wire open line or twisted pair cable. The characteristic impedance of the coaxial cable is given by:

$$Z_0 = \frac{60}{\sqrt{\epsilon_r}} \ln\left(\frac{D}{d}\right) \text{ ohms}$$

Where: D = diameter of the outer conductor
d = diameter of the inner conductor
ϵ_r = relative dielectric constant

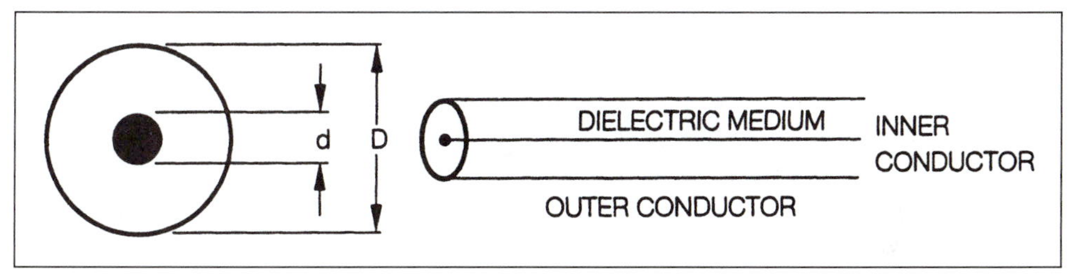

FIGURE 3-7 Coaxial cable

COBOL Common business oriented language. A computer language used for business applications.

code A scheme of symbols and rules to represent information.

code division multiple access (CDMA) A spread spectrum technology that uses frequency hopping and codes to increase spectrum efficiency. The frequency spectrum is divided into 1.25 MHz carrier; conversations are digitized and coded with unique markers, and subscribers are assigned codes—not time slots or frequencies (in contrast to TDMA and FDMA). *Also see* Module 47 *for CDMA, TDMA, and FDMA.*

CODEC A contraction of coder/decoder. A device that converts (using the coder section) an analog signal (e.g., voice) to a digital signal with the help of an analog-to-digital converter (ADC), and also converts (using the decoder section) a digital signal back to analog form using a digital-to-analog converter. At the sending end, a CODEC converts an analog signal into a digital bit stream (PCM), which is transmitted over the digital channel, and at the receiving end another CODEC converts the digital signal (PCM) back to analog form. *Also see* **pulse code modulation (PCM)**. Table 3-5 compares the function of a Codec with a modem and a UART.

TABLE 3-5 Comparison of CODEC and MODEM

DEVICE	FUNCTION	BLOCK DIAGRAM
MODEM	Converts digital signal to analog signal and vice versa	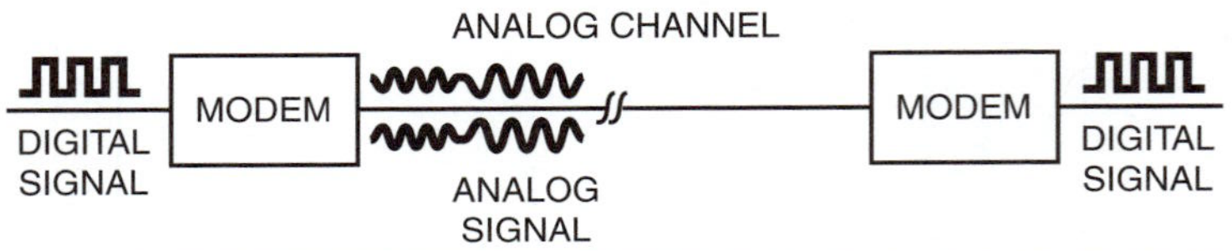 DIGITAL SIGNAL — MODEM — ANALOG SIGNAL — ANALOG CHANNEL — MODEM — DIGITAL SIGNAL
CODEC	Converts analog signal to digital signal and vice versa	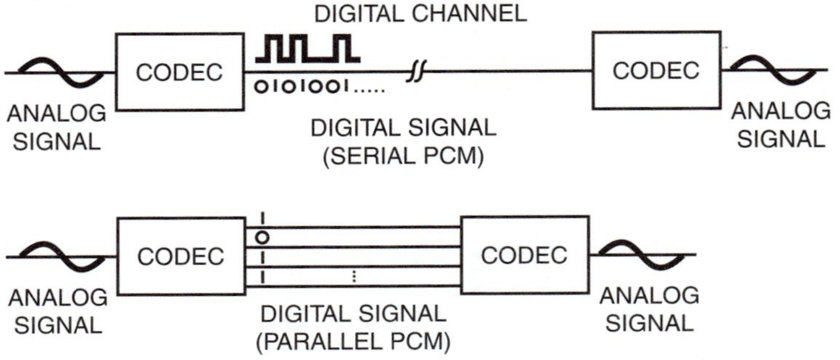 ANALOG SIGNAL — CODEC — 0101001..... DIGITAL SIGNAL (SERIAL PCM) — DIGITAL CHANNEL — CODEC — ANALOG SIGNAL ANALOG SIGNAL — CODEC — DIGITAL SIGNAL (PARALLEL PCM) — CODEC — ANALOG SIGNAL

coherent optical source An optical source that emits optical radiation in which there exists a stable relationship between various points of waves in space (spacial coherence or directionality) or time (temporal coherence or monochromaticity). In contrast to *incoherent optical sources*, a laser is an example of a coherent optical source.

common carrier A business organization that provides regulated telephone, telegraph, telex, and data communications to the public, and is regulated by an appropriate regulating agency.

companding (*compressing/expanding*) A technique used to compress a signal's dynamic range before its transmission and to expand it back to its original range at the receiver, to increase the signal-to-noise ratio (SNR). Companding can be described as a deliberate nonlinear amplitude modulation (AM) that strengthens weak signals and reduces strong signals for transmission. By employing companding in a telephone system, analog channels can maintain acceptable signal-to-distortion levels over a wide dynamic range of input signals. In North America, the μ-law companding algorithm is used, whereas in Europe A-law is employed. *Also see* **A-law, μ-law,** *and* Module 33 *for equations.*

concentrator A line-sharing device whose main function is similar to that of a multiplexer. Table 3-6 compares the characteristics of a concentrator with that of a multiplexer. *See also* **multiplexer**.

TABLE 3-6 Concentrator versus multiplexer

CONCENTRATOR	MULTIPLEXER
• Has processing capability	• Has no processing ability
• Has auxiliary storage	• Has no storage capability
• Multiple inputs/outputs	• Fixed number of inputs and one output line (multiplexer); one input line and fixed number of outputs (demultiplexer)
• Used singly	• Used in pairs (multiplexer/demultiplexer)

connectionless A technique in data communications that does not use a predetermined route, but rather allows different routes for data transmission between networking devices. ATM is an example of connectionless data communication.

connection-oriented A technique in data communications that uses a predetermined routing scheme. Frame relay and X.25 are examples of connection-oriented transmission technologies.

cookie A small text file sent by a website that resides on a user's computer in order to collect information about the user's surfing habits or Internet use. The user has the option of changing the browser settings to accept or not to accept a cookie.

cracker An individual who tries to compromise computer security by gaining unauthorized access to the system for malicious purposes.

CRC Cyclic redundancy check. A technique used for error detection. It can detect over 99 percent of multiple bit errors. Table 3-7 presents a summary of the CRC algorithm.

TABLE 3-7 Cyclic redundancy check

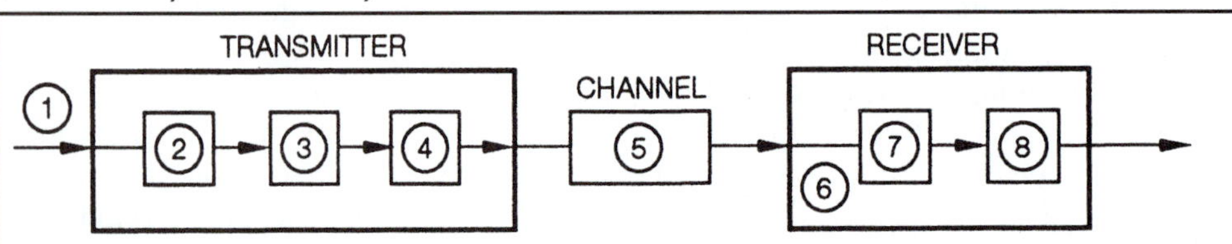

(1) INFORMATION BLOCK TO BE TRANSMITTED, IS REPRESENTED BY A POLYNOMIAL **I (x)** $= x^n + x^{n-1} + x^{n-2} + x^2 + x^1 + x^0$, WHERE x^n REPRESENTS MOST SIGNIFICANT BIT (MSB) AND x^0 REPRESENTS LEAST SIGNIFICANT BIT (LSB).

(2) INFORMATION POLYNOMIAL **I (x)** IS MULTIPLIED BY THE DEGREE(S) OF THE GENERATOR POLYNOMIAL **G (x)**, TO ADD S NUMBER OF CHECK BITS.

$$[\, I(x)\, x^S = \text{SHIFTED POLYNOMIAL} = S(x)\,]$$

(3) THE SHIFTED POLYNOMIAL **S (x)** IS DIVIDED BY THE GENERATOR POLYNOMIAL **G (x)**.

$$\left[\frac{S(x)}{G(x)} = \text{QUOTIENT (Q)} + \text{REMAINDER (R)} \right]$$

(4) THE REMAINDER (R) IS ADDED TO THE SHIFTED POLYNOMIAL **S (x)** TO OBTAIN THE INFORMATION BLOCK T (x) TO BE TRANSMITTED.

$$[\, S(x) + R = T(x)\,]$$

THE POLYNOMIAL **T (x)** IS NOW PERFECTLY DIVISIBLE BY THE GENERATOR POLYNOMIAL **G (x)**.

(5) THE TRANSMITTED INFORMATION BLOCK **T (x)** PROPAGATES THROUGH THE CHANNEL, WHICH MAY ADD NOISE TO THE INFORMATION BLOCK

(6) TRANSMITTED INFORMATION BLOCK IS RECEIVED AT THE RECEIVER AS **R (x)**

(7) THE RECEIVED MESSAGE POLYNOMAL **R (x)** IS DIVIDED BY THE GENERATOR POLYNOMIAL **G (x)**, TO DETERMINE WHETHER THE RECEIVED MESSAGE IS ERROR - FREE OR ERRONEOUS.

$$\frac{R(x)}{G(x)}$$

(8) IF $\frac{R(x)}{G(x)}$ = NO REMAINDER, THE RECEIVED MESSAGE IS ERROR - FREE

IF $\frac{R(x)}{G(x)}$ = A REMAINDER, THE RECEIVED MESSAGE BLOCK IS ERRONEOUS

THE RECEIVER ASKS THE TRANSMITTER FOR THE RETRANSMISSION OF THE PREVIOUS MESSAGE BLOCK.

CRT

1. Cathode ray tube. A television monitor or computer terminal.
2. A vacuum tube used as a display device for televisions, computer terminals, and oscilloscopes. A CRT consists of an electron gun, focusing and accelerating elements, horizontal and vertical deflection plates, and a phosphorous-coated screen (Figure 3-8).

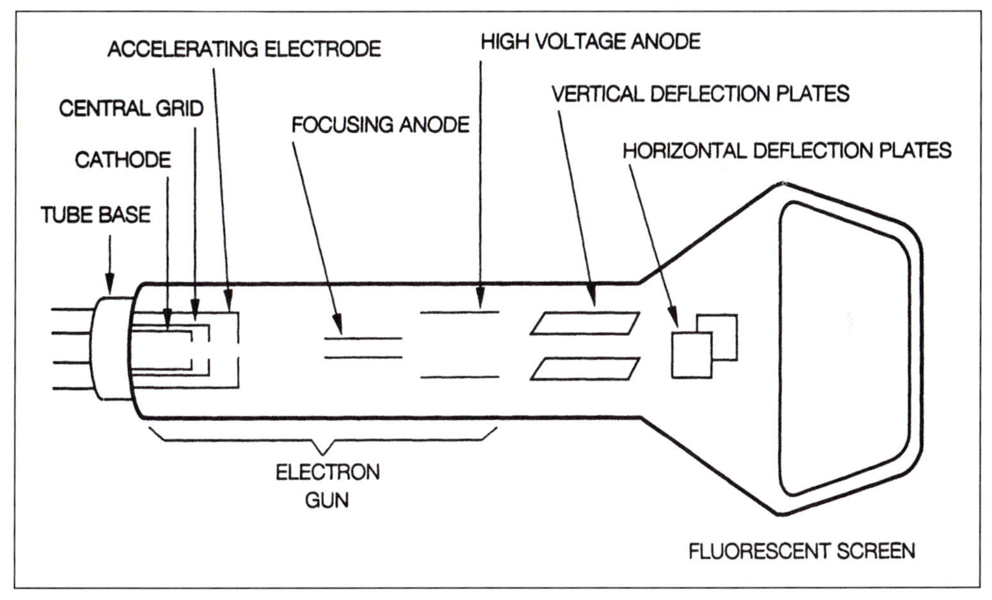

FIGURE 3-8 Cathode ray tube

cryptography The process of converting plain information/text by *encrypting* it at the transmitter and *decrypting* it back to its original form at the receiver with the help of a predetermined mathematical algorithm.

CW

1. Continuous wave
2. Call waiting

MODULE 4 Definitions D

D/A converter An electronic circuit that converts a digital signal to an analog signal (Figure 4-1).

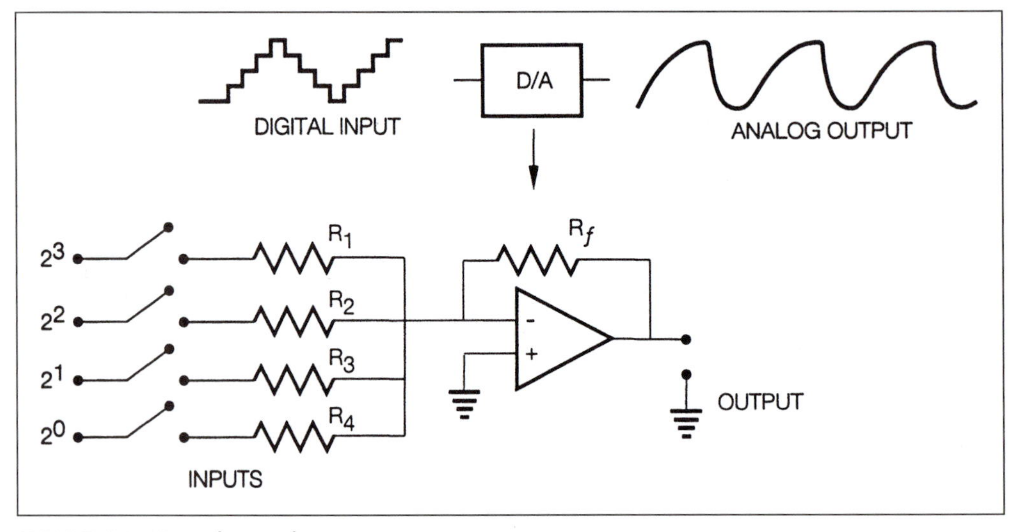

FIGURE 4-1 Digital-to-analog converter

DB-15 Data bus 15, a term used for a type of interface connector with 15 metal pins to connect devices.

DB-x/xx Data bus connector; x/xx indicates the number of pins in the connector.

D-channel The Delta channel in integrated digital services network (ISDN); it is packet-switched and handles out-of-band signaling for the B-channel. When not signaling for the B-channel, it can be used for low-speed data transmission. *Also see* Module 34.

D conditioning Type of line conditioning used to control harmonic distortion and SNR (signal-to-noise ratio) to keep the line characteristics within specific limits.

D layer The lowest layer of the ionosphere. Its intensity is greatest at noon, when maximum solar radiation is absorbed by the upper atmosphere. Like other ionospheric layers, the D layer is also composed of ionized gases. It surrounds the earth at elevations between 60 and 100 kilometers, and reflects electromagnetic signals having frequencies below 5 MHz. The D layer disappears at night (Figure 4-2).

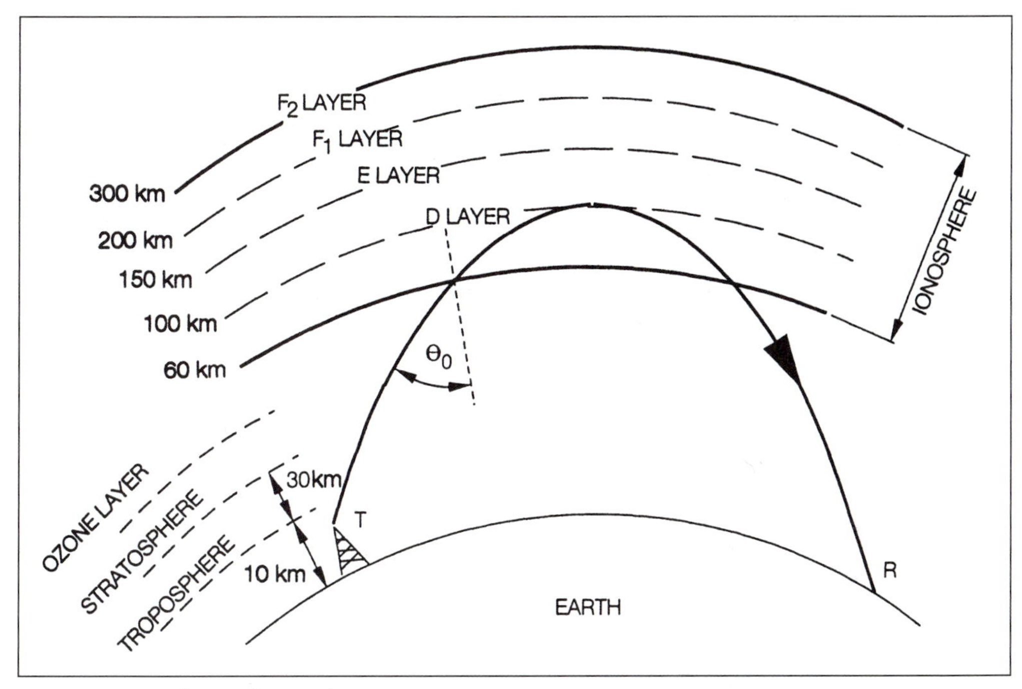

FIGURE 4-2 D-layer of ionosphere

damped wave A wave in which the amplitude of each successive cycle decreases with respect to its predecessor (Figure 4-3).

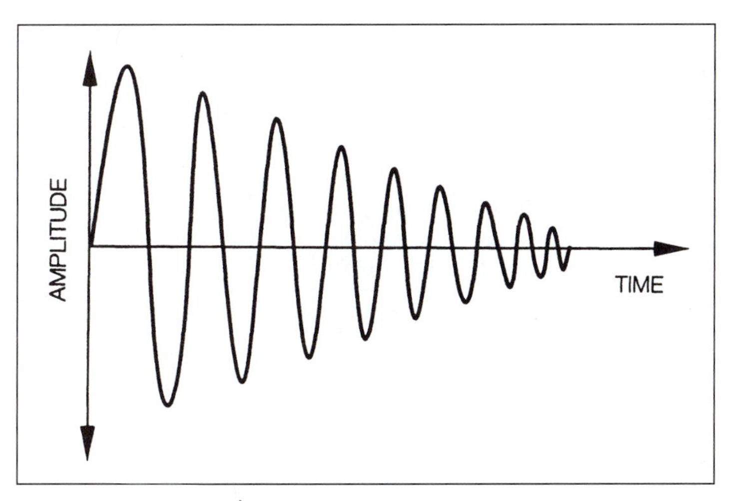

FIGURE 4-3 Damped wave

dark current The small leakage current that flows through a semiconductor photodetector without any incident radiation (Figure 4-4).

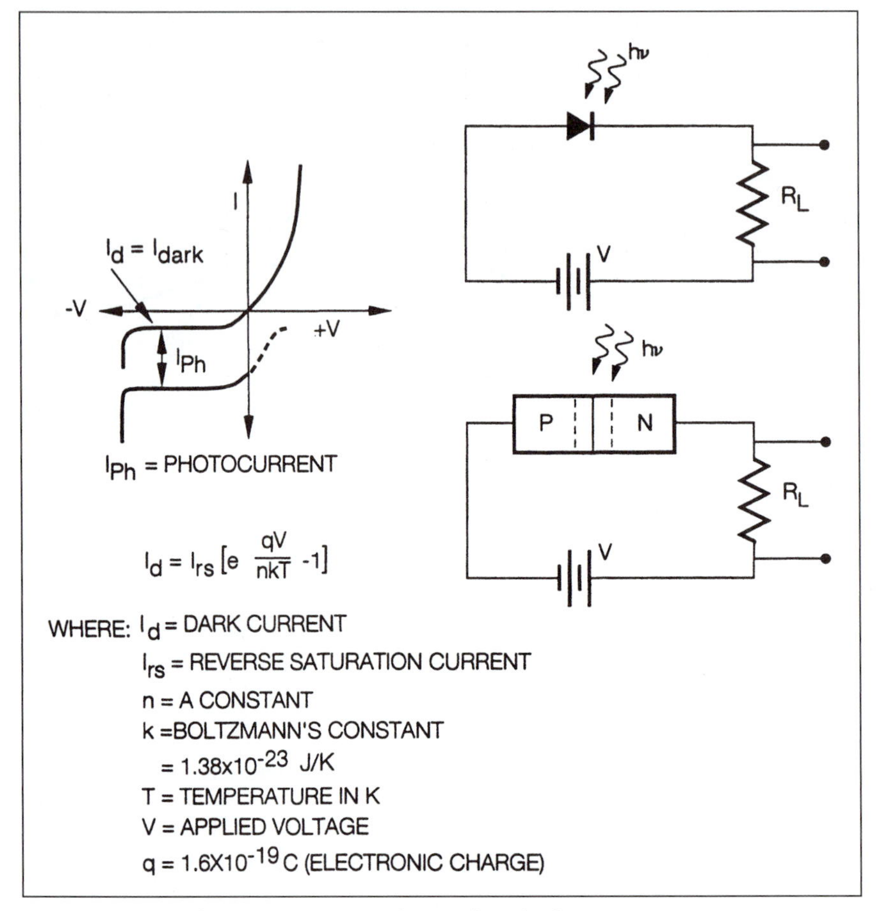

FIGURE 4-4 Dark current in semiconductor photodiode

data Plural of datum. Information in digital or analog form generated by an electronic system or computer. Usually, information generated by a computer is in digital form. Computers generate, store, transmit, and receive information in digital form. Figure 4-5 illustrates various forms of data.

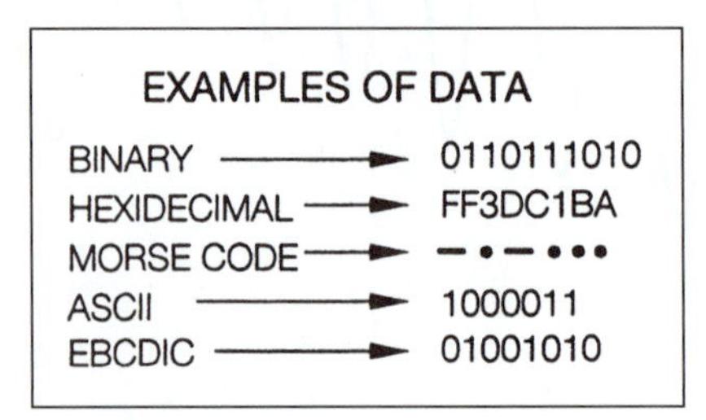

FIGURE 4-5 Data

data access arrangement (DAA) A device used to interface non-telephone-company-provided equipment to DDD (direct distance dial network). DAA provides protection against possible damage resulting from the use of nonstandard equipment.

data base A collection of integrated data files that are self-describing, i.e., they contain information about their structure.

data base management system (DBMS) A complex computer software that creates, maintains, and protects a large, organized, self-describing collection of data and permits various types of access for data retrieval and interrogation of the data base.

data collection The process of bringing data from one or more points to a central point.

data communication Transmission of encoded information from one point to another by means of electrical transmission systems.

data communication equipment (DCE) Communication equipment that provides an interface between data terminal equipment (DTE) and the transmission facility. It provides (1) functions required to establish, maintain, and terminate a connection, (2) the signal conversion, and (3) coding required for communication. Also called *data circuit-terminating equipment. See* **data terminal equipment** *and* Module 39.

data compression A technique in which data is compressed to increase the data transmission rate.

data link A communication facility and associated line termination equipment that permit the transmission of information in data format over the facility.

data link protocol A protocol that controls the manner in which communication devices interact at the data link level. All devices on the data link follow this protocol in controlling the transmission. *Also see* Module 35 *for OSI Model and examples of data link protocols.*

data rate *See* Module 33.

data set A device that performs the modulation/demodulation of a signal transmitted over a telephone line to provide data communication between two communication systems. Also called *modem.*

data terminal equipment (DTE) A part of a computer or data processing machine (comprising the following units: control logic, buffer store, I/O devices) that is capable of transmitting data over a communication facility. A DTE is attached to a DCE to transmit and receive data over a communication circuit. *See* **data communication equipment** *and* Module 39.

datagram A message consisting of a single packet transmitted over a packet-switched network.

DataPhone A service and trademark of AT&T. As a service, it identifies the transmission of data over a telephone network. As a trademark, it identifies the communication equipment supplied by AT&T for data communication.

dbm Decibel with reference to one milliwatt. It is the logarithmic measure of the ratio of the power of a signal to a reference power level of one milliwatt. *Also see* **decibel.**

$$\text{dbm} = 10 \log\left(\frac{\text{P}}{1\text{mW}}\right)$$

dbW The logarithmic measure of the ratio of the power of a signal to a reference power level of one watt. *Also see* **decibel.**

$$\text{dbw} = 10 \log\left(\frac{\text{P}}{1\text{W}}\right)$$

decibel (db or dB) A unit for expressing transmission loss or gain and relative power levels. It is the logarithmic measure of the ratio of output power to input power:

$$db = 10 \log\left(\frac{P_o}{P_i}\right)$$

Where: P_o = output power
P_i = input power

For voltage levels, db is given by:

$$db = 20 \log\left(\frac{V_0}{V_i}\right)$$

Where: V_o = output voltage
V_i = input voltage

decimetric waves An ultrahigh-frequency band of electromagnetic waves having wavelengths in the range of 0.1 to 1 meter.

decoder An electronic circuit that decodes encoded information from n-input lines to a maximum of 2^n unique output lines. Figure 4-6 illustrates a three-to-eight line decoder.

INPUTS			OUTPUTS							
X	Y	Z	D_0	D_1	D_2	D_3	D_4	D_5	D_6	D_7
0	0	0	1	0	0	0	0	0	0	0
0	0	1	0	1	0	0	0	0	0	0
0	1	0	0	0	1	0	0	0	0	0
0	1	1	0	0	0	1	0	0	0	0
1	0	0	0	0	0	0	1	0	0	0
1	0	1	0	0	0	0	0	1	0	0
1	1	0	0	0	0	0	0	0	1	0
1	1	1	0	0	0	0	0	0	0	1

FIGURE 4-6 Decoder

DECNET A peer-to-peer network technology, developed by Digital Equipment Corporation, that represents the implementation of Digital Network Architecture (DNA).

dedicated A communication channel/link or service leased from a carrier and continuously available.

de-emphasis The technique used in FM receivers to bring back the pre-emphasized signal to its original amplitude by the same amount during pre-emphasis. *See also* **pre-emphasis**.

default A setting, option, value, or choice that a device or a software program selects if the user does not make any selections.

delay distortion Distortion due to nonuniform speeds of various frequency components of a signal through a transmission channel.

delta modulation A digital modulation technique in which a train of binary pulses is transmitted and their polarity indicates whether the signal at the output of the demodulator should rise or fall at each pulse. A delta modulation system consists of an encoder and a decoder.

demodulation The process in which a low-frequency information signal (voice, data, video) is retrieved from a high-frequency modulated carrier signal. It is the opposite of *modulation*. Figure 4-7 illustrates the process of demodulation for an AM signal.

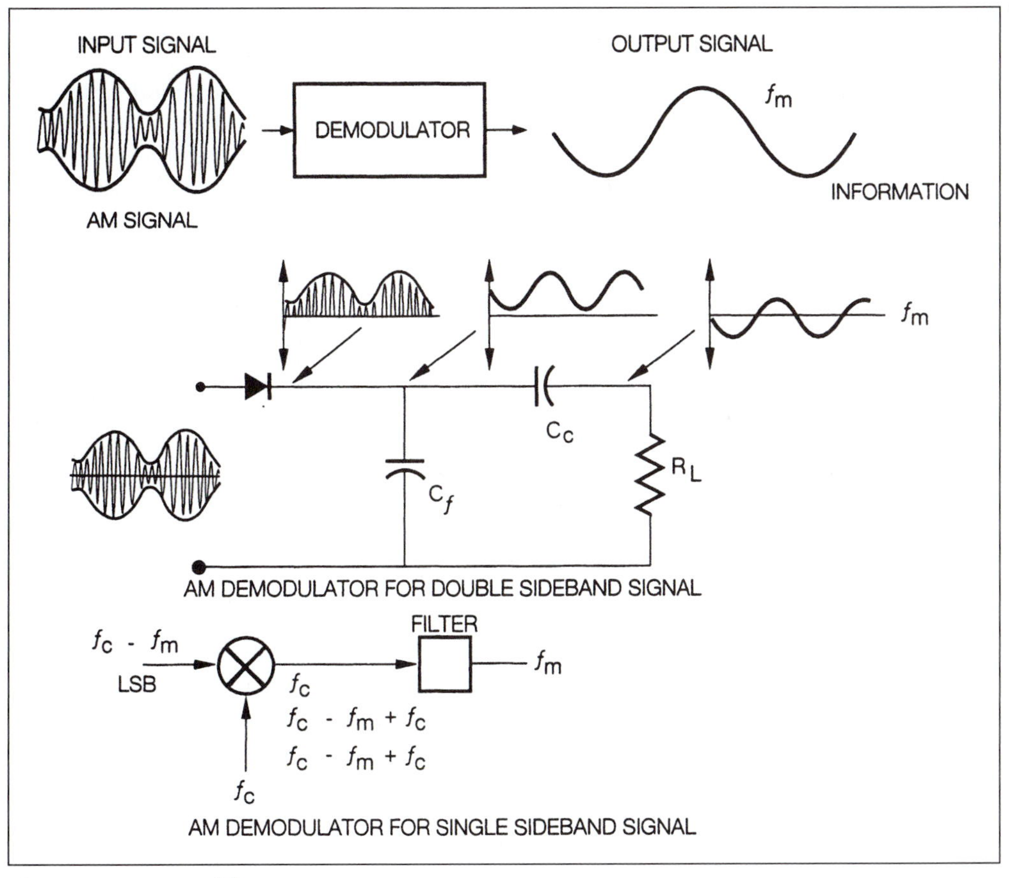

FIGURE 4-7 Demodulation

demultiplexer An electronic circuit that receives encoded information on a single line and transmits this information on one of 2^n possible output lines. The selection of a specific output line is controlled by the value of the selection lines. Figure 4-8 illustrates a 1×4 demultiplexer.

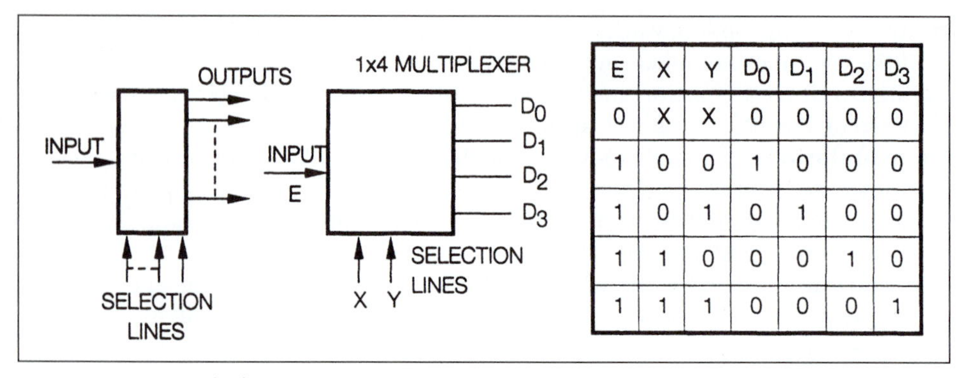

E	X	Y	D_0	D_1	D_2	D_3
0	X	X	0	0	0	0
1	0	0	1	0	0	0
1	0	1	0	1	0	0
1	1	0	0	0	1	0
1	1	1	0	0	0	1

FIGURE 4-8 Demultiplexer

denial of service (DoS) An attack on a system that makes it unavailable for users.

detector

1. A circuit used in radio and television receivers that retrieves (demodulates) audio or video signals from the modulated radio frequency signal. Also called *demodulator*. Figure 4-7 illustrates detectors used for AM reception.
2. A device that converts incident optical radiation into an electrical signal. *Also see* Module 27 *for optical detectors.*

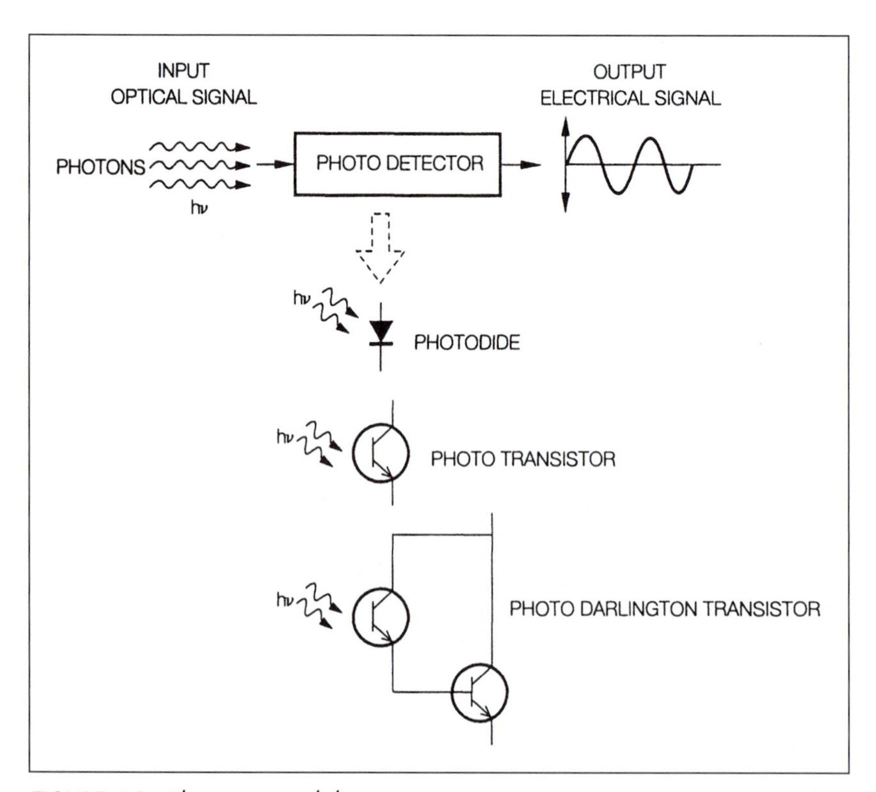

FIGURE 4-9 Photo or optical detectors

dial up line A communication link established by a switched circuit connection using the telephone network.

dialog box A small information window that appears on a computer screen and asks a user to make selections or provide information. It may also contain check boxes to solicit user selection or options regarding a specific task.

dibit A pair of bits used in four-phase modulation, such as differential phase shift keying (DPSK). The possible values of dibits are 00,01,10, and 11. Each dibit value represents a unique carrier phase shift.

Differential Manchester Coding *See* **digital signals**.

differential phase shift keying (DPSK) A modulation scheme in which the phase of the carrier is shifted in accordance with the changes in the binary signal. *Also see* **modulation**.

diffusion The process of scattering light by reflection or transmission.

digital certificate A digital certificate issued by the certificate authority that establishes the sender's identity. Also called *electronic certificate*, it is attached with the message so that the receiver knows the sender is who he/she/it claims to be.

digital data Information represented by a sequence of discrete elements. A single element of digital data is called a *bit*.

digital modulation A form of modulation using a digital pulse train is used as the carrier (also called *pulse modulation*). Its characteristics, such as amplitude, time, or position, are varied in accordance with changes in the analog information signal. It is used for the digital transmission of analog signals. Most commonly used pulse modulation techniques are pulse amplitude modulation (PAM), pulse width modulation (PWM) or pulse time modulation (PTM), and pulse code modulation (PCM). In PAM, the amplitude of the pulse carrier is varied in accordance with the changes in the successive samples of the information signal. In PWM or PTM, the time-dependent quantities (pulse width or duration, pulse position) of the pulse carrier are varied in accordance with the changes in the successive samples of the information signal. In PCM, an analog signal is first sampled at regular intervals and converted into a PAM signal, i.e., into a pulse train where amplitude of each pulse is proportional to the amplitude of the information signal at the sampling time. The PAM signal is then quantized and converted into a series of coded pulses whose bit combination reflects the quantized amplitude value of the PAM signal. PCM is widely used in telecommunication systems, especially in telephone networks. *Also see* **modulation** *and* Table 13-5.

Digital Network Architecture (DNA) The eight-layer network architecture model developed by Digital Equipment Corporation. It is similar to the OSI model at lower layers but different at the top three layers.

digital signal A signal whose amplitude variations with respect to time are not continuous but discrete. In contrast to an *analog signal*, which is continuous with respect to time. Figure 4-10 illustrates a digital signal.

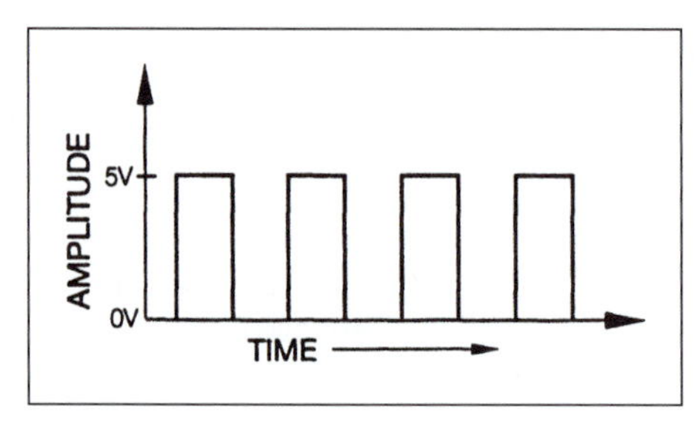

FIGURE 4-10 Digital signal

digital signals Figure 4-11 illustrates unipolar and bipolar digital signals. Table 4-1 lists various digital signal formats used in telecommunications. *Also see* **digital signal** *and* **analog signals**.

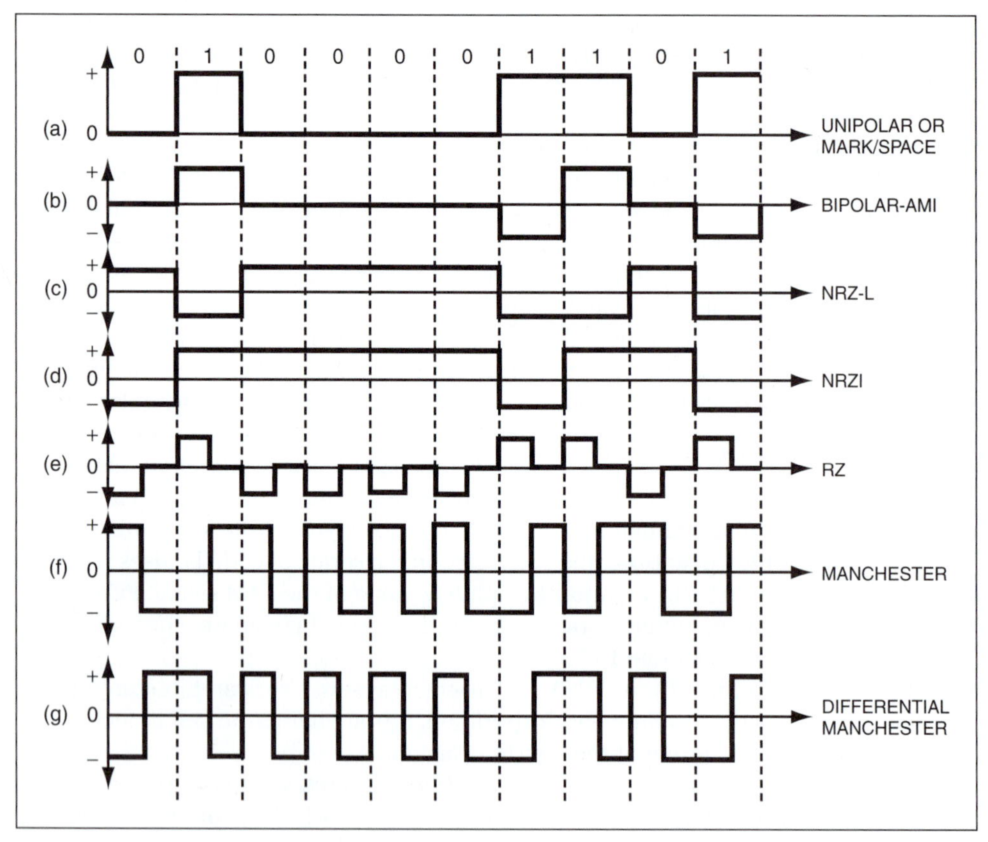

FIGURE 4-11 Digital signal formats

TABLE 4-1 Digital signal formats

TYPE	DESCRIPTION	EXAMPLE
Unipolar or Mark/space	High level (1) is represented by a flow of current value and low level (0) is represented by absence of current flow.	Figure 4-11(a)
Bipolar-AMI	A high level (1) is represented by a positive or a negative voltage value. The successive high levels alternate for polarity. The low level (0) is represented by zero voltage value.	Figure 4-11(b)
NRZ-L (non return-to-zero-level)	A high level (0) is represented by a negative voltage and a low level (0) is represented by a positive voltage.	Figure 4-11(c)
NRZI (non return-to-zero, Invert on ones)	A high level (1) is represented by a voltage transition (low-to-high or high-to-low) at the beginning of a bit time interval. A low level (0) is represented by no transition.	Figure 4-11(d)
RZ (return to zero)	Each encoded bit regardless of its value has a transition during the bit interval. A high level (1) is indicated by positive voltage for the first half of the bit interval and zero level for the second half. A low level (0) is represented by a negative voltage for the first half of the bit duration and zero level for the second half.	Figure 4-11(e)
Manchester	A high level (1) is represented by a transition from low to high in the middle of the bit interval. And a low level (0) is represented by a transition from high to low in the middle of the bit interval.	Figure 4-11(f)
Differential Manchester	There is always a transition in the middle of the bit duration. A high level (1) is represented by absence of a transition at the beginning of bit interval. And a low level (0) is represented by the presence of a transition at the beginning of a bit interval.	Figure 4-11(g)

digital subscriber line (DSL) A dedicated high-speed connection that uses advanced modulation techniques such as QAM to achieve high data rates using twisted pair wires. See Module 36.

DIMM (Dual Inline Memory Module) A small circuit board that contains memory chips in a computer. DIMMs can hold 16, 32, 64, 128, 256, and 512 megabytes (MB) or 1 gigabyte (GB) of random access memory (RAM) on a single module.

DIP (dual inline package) switch A switch that can be turned on or off in a device or circuit board to hold configuration or setup information.

direct broadcast satellite (DBS) A type of satellite that operates in the range of 12.2–12.7 GHz and provides television programming directly to the consumer.

direct distance dialing (DDD) A telephone company service that allows direct dialing for long-distance calls.

direct wave An electromagnetic wave that reaches the receiver directly from the transmitter through air/space without getting reflected by the earth or ionosphere (Figure 4-12).

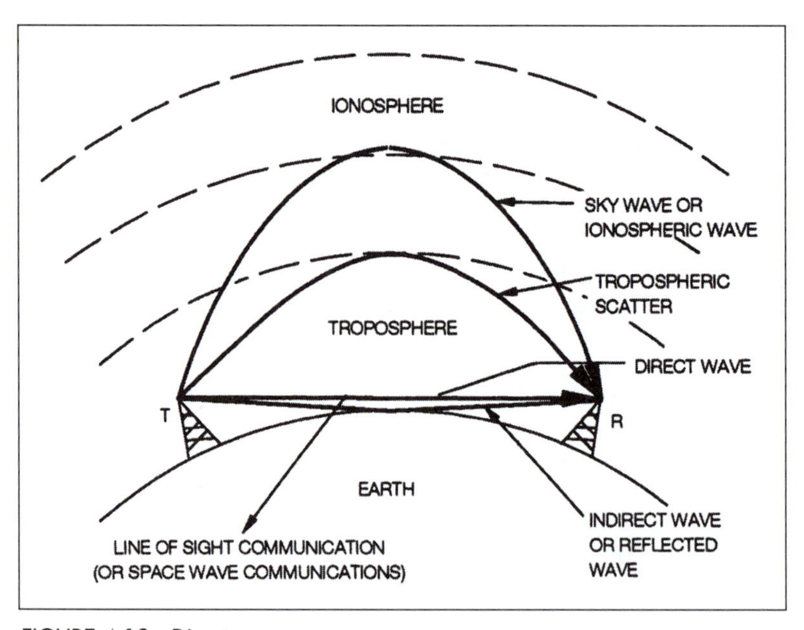

FIGURE 4-12 Direct wave

directional antenna A type of antenna whose radiation and reception characteristics are highly directional.

dish antenna A dish-shaped directional antenna used for microwave and satellite communications. Figure 4-13 shows a dish antenna and its characteristics. Also called a *parabolic reflector*.

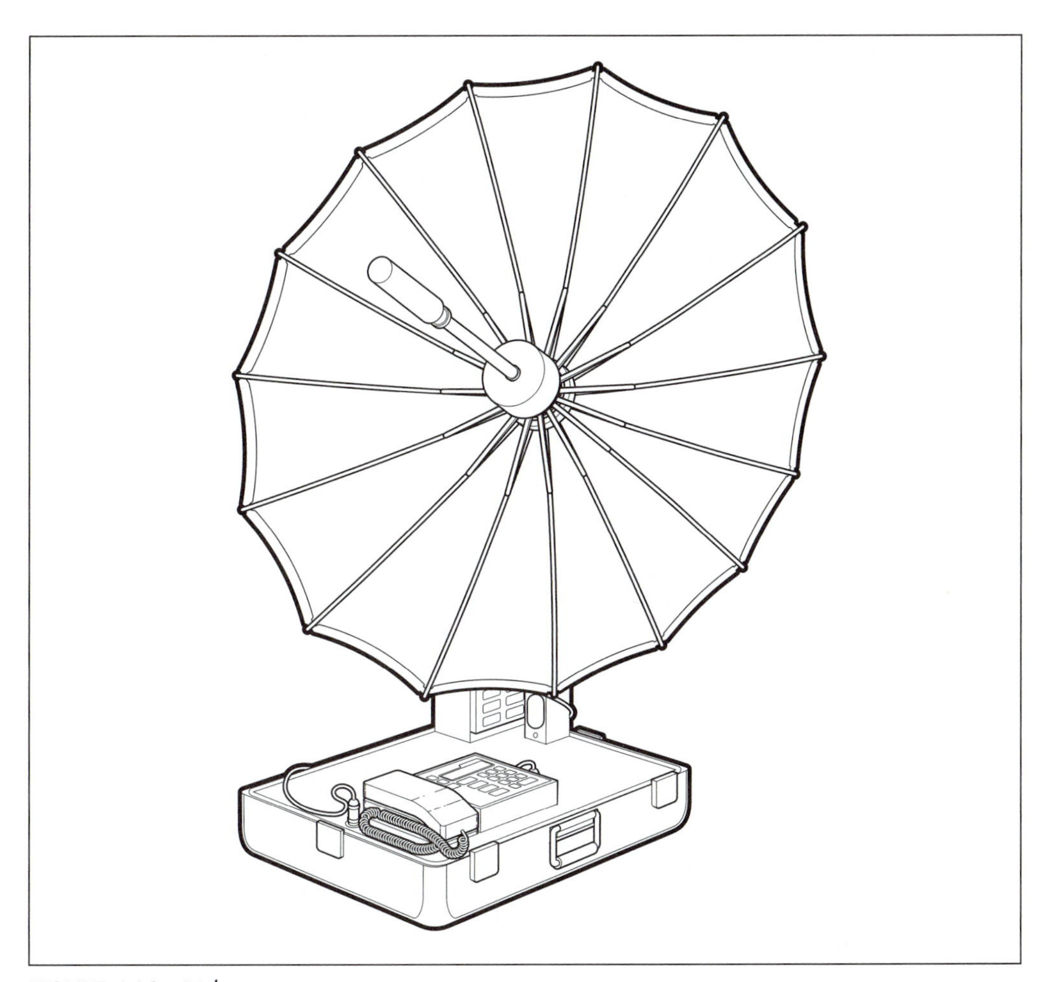

FIGURE 4-13 Dish antenna

disk duplexing A network server that has multiple drives connected to separate disk controller cards in a computer system.

disk mirroring A network server that has multiple drives connected to the same disk controller cards in a computer system.

dispersion In fiber optics communication systems, pulse spreading caused by various parameters such as number of modes, frequency-dependent velocity of modes, and waveguide dimensions. Table 4-2 compares various types of dispersion. Figure 4-14 illustrates dispersion.

TABLE 4-2 Dispersion

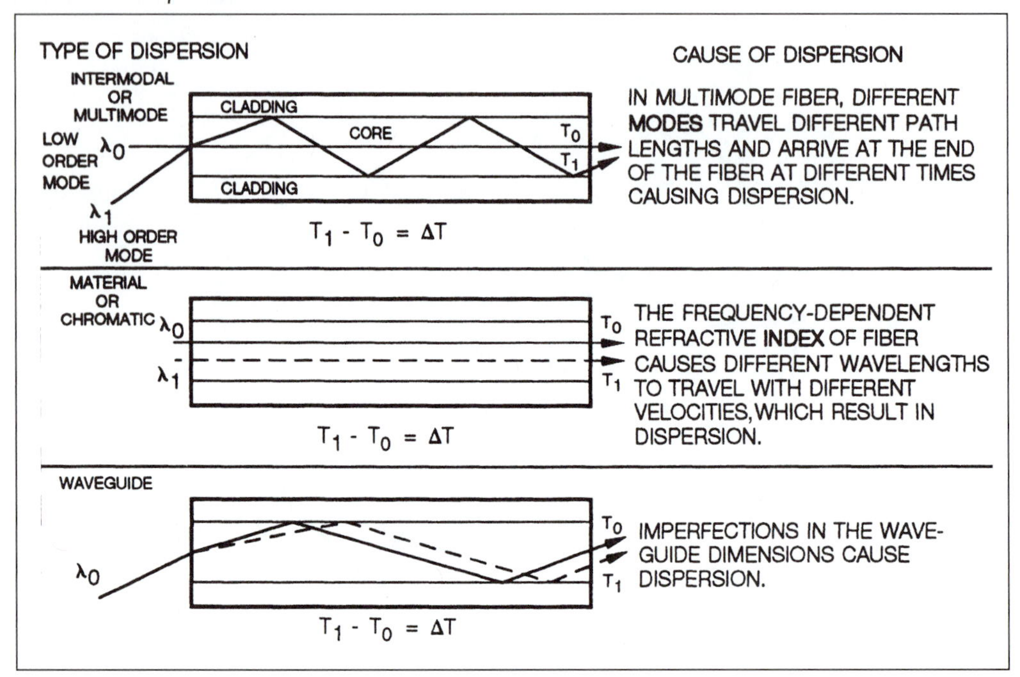

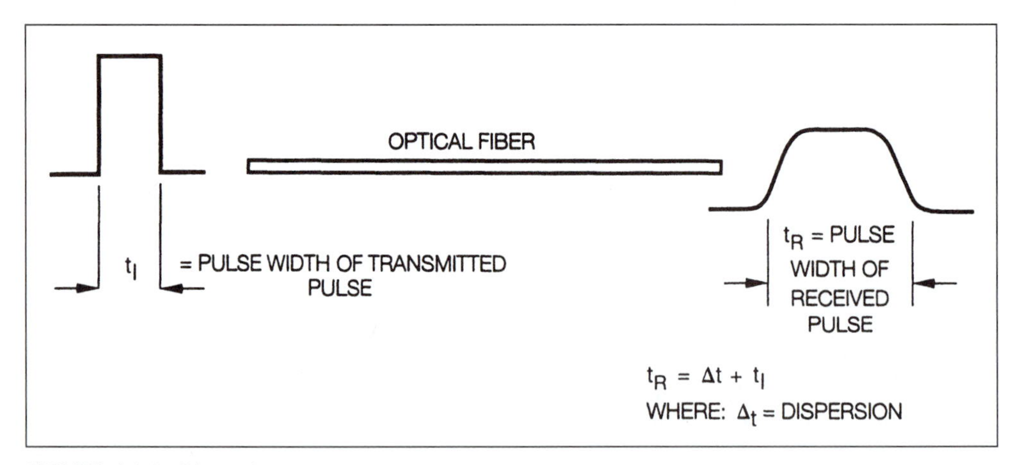

FIGURE 4-14 Dispersion

display Equipment or device that translates the output of an electronic circuit or system into readable form.

distance learning An instructional mode through which students can receive instruction (synchronously or asynchronously) via telecommunication technologies.

distortion The unwanted changes in wave shape that take place during the transmission of a signal and the reception of a signal between two points.

distributed data processing The movement of data processing functions from a central processing facility to separate locations equipped with independent data processing systems.

domain name An address (URL) that identifies a website on the Internet. For example, http://www.un.org is the domain name for the United Nations organization. Table 4-3 illustrates the breakdown of an Internet domain name/Uniform resource locator (URL). A top-level domain (TLD) identifies either the generic part of the domain name (in an Internet Address/uniform resource locator (URL) or the country code. Table 4-4 lists various generic top-level domains (gTLD), and Table 4-5 presents a summary of country codes for generic top level domains (ccTLD) for selected countries. The domain name system (DNS) converts a domain name to a 32-bit IP address. *See* **Domain Name System (DNS)** *and* Module 37.

TABLE 4-3 Breakdown of an Internet Domain Name/Uniform Resource Locator (URL)

Generic format of an Internet Domain Name or URL:
protocol://subdomain or third level domain.domain(also called second level domain).top-level domain (TLD=gTLD/ccTLD]
[subdomain, domain, and gTLD/ccTLD entities are separated by a dot (.)]

EXAMPLES

	Protocol	Third level domain name (host server)	Subdomain name	Second level domain name	Top level domain (TLD) name	
	http:// (hyper-text transport protocol)	Host server		Name of entity	gTLD (see Table 4-4)	ccTLD (see Table 4-5)
Example	http://	www		ups	com	au
	http://	www		harvard	edu	
	http://			dawn	com	
	http://	www		china	org	cn
	http://	www	dpg (campus name)	deVry	edu	
	http://'	www2	nature	nps	gov	
	http://	www		ftp-sites	org	
	ftp:// (file transfer protocol)	ftp 1	us	IreeBSD	org	

TABLE 4-4 Description of generic top-level domains (gTLD)

TOP LEVEL DOMAIN (TLD)	DESCRIPTION
com	Commercial entity
edu	Educational institution
gov	Government entity
int	International organization
mil	Military
net	Network
org	Organization

TABLE 4-5 Description of country code generic top-level domains (ccTLD) for selected countries

COUNTRY CODE	COUNTRY	COUNTRY CODE	COUNTRY
ae	United Arab Emirates (U.A.E)	bz	Belize
af	Afghanistan	ca	Canada
al	Albania	cf	Central African Republic
am	Armenia	co	Congo
ao	Angola	cm	Cameroon
ar	Argentina	cn	China
at	Austria	co	Colombia
au	Australia	cr	Costa Rica
az	Azerbaijan	cu	Cuba
ba	Bosnia and Herzegovina	cz	Czech Republic
bb	Barbados	de	Germany
bd	Bangladesh	dk	Denmark
be	Belgium	dj	Djibouti
bf	Burkina Faso	ec	Ecuador
bg	Bulgaria	ee	Estonia
bh	Bahrain	eg	Egypt
bi	Burundi	er	Eritrea
bj	Benin	es	Spain
bm	Bermuda	ec	Ecuador
bn	Brunei	fi	Finland
bo	Bolivia	fj	Fiji
br	Brazil	fr	France
bs	Bahamas	gh	Ghana
bt	Bhutan	gl	Greenland
bw	Botswana	gm	Gambia
by	Belarus	gr	Greece
		hk	Hong Kong

(continues)

TABLE 4-5 (continued)

COUNTRY CODE	COUNTRY	COUNTRY CODE	COUNTRY
hu	Hungary	pa	Panama
id	Indonesia	pk	Pakistan
ir	Ireland	pl	Poland
il	Israel	qa	Qatar
in	India	ru	Russia
iq	Iraq	sa	Saudi Arabia
ir	Iran	sd	Sudan
it	Italy	se	Sweden
jm	Jamaica	sn	Senegal
jo	Jordan	so	Somalia
ke	Kenya	sv	Salvador
kg	Kyrgyzstan	sy	Syria
kr	Korea (south)	th	Thailand
kp	Korea (north)	tj	Tajikistan
la	Laos	tm	Turkmenistan
lb	Lebanon	tr	Turkey
lt	Lithuania	tw	Taiwan
lu	Luxembourg	tz	Tanzania
lk	Sri Lanka	ua	Ukraine
ma	Morocco	ug	Uganda
mg	Madagascar	uk	United Kingdom
ml	Mali		
mm	Myanmar	us	United States
mx	Mexico	uy	Uruguay
my	Malaysia	uz	Uzbekistan
na	Namibia	va	Vatican
ne	Niger	ve	Venezuela
ng	Nigeria	vn	Viet Nam
no	Norway	ye	Yemen
np	Nepal	zr	Zaire
nz	New Zealand	zm	Zimbabwe
om	Oman		

domain name system or domain name service (DNS) The software on a computer that provides a name service for networked-connected computers. It translates domain names to or from an Internet Protocol (IP) address.

down converter That part of a satellite communication system that converts incoming high-frequency signals to lower frequency signals.

down link That part of a satellite link in which signals transmitted from the satellite are received by earth stations (Figure 4-15).

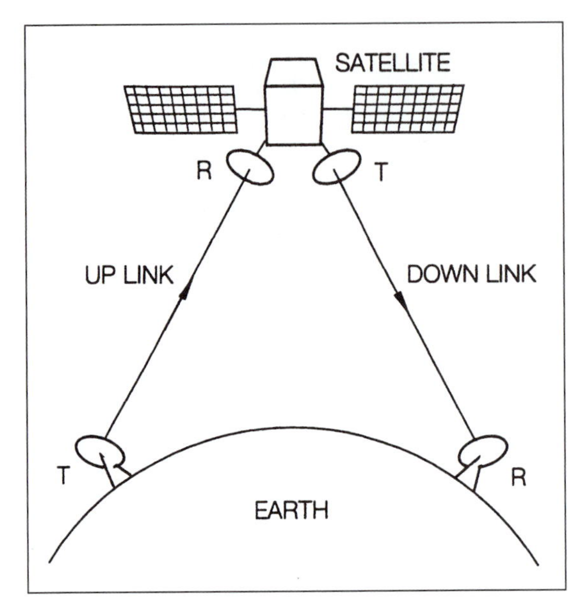

FIGURE 4-15 Satellite up and down links

downstream The data traffic that flows from a carrier's point-of-presence (POP) to a customer's equipment/premises. In symmetric networks, downstream and upstream data speeds are equal, whereas in asymmetric networks downstream data speed is higher than the upstream.

down time The duration of time during which a device, component, or system is inoperable.

double heterojunction diode A laser having two heterojunctions that allow full carrier confinement and improved control of the recombination of charge carriers. It is used in fiber optics communication systems.

drop A communication line that connects a user to a communication network. Also called *drop line.*

DS The classification of digital signal speeds of lines and trunks. Table 4-6 presents a summary of DS hierarchy. *See also* **T-1** *and* **T-carrier**.

TABLE 4-6 DS hierarchy

TYPE OF DIGITAL SERVICE	CHANNEL CAPACITY	TRANSMISSION FACILITY
DS-0: Digital Service level 0	64 kbps	
DS-1: Digital Service level 1	1.544 Mbps in North America, 2.048 Mbps in Europe	T-1
DS-1C: Digital Service level 1C	3.152 Mbps in North America	T1C, T1D
DS-2: Digital Service level 2	6.312 Mbps	T-2

(continues)

TABLE 4-6 (continued)

DS-3: Digital Service level 3	44.736 Mbps	T-3
DS-4: Digital Service level 4	274.176 Mbps	T-4

DS-0 Digital signal level 0. In a T-carrier system, it refers to one data or voice channel (64 kbps). *Also see* **T-carrier**.

DSBSC Double side band suppressed carrier. A type of amplitude modulated signal in which the carrier is suppressed and the transmitted signal consists of upper and lower side bands.

DSL access multiplexer (DSLAM) A multiplexer, located at the carrier's premises, which combines a number of DSL channels for transmission on a carrier link or high-speed backbone. *Also see* Module 36 *for Digital subscriber line (DSL).*

DSL modem Digital subscriber line *mo*dulator/*dem*odulator. A device that modulates an outgoing DSL signal and demodulates an incoming DSL signal, and thus provides connectivity to data equipment such as computers and telephones. *Also see* Module 36 *for Digital subscriber line (DSL).*

dual tone multifrequency (DTMF) In a telephone system, a dialing scheme that uses two simultaneous voice band tones for dialing. Also called *touch tone.* Figure 4-16 illustrates DTMF.

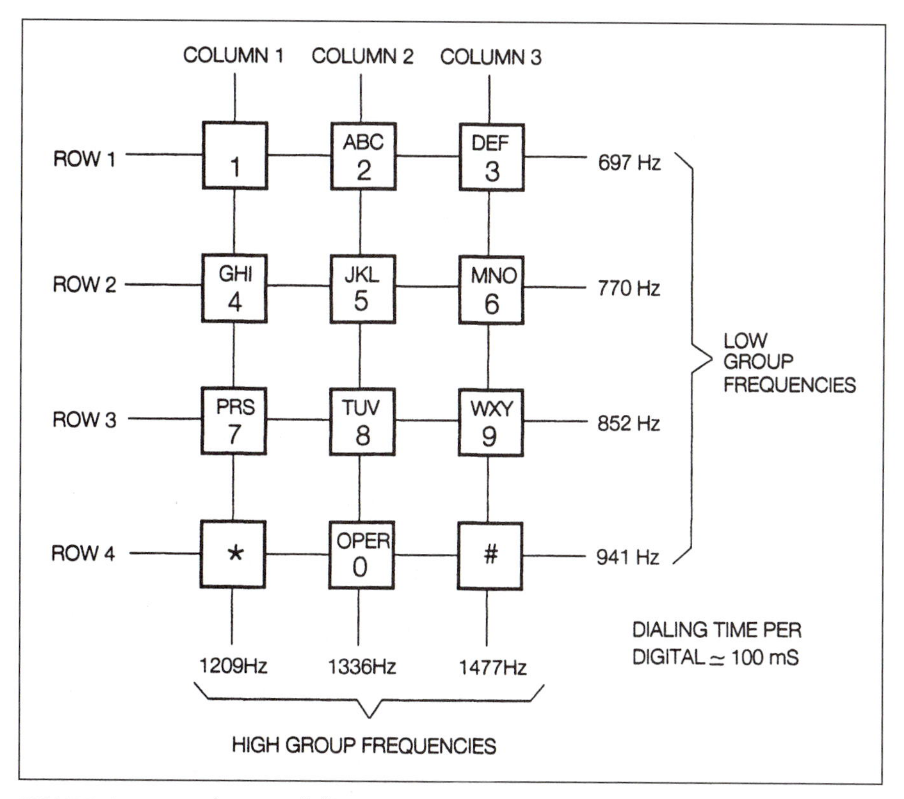

FIGURE 4-16 Dual tone multifrequency

dumb terminal A data terminal that has no processing capabilities and is used only for data input and output. Contrast to *intelligent terminal,* which has processing capabilities.

duplex Simultaneous two-way independent transmission in both directions (Figure 4-17). Also called *full duplex (FDX). See also* **simplex** *and* **half duplex**.

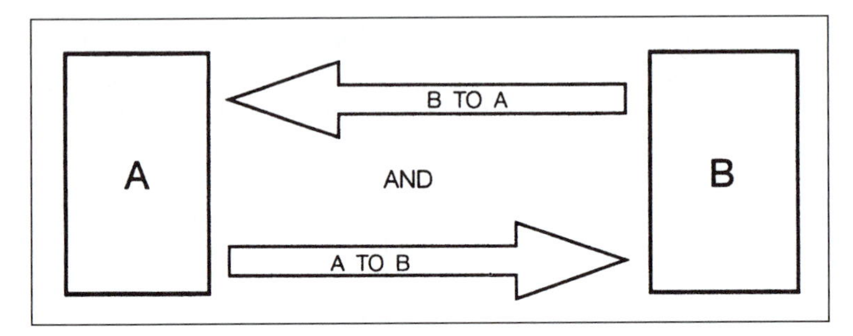

FIGURE 4-17 Full-duplex transmission

duty cycle The ratio of on time to time period (on time + off time) of an electrical signal (Figure 4-18).

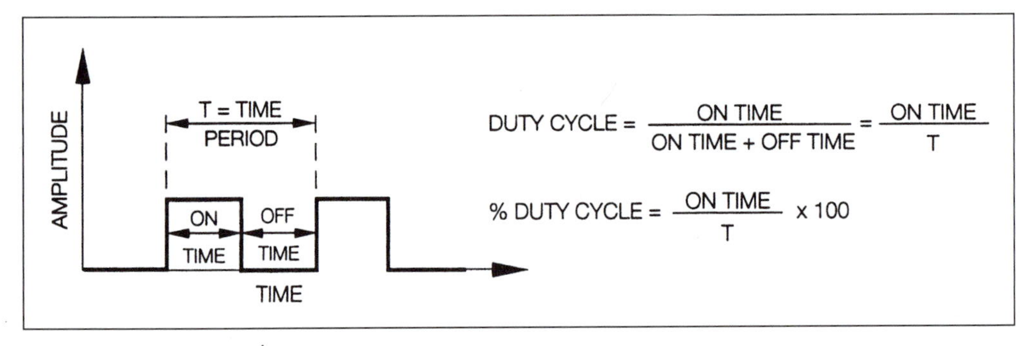

FIGURE 4-18 Duty cycle

DVD Digital versatile disk or digital video disk. A computer storage medium, the successor to the CD-ROM that can store 4.7–17 GB of data.

DWDM Dense wavelength division multiplexing. A type of wavelength division multiplexing technique that uses multiple wavelengths (colors) to transmit signals in a single fiber. The use of DWDM increases the capacity of the optical fiber. *Also see* **wavelength division multiplexing (WDM)**.

Dynamic Host Configuration Protocol (DHCP) A protocol that automatically provides computers with unique IP addresses.

MODULE

5 Definitions E

E-1, E-2, E-3, E-4, E-5 *See* **E-Line**.

E-Line European version of T-carrier system.

TABLE 5-1 Classification of E-Lines

E-LINE	DATA RATE (Mbps)	NUMBER OF VOICE CHANNELS
E-1	2.048	30
E-2	8.448	120
E-3	34.368	480
E-4	139.264	1920
E-5	565.148	7680

e-mail The most common application of the Internet. Allows users to send and receive messages in text or graphic form.

earth station The part of a satellite system that is located on the earth's surface. An earth station can transmit signals to or receive signals from a satellite in space (Figure 5-1). *See also* Module 5-1.

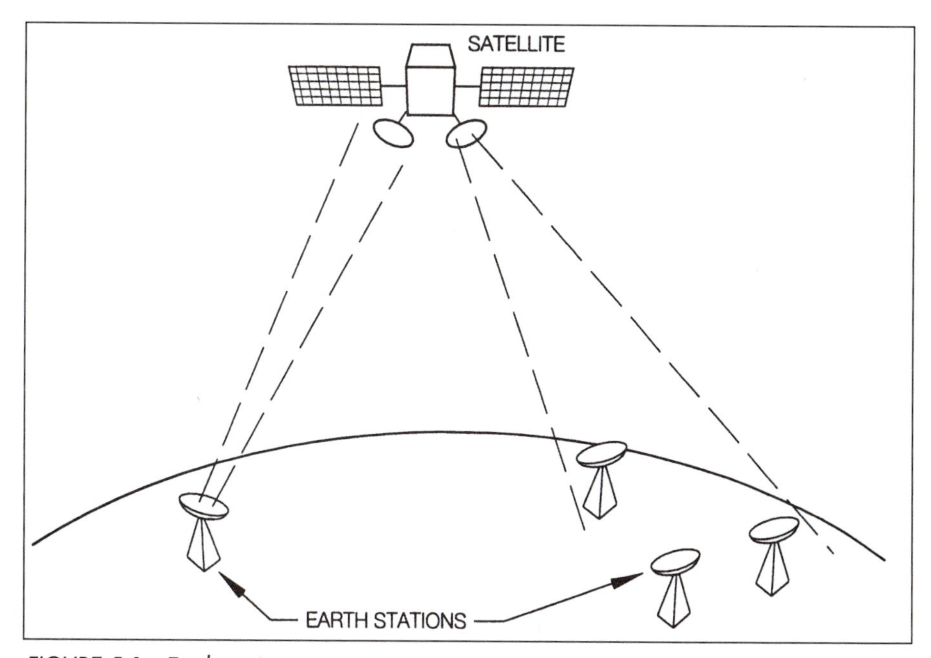

FIGURE 5-1 Earth station

EBCDIC An acronym used for extended binary coded decimal interchange code. It is an 8-bit code used for the representation of alphanumeric characters in data transmission. *See* Module 40.

Examples:

		b_7	b_6	b_5	b_4	b_3	b_2	b_1	b_0
A	=	1	1	0	0	1	0	0	0
M	=	1	1	0	1	0	0	1	0
1	=	1	1	1	1	1	0	0	0

echo

1. A part of a radiated electromagnetic signal that gets reflected with sufficient magnitude and delay by an object at a point to the source.
2. The reflection of a transmitted signal on a transmission line due to an impedance mismatch. Figure 5-2 illustrates echo on a transmission line.

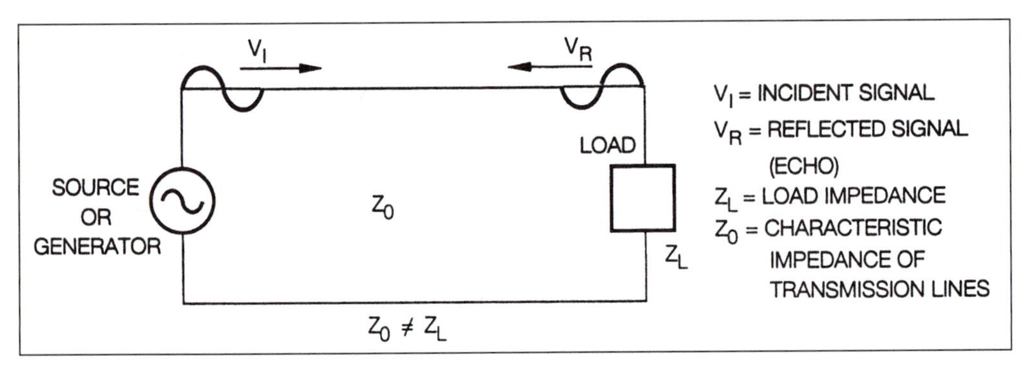

FIGURE 5-2 Echo

echo check An error checking method used in a data communication system in which the receiver returns the data received to the transmitter for the purpose of comparison.

echo suppressor A device primarily used in voice communication circuits to attenuate the reflection of the transmitted signal in order to provide an echo-free transmission.

echoplex An error checking method that uses information feedback for error control in data communication networks.

edit The process of revising the contents of a data file.

effective isotropic radiated power (EIRP) The product of the power supplied to an antenna and its gain in a given direction. *See also* Module 33.

EIA The Electronic Industry Association, a trade association of electronics manufacturers of North America that prepares comprehensive standards for the industry. (*See* Module 39.)

EIA RS-232-C (TIA/EIA-232-C) The Electronics Industry Association's physical level standard that defines a 25-pin interface that can be used to connect two communication stations. This interface allows serial transmission speeds up to 20 kbps at a maximum distance of 50 feet. (*See* Module 39.)

electromagnetic interference (EMI) Random or periodic disturbance, caused in a signal during its transmission or reception by electromagnetic waves (external noise sources) whose sources of origin could be man-made (ignition systems, microwave ovens, switching of high-current loads, etc.) or nonman-made (atmospheric noise, solar noise, cosmic noise, etc.).

electromagnetic spectrum The continuum of electromagnetic signals from longest wavelengths (lowest frequencies) to shortest wavelengths (highest frequencies). Figure 5-3 illustrates the electromagnetic spectrum and lists various applications of different frequency bands. *See also* Module 29.

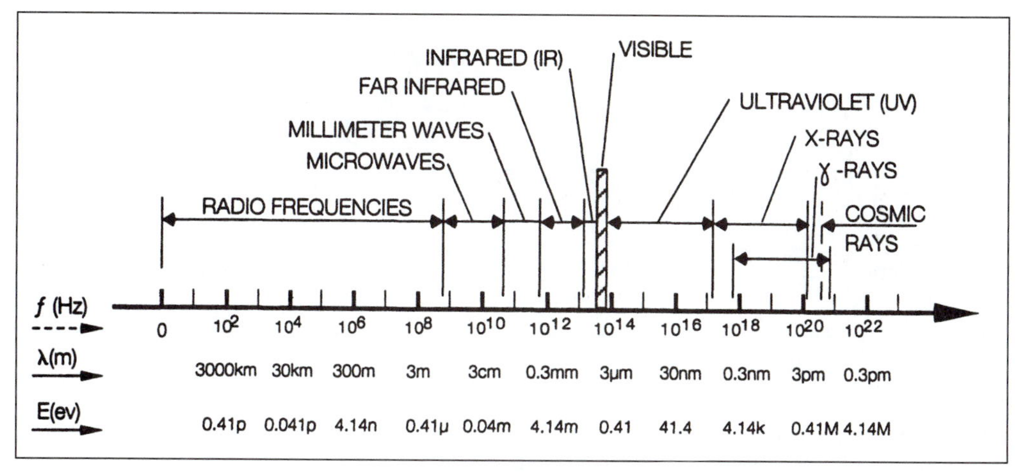

FIGURE 5-3 Electromagnetic spectrum

electromagnetic wave Electric (E) and magnetic (H) fields propagated by regular variations in these fields. An electromagnetic wave travels through air and free space at a speed of 3×10^8 meters per second or 186,000 miles per second. The speed of an electromagnetic wave in any medium can be determined by:

$$v = \frac{1}{\sqrt{\epsilon\mu}} \text{ m/s}$$

Where: ϵ = permittivity of the medium
μ = permeability of the medium

or

$$v = \frac{c}{\sqrt{\epsilon_r}}$$

Where: c = speed of electromagnetic signal in free space/air = 3×10^8 m/s
ϵ_r = relative dielectric constant of the medium

or

$$v = \frac{c}{n}$$

Where: n = refractive index of the medium

Figure 5-4 illustrates an electromagnetic wave. *See also* Module 29.

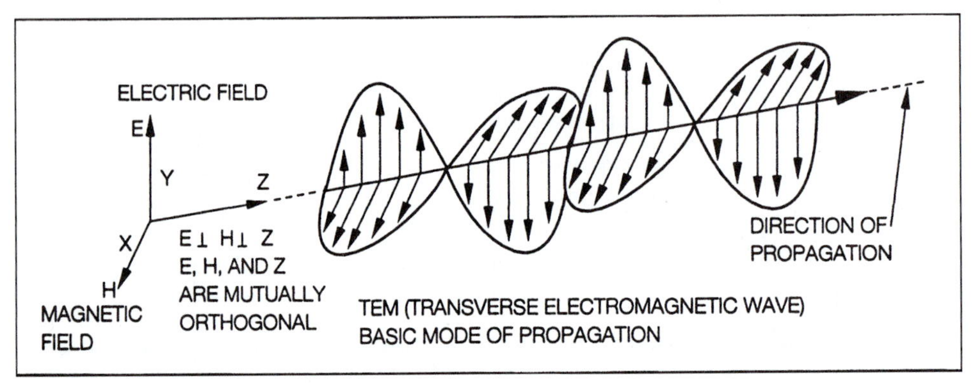

FIGURE 5-4 Electromagnetic wave

electronic switching Circuit switching equipment in which connects are made electronically rather than electromechanically.

end of text (ETX) A control character used in data communication that indicates the end of a message.

end of transmission (EOT) A control character used in data communication that indicates the end of a transmission.

end of transmission block (ETB) A control character used in data communications that indicates the end of the transmission of a block of data.

encryption The process of encoding information in which plain text is converted to coded information before transmission for security purposes.

encryption equipment Devices that allow data to be coded at the transmitting end and decoded at the receiving end.

equalization Compensation for attenuated increase, with frequency used to produce a flat frequency response.

equalizer A combination of reactive and resistive components inserted in a transmission line or amplifier circuit to improve frequency response.

equatorial orbit The orbit of a geosynchronous or geostationary satellite located above the equator.

ergonomics The study of how people adjust to their work environment.

erlang A unit of traffic intensity measurement. One erlang refers to one fully occupied traffic path or multiple paths carrying a combined load of one call-hour per hour. One erlang equals 36 CCS (centum call seconds) or 60 MOU (minutes of use). *Also see traffic formulas in* Module 33.

error A discrepancy between transmitted and received information.

error control An arrangement that detects the presence of errors in data. Control information is appended to the data to detect errors either by performing operations on the received data or by retransmitting information from the source.

ESS Electronic switching system. A computer-controlled system used by a public switched telephone network (PSTN) to process and route telephone calls. Table 5-2 lists the evolution of ESS switches at Bell Labs.

TABLE 5-2 Evolution of ESS

1 ESS (1965)	– Capability to handle 10,000–70,000 subscribers
	– Used stored program control (SPC) to perform switching
	– Used diode-transistor logic (DTL)
2 ESS	– Capacity to handle 1,000–10,000 lines
	– Used resistor-transistor logic (RTL)
3 ESS (1970s)	– Capacity to handle 100–1,000 lines
	– Used large scale integrated (LSI) technology ROM
4 ESS (1976)	– Capacity to handle 10,000 + trunks
	– Used time division multiplexing (TDM) and space division
	multiplexing (SDM) techniques
5 ESS	– Capacity to handle 1,000–100,000 subscribers
	– Uses software based on C-Language and UNIX operating system
	– Offers custom calling features such as call forwarding, call waiting, speed call, three-way calling, caller ID, etc.

Ethernet A local area network architecture standard developed by Xerox Corporation. It is characterized by a 10 Mbps baseband transmission. IEEE 802.3 protocol refers to Ethernet.

Euronet European packet distribution network.

European Telecommunications Standards Institute (ETSI) The regulatory body that establishes standards in Europe. [http://www.etsi.org]

exchange An area serviced by a communications common carrier at the exchange rate and under regulations applicable as prescribed in the carrier's tariffs. The service area has one or more central offices that house the equipment used for providing communication service. Also called *central office (CO).*

exchange, private automatic (PAX) A dial telephone exchange that provides private telephone service to a company and does not permit calls to be transmitted to or from the public telephone network.

exchange, private automatic branch (PABX) A private automatic telephone switching system connected to the public telephone network.

exchange, private branch (PBX) A private automatic or manual switching system connected to the public telephone system.

exchange termination (ET) In integrated services digital network (ISDN), the ISDN loop termination interface to the carrier's local exchange switch. *See also* Module 34.

exchange, trunk An exchange devoted to interconnecting trunks.

expansion slots In computers, internal slots for adding additional PC boards (hardware).

MODULE

6 Definitions F

facility A transmission link between two locations.

facsimile (FAX) An image transmitted via a telephone network or by radio. The image is scanned and converted into codes that are transmitted on the line then received and reconstructed into an exact duplicate on paper. Figure 6-1 illustrates the concept of FAX.

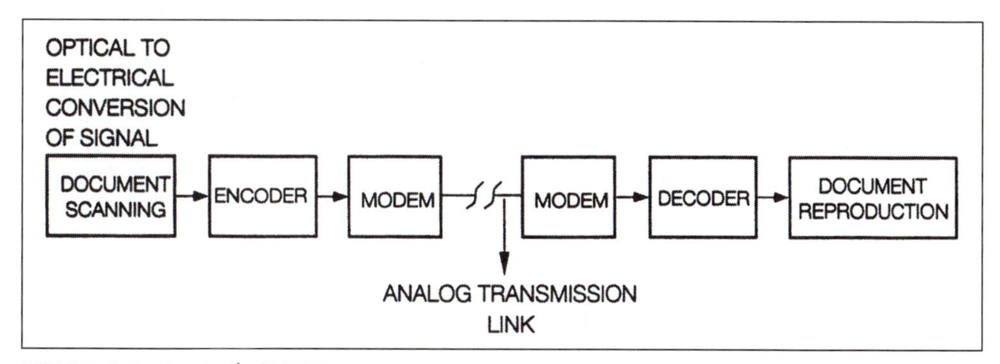

FIGURE 6-1 Facsimile (FAX) Transmission

fading Decrease in signal strength of an electromagnetic signal due to increased distance from the transmitting antenna, obstruction of the receiving antenna, interference from other electromagnetic signals, or atmospheric conditions.

fail safe A backup system used in the event of a failure of the primary system.

Fast Ethernet A type of Ethernet LAN configuration capable of transmitting data at 100 Mbps. 100BaseT and 100BaseFX are examples of Fast Ethernet. *Also see* Table 0-1 *for Gigabit Ethernet specifications and* Table 0-2 *for IEEE 802.3 specifications.*

FAT File allocation table. A simple and reliable file system used by DOS (disk operating system) and supported by all DOS—and Windows—based operating systems. *See also* **file system**.

FCC The Federal Communications Commission. A board of five commissioners appointed by the President of the United States under the Communications Act of 1934. The commission has the power to establish and enforce regulations regarding all interstate and foreign electrical communication systems originating in the United States. [http://www.fcc.gov]

fiber cladding The light conducting material that surrounds the core of an optical fiber (Figure 6-2). It has a lower refractive index than the core material. *See also* Module 27.

fiber core The central portion of an optical fiber through which light propagates from one end to another (Figure 6-2). It has a higher refractive index than the cladding material that surrounds it. *See also* Module 27.

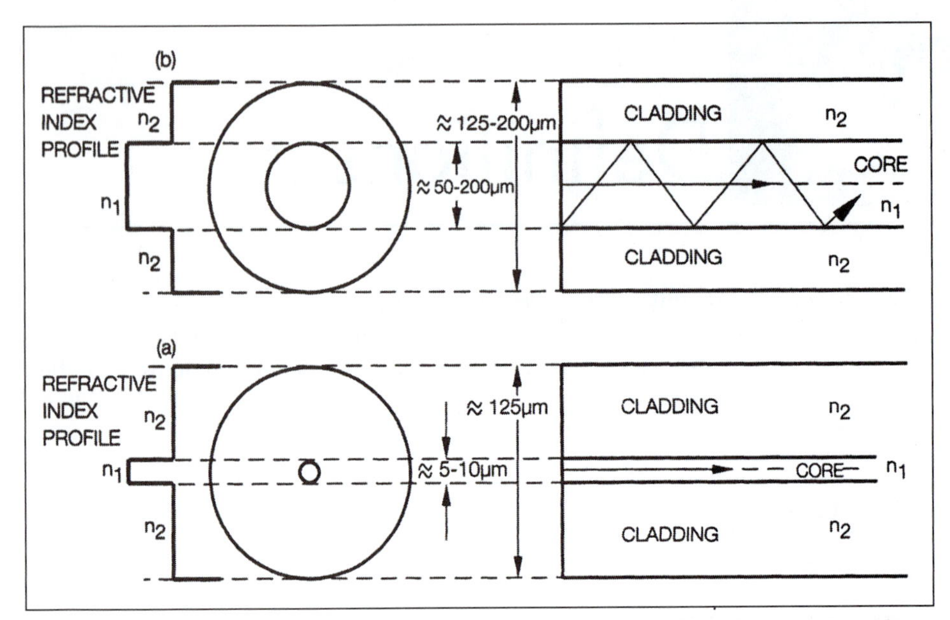

FIGURE 6-2 Fiber core and cladding: (a) monomode optical fiber; (b) multimode optical fiber

Fiber Distributed Data Interface (FDDI) An optical fiber LAN technology that uses token-passing dual counter-rotating ring topology. Each ring supports a data rate of 100 Mbps. FDDI uses multimode fiber and light emitting diodes, and supports two types of nodes/devices: a single attach station connected to one ring, and a dual attach station connected to both rings. FDDI is used as a backbone to connect LANs (Figure 6-3).

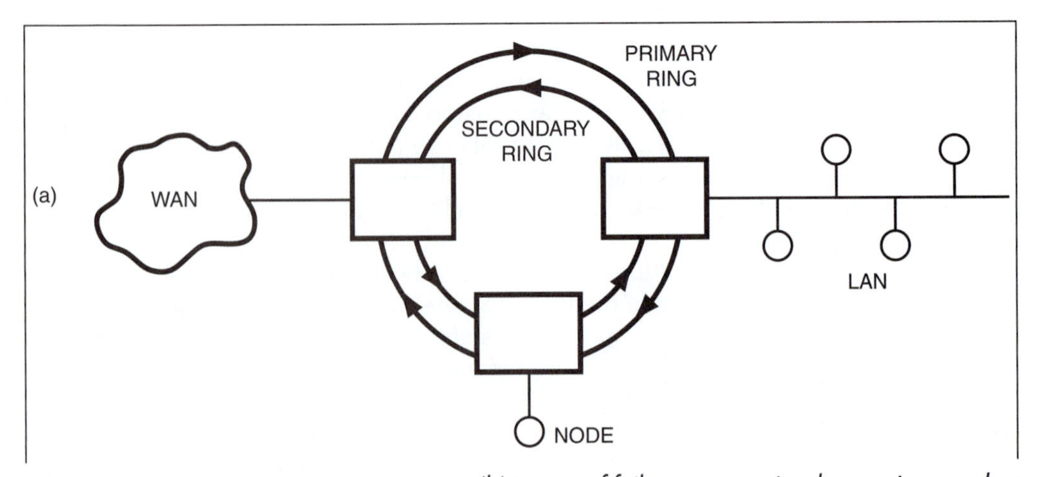

FIGURE 6-3 (a) FDDI: counter-rotating rings; (b) in case of failure at one point, the two rings can be configured to form a single ring; (c) FDDI Frame Format *(continues)*

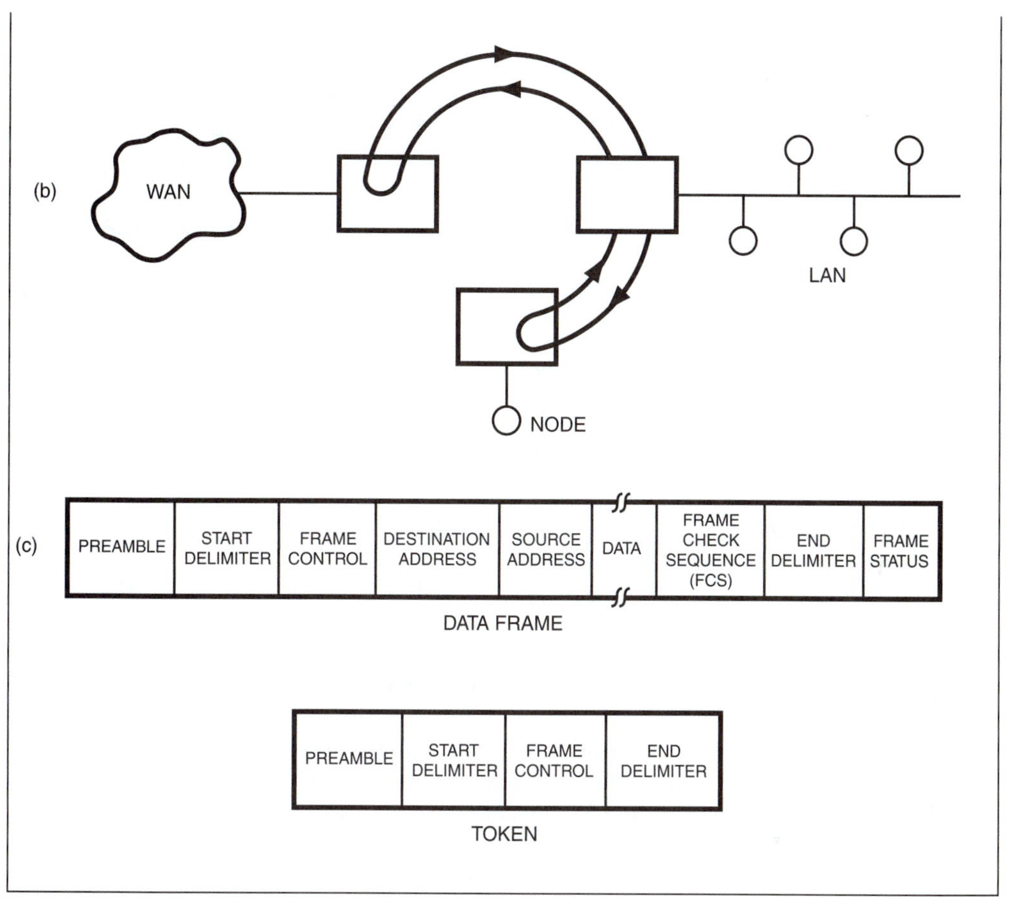

FIGURE 6-3 (continued)

fiber optics Transmission technology in which information is transmitted over a very thin strand of glass using a modulated optical beam. A typical fiber optics communication system consists of an optical source, optical fiber, and an optical detector. Figure 6-4 illustrates a fiber optics communication system. *See also* Module 27.

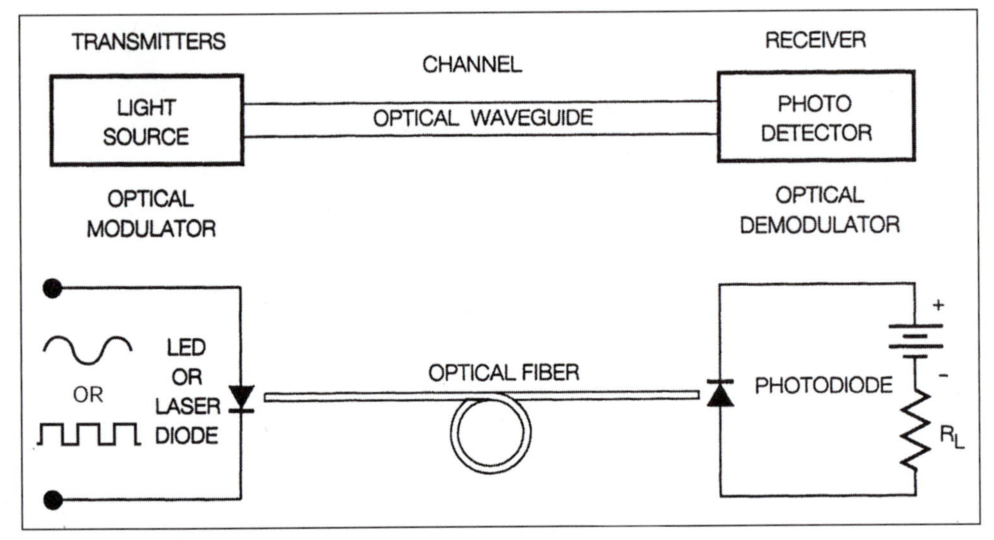

FIGURE 6-4 Fiber optics communication system

fiber to the curb (FTTC) Deployment of fiber optic cable from a telephone company's central office to the curb near a subscriber's home or business premises. From the curb signals are transmitted to the subscriber's home or the business using another medium such as coaxial cable. FTTC's architecture provides high-speed data transmission capability, and is less expensive to implement than fiber to the home (FTTH) (Figure 6-5). *See also* **fiber to the home (FTTH)**.

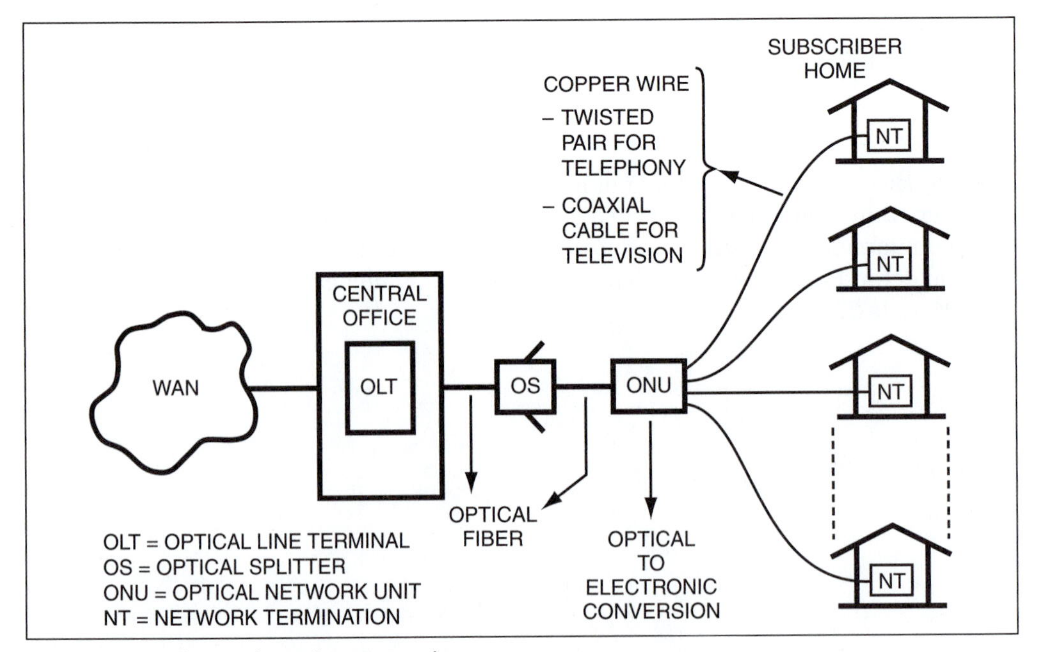

FIGURE 6-5 Fiber-to-the-curb (FTTC) architecture

fiber to the home (FTTH) Deployment of fiber optic cable from a telephone company's central office to the subscriber's home. Its architecture provides high-speed data transmission capability (Figure 6-6). *See also* **fiber to the curb (FTTC)**.

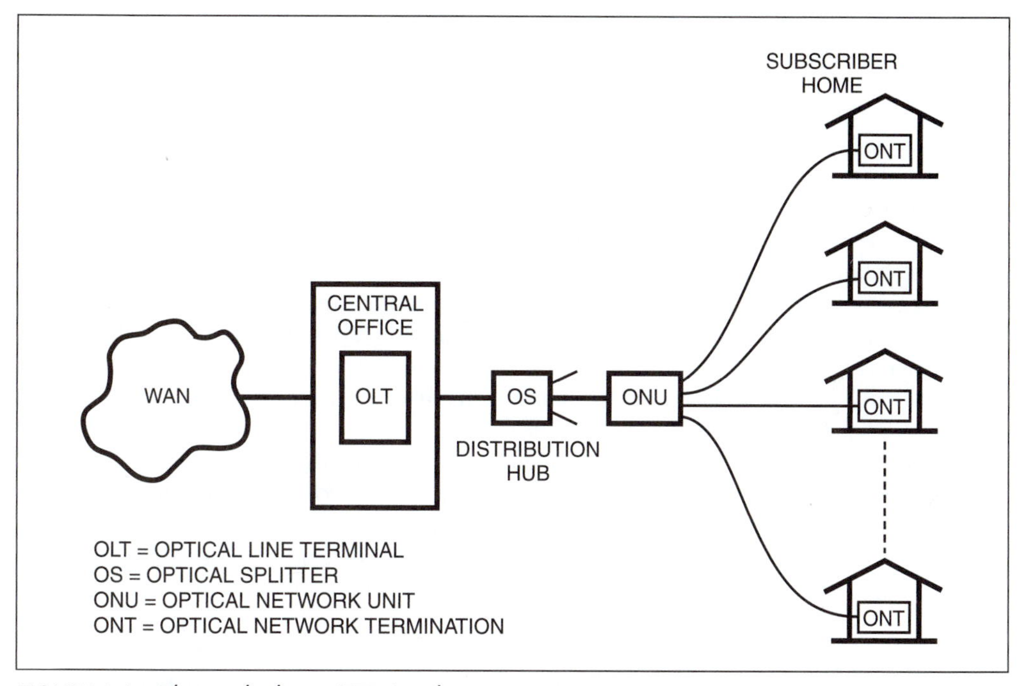

FIGURE 6-6 Fiber-to-the-home (FTTH) architecture

file format The layout of organizing data in a file so that a software program can recognize and access it. Table 6-1 lists some of the commonly used file formats.

TABLE 6-1 Commonly used file formats

FILE FORMAT	DESCRIPTIONS
.bmp	Bitmap: Used for saving graphics.
.doc	Document: Used for saving text in MS Word and WordPerfect.
.gif	Graphics Interchange Format: Used for saving graphics.
.jpeg	Joint Photographic Expert Group: Used for saving photos and graphics.
.pdf	Portable Document Format: Used for saving documents and graphic in a layout so that they can be viewed using any computer screen or printer in manner they were intended to appear.
.tiff	Tagged Image File Format: Used for saving graphics in high resolution.

file system An operating system's method of managing, organizing, and accessing files by using the logical structuring of a computer's hard disk.

filter An electronic circuit or system that allows a single frequency or a band of frequencies to pass through and attenuate all other frequencies.

firewall Computer hardware and/or software that acts as a buffer between an internal network and the Internet to provide security against unauthorized access.

FireWire Also known as IEEE 1394, a scalable, flexible, and low-cost serial digital interface that connects consumer electronics devices such as video cameras, digital cameras, etc., with personal computers. IEEE 1394 supports multiple speeds (100, 200, and 400 Mbps). Most 1394 devices use a 6-pin connector, but some devices, such as digital camcorders, use a smaller 4-pin connector. IEEE 1394 is also known as Apple's FireWire and Sony's i.LINK.

firmware Software that is permanently stored in a computer's memory.

first in first out (FIFO) A type of memory from which data is retrieved in the same order as it was stored.

fixed routing A transmission process in which a user gets access to a desired station through a defined route.

flag A bit pattern used to indicate the occurrence of a specific state.

flow control A mechanism used in data communications for regulating the flow of data in a network.

foreign exchange service (FX) A service that connects a subscriber's telephone to a central office other than the one that serves the subscriber's location.

format An accepted standard form for the representation of information.

forward error correction (FEC) An error correction technique in which enough redundancy is built into each message so that the receiver can detect as well as correct errors that occur during transmission. Hamming code is an example of FEC.

Fourier series A mathematical series (developed by French mathematician Jean-Baptiste Fourier) by which most periodic functions can be represented by a series expansion of sine and cosine components. The Fourier series representation of a periodic function is given by:

$$x(t) = \frac{a_o}{z} + \sum_{n=1}^{\infty} (a_n \cos 2\pi ft + b \sin 2\pi ft)$$

Where

$$a_n = \frac{2}{T} \int_{-T/2}^{T/2} x(t) \cos 2\pi ft \, dt$$

$$b_n = \frac{2}{T} \int_{-T/2}^{T/2} x(t) \sin 2\pi ft \, dt$$

f = fundamental frequency (Hz)
T = time period of a signal

Fourier transform A mathematical transform that decomposes a time function into its frequency components. The frequency representation of a signal is required in many applications, such as communications (efficient transmission of data), medical (multidimensional convolution in tomography), audio (speech analysis), mechanical (vibration analysis of rotational machinery), and defense (radar and sonar target identification and tracking). The Fourier transform of a time domain signal x(t) is given by:

$$X(f) = \int_{-\infty}^{\infty} x(t)e^{-2\pi ft} \, dt$$

Where X(f) = frequency domain representation of the signal
x(t) = time domain representation of the signal
f = frequency

four-wire circuit A circuit consisting of two pairs of conductors, used in a telephone system for providing full duplex operation. One pair is used for transmission and the other for reception of a signal.

four-wire terminating circuit A hybrid circuit that allows the connection of four-wire circuits with two-wire circuits.

fractional T-1 A leased connection that provides data speeds of 128 kbps, 256 kbps, 512 kbps, or 768 kbps.

frame

1. Information contained in a specific time interval.
2. In data communications, a group of bits whose position is predefined in a data block for control purposes.

frame relay A packet-switched transmission technique that supports higher data rates (45 Mbps). It is the digital version of X.25.

framing error A type of error in data communication occurring when the receiver incorrectly interprets the start and stop bits within a frame.

free space optics (FSO) An optical link used to transmit data between remote sites. Also called *free space optical communication (FSOC)* (Figure 6-7).

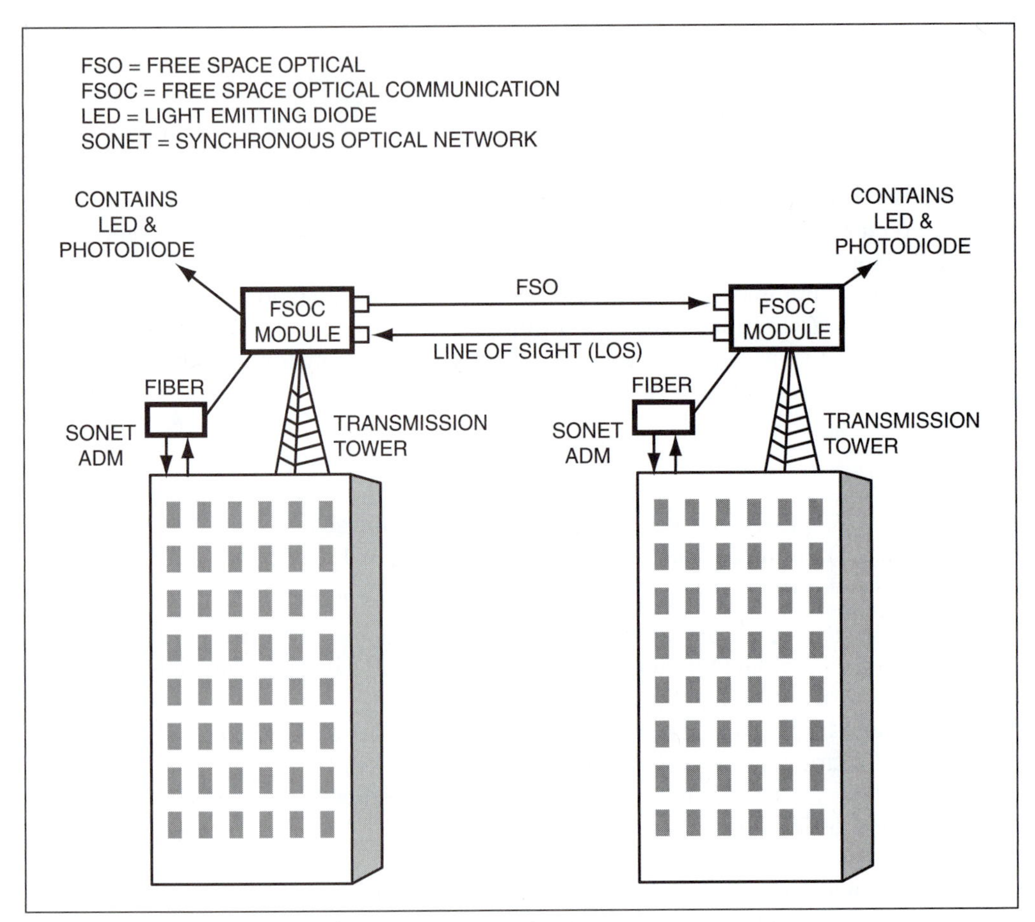

FIGURE 6-7 Free space optical communication architecture

freeware Software that is available at no cost on the Internet for users to download, in contrast to shareware (for which a user pays a small fee).

frequency Number of oscillations, cycles, or events per unit of time (Figure 6-8). It is measured in hertz (Hz), which is defined as the number of cycles per second (hertz = cycles/second).

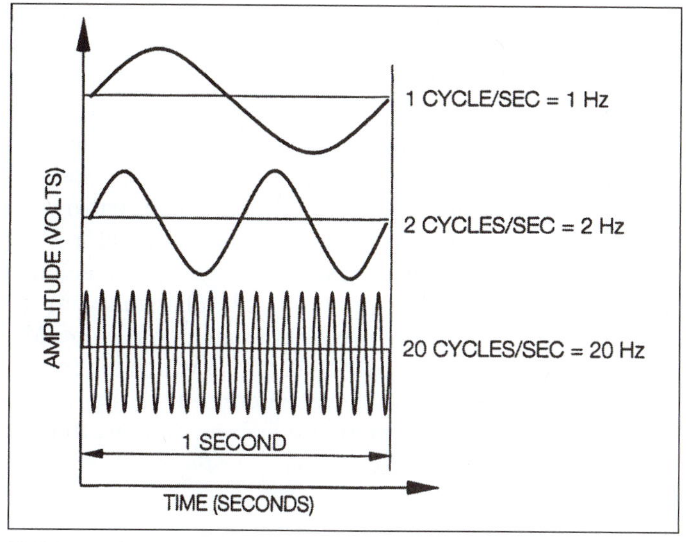

FIGURE 6-8 Frequency

frequency-division multiplexing (FDM) A multiplexing technique in which the range of transmission frequencies is divided into narrow bands, each representing a separate channel. This multiplexing technique allows the transmission of a large number of channels within a specific range of frequencies over a transmission medium. Figure 6-9 illustrates FDM. *See also* **time-division multiplexing** *and* **multiplexing**.

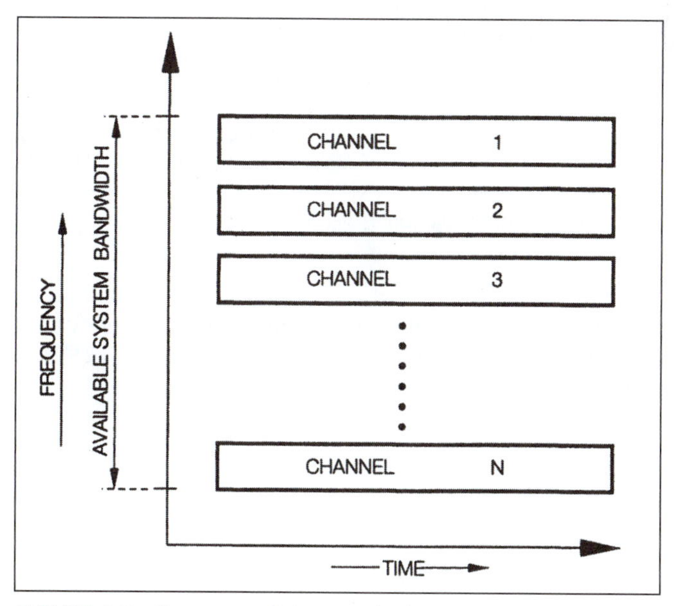

FIGURE 6-9 Frequency-division multiplexing

frequency division multiple access (FDMA) A frequency division multiplexing technique, used in cellular communications, in which the frequency spectrum is divided into dedicated 30 kHz channels. Each channel is dedicated to a single subscriber. Voice is modulated in the digital realm. *See also* Module 47.

frequency hopping A modulation technique used in spread spectrum technology, in which a signal at the transmitter and the receiver hops to preassigned values determined by a mathematical algorithm. The signal itself is narrowband, but because of continuous frequency shifts or skips it behaves like a spread spectrum signal.

frequency hopping spread spectrum (FHSS) technology A technique for generating spread spectrum signals by "hopping"—rapidly changing values based on pseudo-random noise (PN) code—the carrier frequency. *Also see* Module 47.

frequency modulation (FM) A type of modulation in which the frequency of a high-frequency carrier signal is varied in accordance with the variations in the amplitude of a low-frequency information signal (modulating frequency). Figure 6-10 illustrates frequency modulation. (*For the mathematical expression for an FM wave, see* Module 33.)

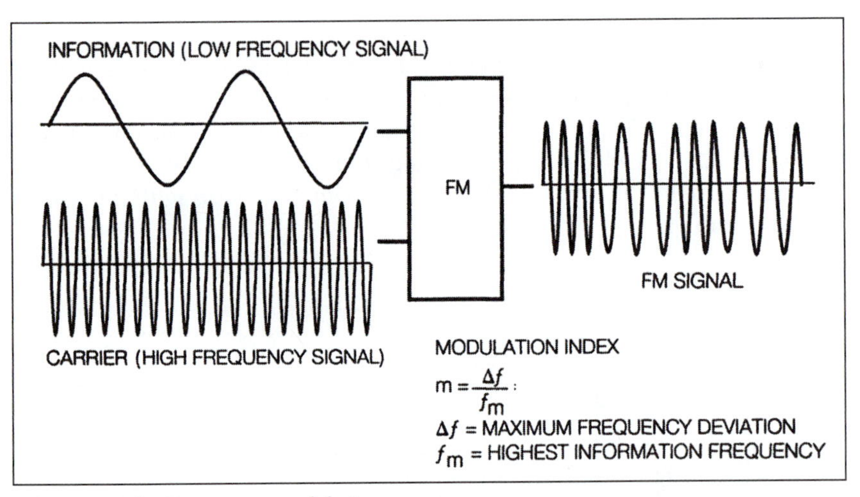

FIGURE 6-10 Frequency modulation

frequency response A plot of amplitude or gain versus frequency that shows how a device, circuit, or system responds to a range of frequencies (Figure 6-11). It is a measure of the effective bandwidth.

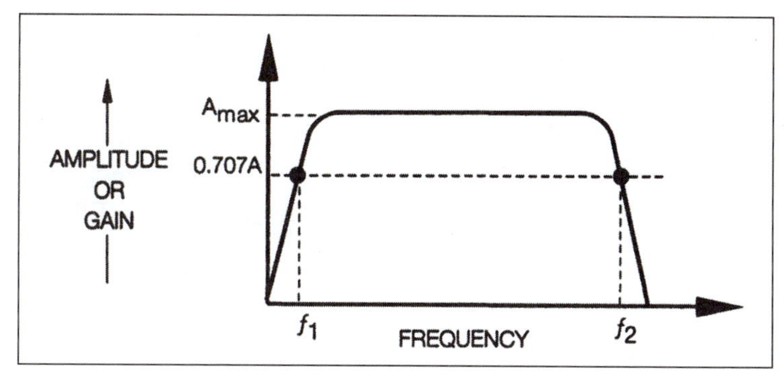

FIGURE 6-11 Frequency response

frequency shift keying (FSK) A type of RF modulation technique in which the frequency of an analog signal is varied in accordance with the variations in the amplitude of a digital signal. FSK is employed in modems. Figure 6-12 illustrates (a) FSK and (b) the frequency spectrum of FSK. *See also* **modulation**.

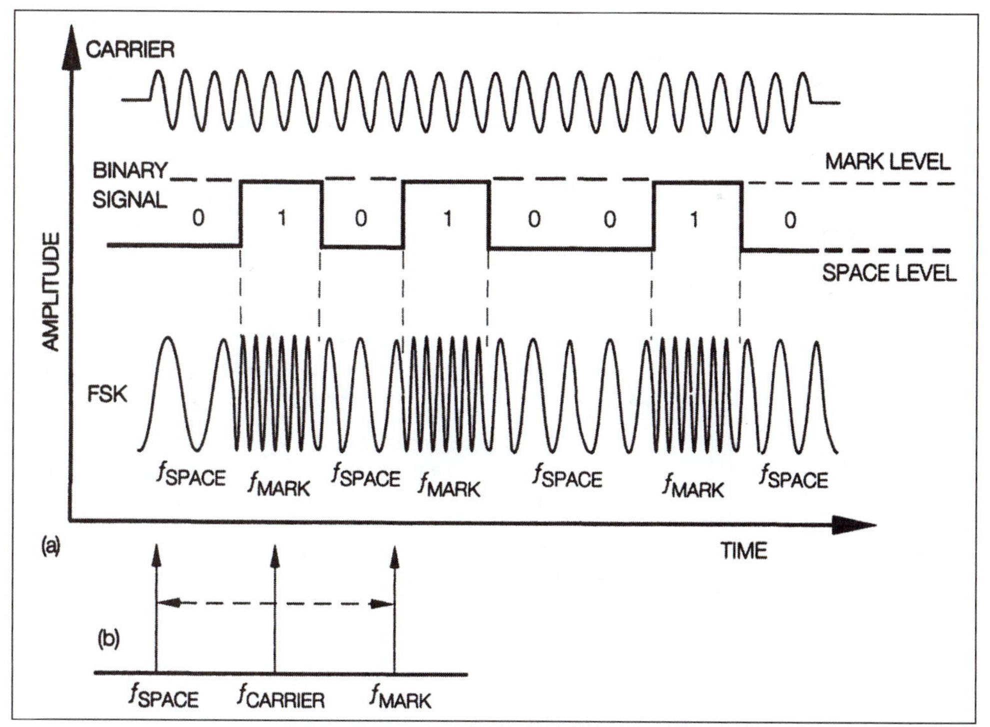

FIGURE 6-12 Frequency shift keying (a) FSK; (b) frequency spectrum of FSK

Fresnel reflection The reflection of a part of incident light at the interface of two media that have different refractive indices (Figure 6-13).

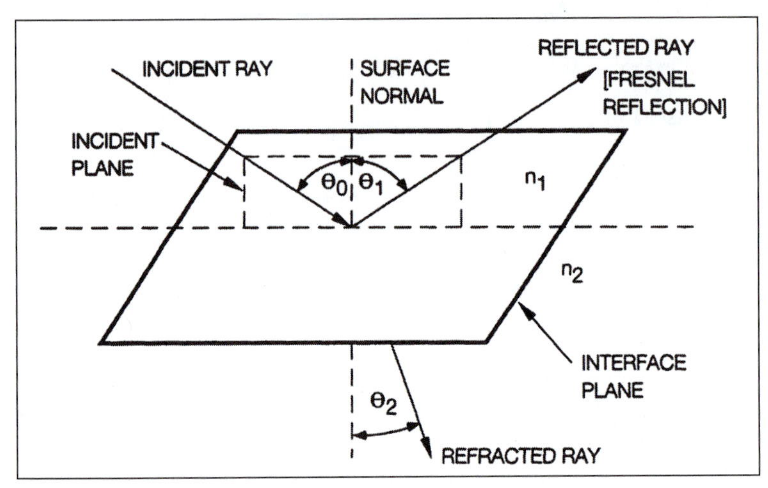

FIGURE 6-13 Fresnel reflection

front end processor (FEP) A stored program device or network processor that provides communication management service to a central computer to which it is connected. The heart of a front end processor is a software program called *Network Control Program*, which is responsible for functions like code conversion, management of multidrop lines, traffic handling, data format conversion, monitoring of network performance, error checking, etc. Figure 6-14 illustrates FEP.

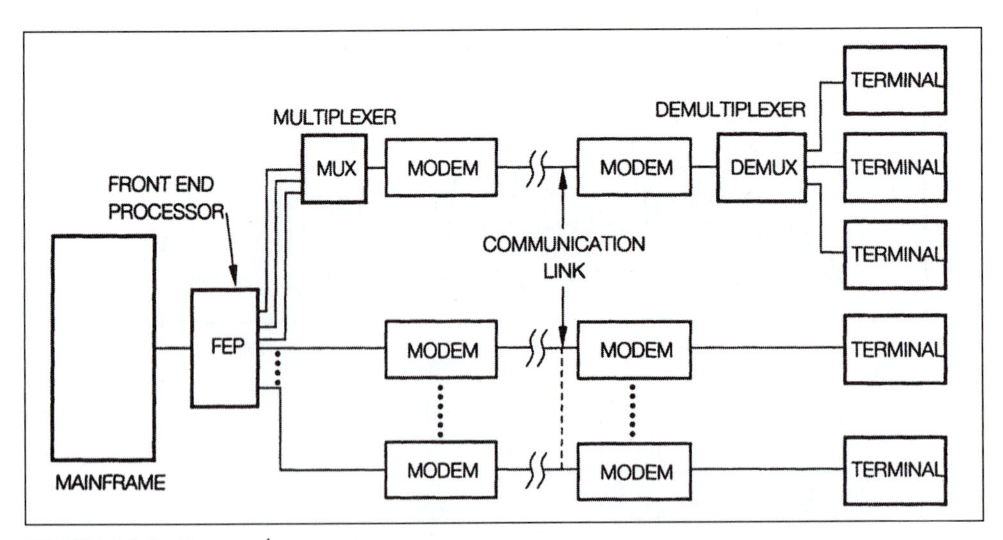

FIGURE 6-14 Front end processor

full duplex (FDX) A type of communication technique in which simultaneous two-way independent transmission takes place between two points. *See also* **duplex, simplex,** *and* **half duplex**.

MODULE 7 Definitions G

gain An increase in a signal's current, voltage, or power level due to the amplification process. It is represented by a ratio of the signal's output strength to input strength. A ratio greater than one represents amplification and less than one attenuation (Figure 7-1). It is expressed as a unitless quantity or in terms of decibels.

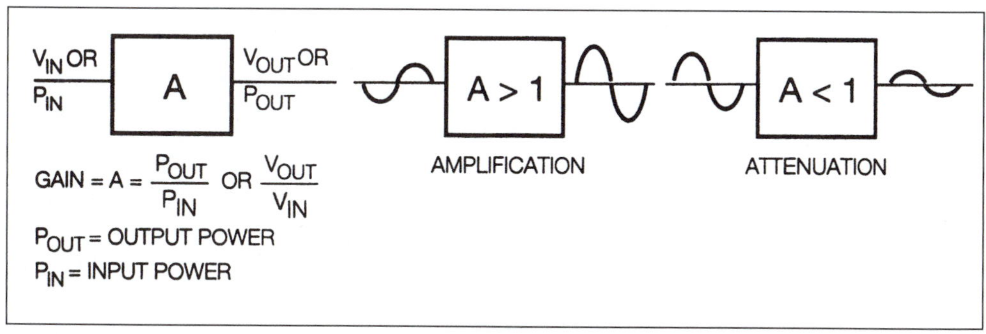

FIGURE 7-1 Gain

gain bandwidth product The product of the gain of a device and its bandwidth.

gateway A networking device that allows connection between two dissimilar computer networks. It performs the protocol conversion/translation at the transport and application layers of a network (Figure 7-2). *Also see* Module 48 *for computer networks.*

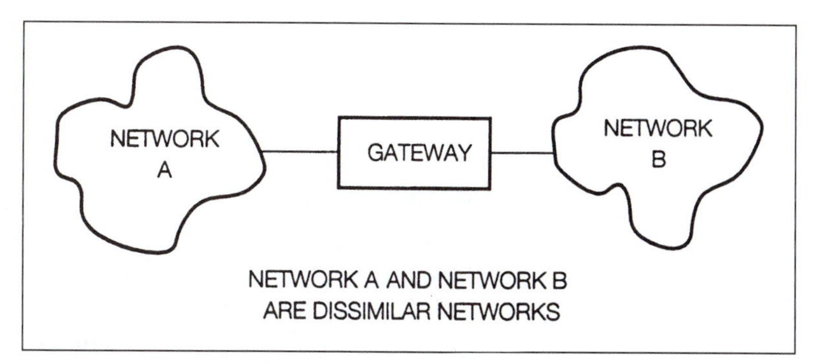

FIGURE 7-2 Gateway

Gaussian noise A type of noise generated by the thermal motion of electrons in a medium at a temperature above absolute zero. Also known as *white noise* and *thermal noise. Also see* **noise**.

generating polynomial The polynomial used as the divisor in computing an error check character in the CRC error detection code. *See also* **CRC**.

geometric optics A branch of optics that treats the propagation of light in terms of rays. Light rays follow defined paths in a straight line or curved lines when propagated through optical elements such as lenses, prisms, and other optical media that exhibit reflection, refraction, and transmission according to geometric formulas.

geosynchronous orbit An earth orbit, about 36,000 km above the equator, in which a satellite circles the earth in the same time it takes the earth to rotate on its axis.

geosynchronous satellite A satellite positioned about 36,000 km above the equator (Figure 7-3). The satellite's equatorial orbit is synchronous with the earth and the satellite appears to be stationary. It is used for long-distance communications. Also called *geostationary satellite*. (*See* Module 46.)

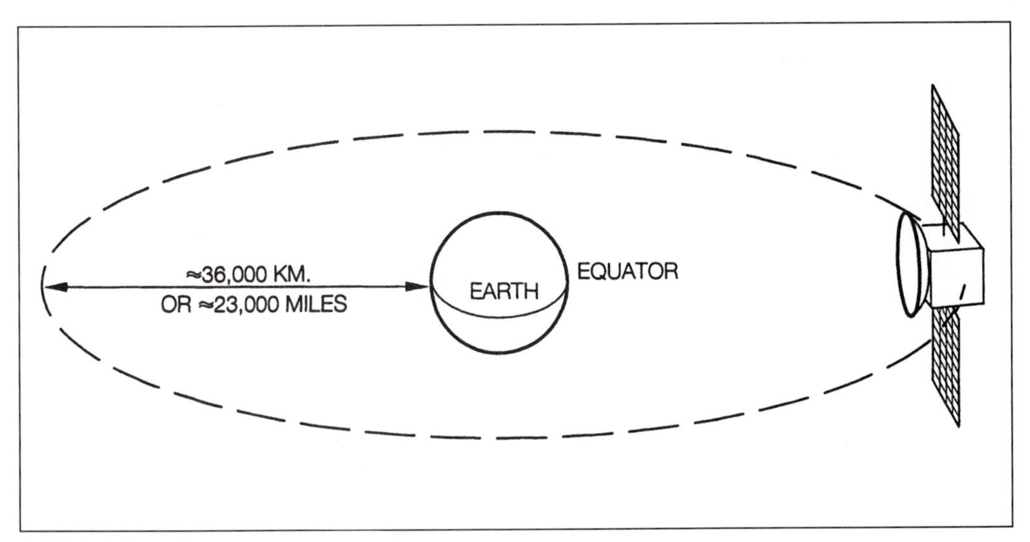

FIGURE 7-3 Geosynchronous satellite

Gigabit Ethernet A LAN configuration that transmits data at 1 Gbps or 10 Gbps. *Also see* Table 0-1 *for Gigabit Ethernet standards.*

gigahertz (GHz) One billion cycles per second.

$$1\ \text{GHz} = 1{,}000{,}000{,}000 = 1 \times 10^{9}\ \text{Hz}$$

$$1\ \text{GHz} = 1{,}000\ \text{MHz}$$

glare A condition in two-way circuits, such as PBX trunks, when an incoming call meets an outgoing call causing a blockage by connecting the wrong parties (callers). Ground start helps prevent glare. *See also* **ground start**.

Global System for Mobile Communications (GSM) A cellular communication standard developed in Europe (Groupe Speciale Mobile Communications, translated in English as Global System for Mobile Communications) but used worldwide. This standard offers three times the capacity of AMPS. Also called European TDMA. The U.S. version is called PCS-1900. *Also see* Module 47 *for wireless technologies and standards.*

grade of service The probability of a telephone call being blocked and receiving a network busy signal on the telephone circuit.

graded index fiber A type of optical fiber in which the refractive index of the core decreases as the radial distance from the central axis is increased. Figure 7-4 illustrates graded index fiber and various ray paths. *Also see* **optical fiber** *and* Module 27.

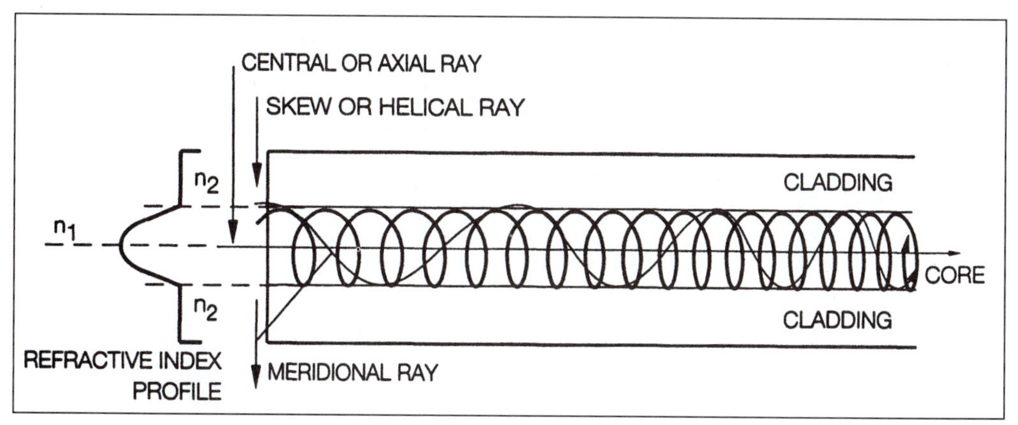

FIGURE 7-4 Graded index fiber

ground (GND)

1. A point considered to be at zero potential and used as a reference point for all other potentials.
2. A common return path for electric current in circuits.

ground start A type of signaling technique used between two communications systems where one system momentarily grounds the PBX trunk loop to get a dial tone. Ground start is used to prevent incoming calls from meeting outgoing calls and causing a call failure in the private branch exchange. *See also* **exchange, private branch (PBX)** *and* **glare**.

ground wave A low-frequency (long wavelength), usually less than 30 MHz, radio signal that travels along the earth's surface (Figure 7-5). *See also* **sky wave** *and* **space wave**.

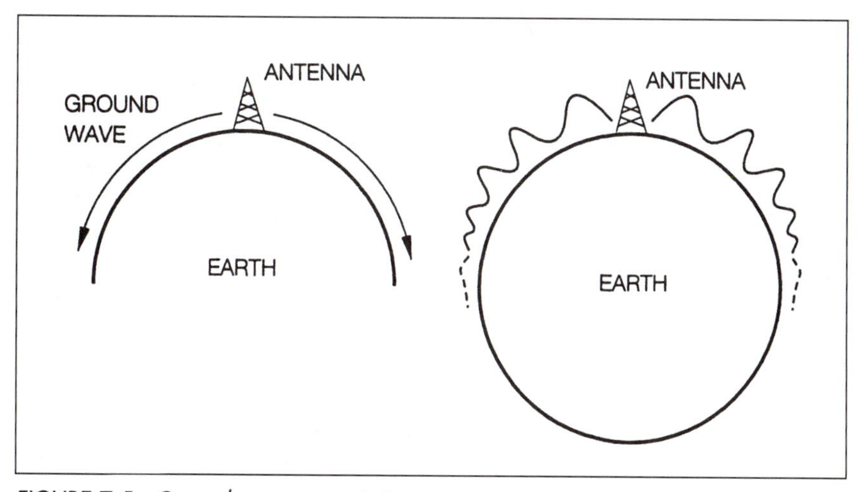

FIGURE 7-5 Ground wave transmission

group Formed by combining twelve telephone channels using different carrier frequencies in broadband transmission (Figure 7-6). *See also* **super group, master group,** *and* **jumbo group**.

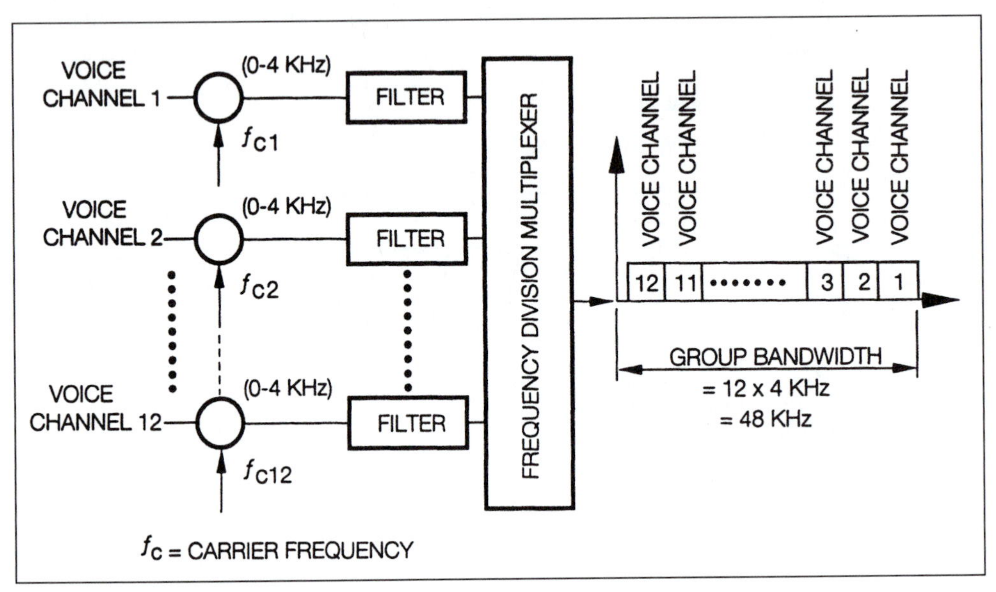

FIGURE 7-6 Formation of a group

groupware Software programs that enable people to work together jointly.

guard band A band of frequencies that separates two channels to prevent interchannel interference (Figure 7-7).

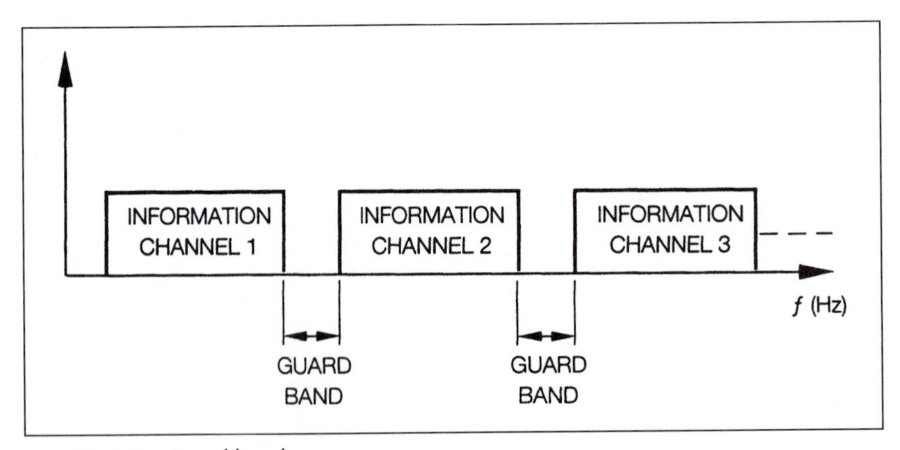

FIGURE 7-7 Guard band

Gunn diode A semiconductor diode invented by J. B. Gunn and used in microwave circuits. It oscillates when electrons are forced to transfer between two mobility states caused by fields of several kilovolts per centimeter.

MODULE

8 Definitions H

H channel The ISDN packet-switched channel that carries user data at different speeds (e.g., H0 = 384 kbps, H11 = 1536 kbps, and H12 = 1920 kbps). See Table 8-1. *Also see* Module 34.

TABLE 8-1 ISDN channels

ISDN CHANNEL	BIT RATE	INTERFACE
D	16 kbps 64 kbps	Basic Rate Interface (BRI) Primary Rate Interface (PRI)
B	64 kbps	Basic Rate Interface (BRI) Primary Rate Interface (PRI)
H0	384 kbps	Primary Rate Interface (PRI)
H11	1536 kbps	Primary Rate Interface (PRI)
H12	1920 kbps	Primary Rate Interface (PRI)

hacker A computer user who gains access without authorization to someone else's computer system for personal enjoyment or satisfaction.

half duplex A circuit or a transmission link that allows the flow of information in either direction, but only in one direction at a time (Figure 8-1).

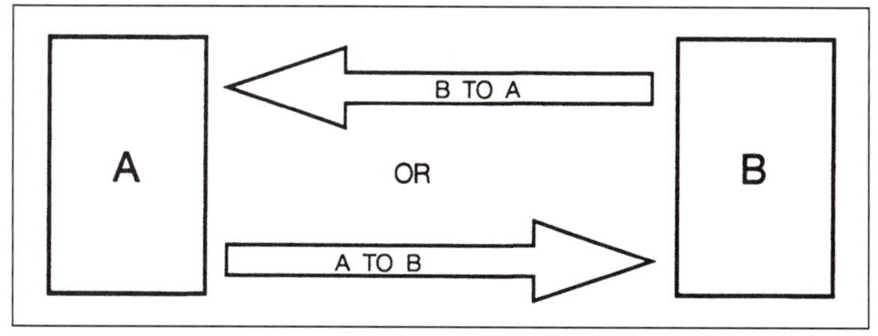

FIGURE 8-1 Half-duplex transmission

handset That part of a telephone set that houses the transmitter and receiver, and is held in one's hand when dialing or receiving a call.

handshaking Exchange of predetermined information for control between two devices to establish a connection.

hard disk A storage device in which magnetic material is deposited on a metallic disk. Contrast to *floppy* or *flexible disk* or *diskette*.

hardware The physical components of a system, in contrast to *software*.

hard wired A communication path that permanently connects two devices or circuits.

harmonic An oscillation whose frequency is an integral multiple of the fundamental frequency.

harmonic distortion Distortion in the waveshape of a signal caused by a nonlinear frequency response of a device or transmission channel.

Hartley oscillator A frequency generator that utilizes inductive feedback to maintain oscillation. (Table 15-2 illustrates four oscillators, including a Hartley.) *See also* **oscillator**.

hashing A process used in encryption algorithms in which a long string of characters is changed into a shorter fixed-length string or value or key that represents the original string.

HD or HDX *See* **half duplex**.

HDLC *See* **high-level data link control**.

head end A central point in a broadband network that transmits and receives between users. It transmits signals using one set of frequencies and receives signals using another set of frequencies.

header The initial part of a data block, consisting of information related to address, destination, control codes, etc. It is not a part of the text.

hertz (Hz) A unit for measurement of frequency of electromagnetic waves. Named after Heinrich Hertz, one hertz equals one cycle per second. The commonly used prefixes in frequency measurement are:

one kilohertz	1 kHz = 1,000 Hz
one megahertz	1 MHz = 1,000,000 Hz
one gigahertz	1 GHz = 1,000,000,000 Hz
one terahertz	1 THz = 1,000,000,000,000 Hz

heterodyning Heterodyning is a technique of combining two signals (an incoming radio signal and a local oscillator signal) to transfer the information signal from its carrier to a fixed local intermediate frequency (IF) in a receiver. Regardless of the changes in the incoming or received signal frequency, the IF frequency remains the same, thus providing an efficient way to tune to a wide range of broadcast channels. See Figure 8-2. *Also see* **beat frequency**.

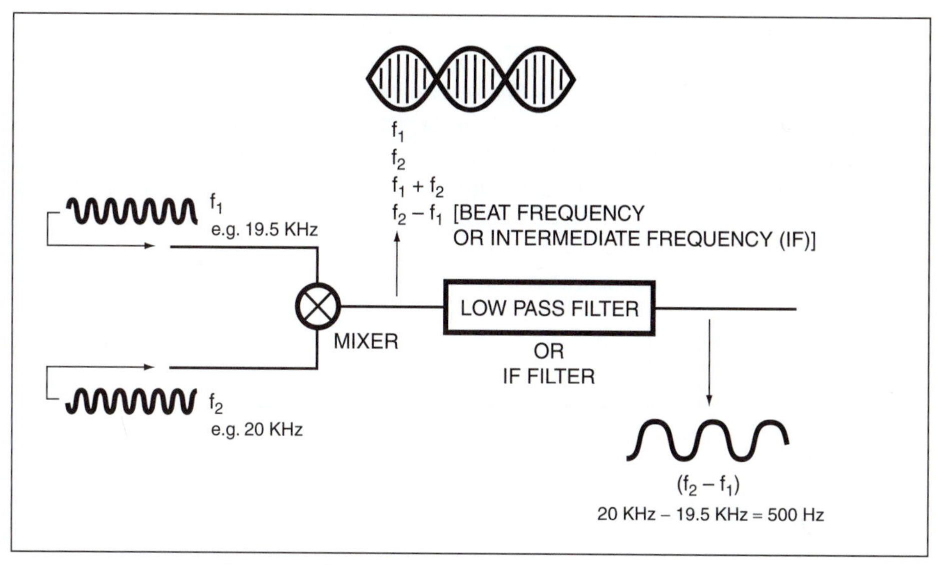

FIGURE 8-2 Heterodyning and generation of beat frequency or intermediate frequency (IF)

heterojunction A junction between two semiconductor materials in the same device having different bandgap energies. Heterojunction optical emitters are used in fiber optics communications.

hexadecimal A number system that represents quantities by base 16. Four-bit binary numbers are represented by integers 0 (zero) through 9 and A, B, C, D, E, and F. Table 8-2 compares hexadecimal numbers with binary and decimal equivalents.

TABLE 8-2 Hexadecimal numbers

DECIMAL	BINARY	HEXADECIMAL
0	0000	0
1	0001	1
2	0010	2
3	0011	3
4	0100	4
5	0101	5
6	0110	6
7	0111	7
8	1000	8
9	1001	9
10	1010	A
11	1011	B
12	1100	C
13	1101	D
14	1110	E
15	1111	F

hierarchical network A network topology in which nodes are arranged in a hierarchical order and processing of information can take place at several levels.

high-definition television (HDTV) *See* Module 41.

high-level data link control (HDLC) A bit-oriented protocol for transmitting data over a link. Predecessor to SDLC (used for point-to-point communication), HDLC can be used in point-to-point as well as point-to-multipoint communication (Figure 8-3).

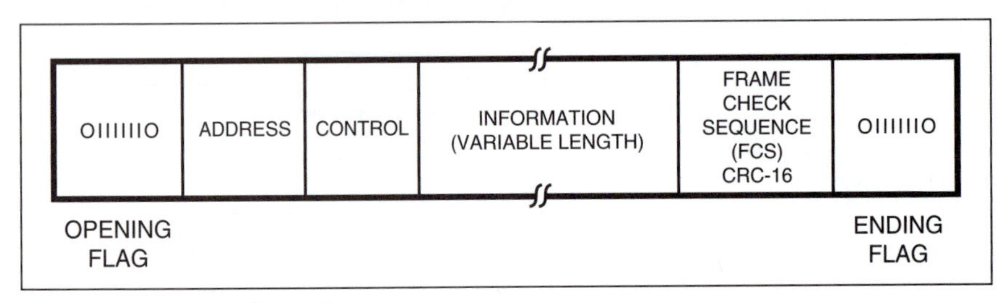

FIGURE 8-3 HDLC frame format

high-order mode The longest path for a light ray to travel in a multimode optical fiber. The difference in the path lengths for high-order and low-order modes causes dispersion of the signal. *See also* **dispersion**.

holding computer The duration of time that a communication channel is in use.

Hollerith code An 80-character code used to represent data in the form of two or three out of twelve punched-hole patterns on cards.

homojunction A PN junction formed in a single type of semiconductor material.

hookswitch A switch used to hang up the handset on the telephone set in order to turn off the loop current.

host computer A computer in a network that acts as a repository for offering services (e.g., WWW, SMTP, HTTP, USENET, etc.) to other computers connected to the network.

hub A device, used in computer networks, to connect cables from multiple computers to form a signal network connection. A *passive hub* just repeats the signal, whereas an *active hub* (multiport repeaters) extends the range of signal transmission. An *intelligent hub* also performs remote management and diagnostic functions about the network.

hunt group A group of lines organized in such a way that if the first line is busy, the next line is hunted and so on until an unused line is found so that a connection can be established.

hybrid circuit In telephony, a circuit used to interface a four-wire circuit to a two-wire circuit. This allows full duplex transmission of information. A hybrid circuit consists of two interconnected transformers, as shown in Figure 8-4.

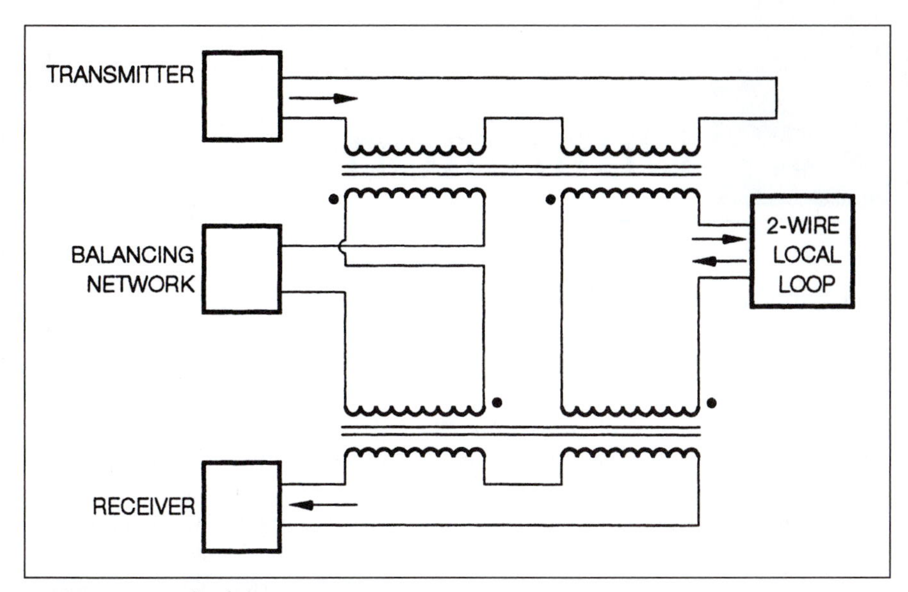

FIGURE 8-4 Hybrid circuit

hybrid/fiber coax (HFC) The use of fiber optic and coaxial transmission media in a telecommunications network to provide high-speed data channels to support interactive video, data, and voice services.

hypertext markup language (HTML) A computer language used for creating and displaying a web page that may contain text and graphics.

hypertext transfer protocol (HTTP) A protocol used by web servers and clients to communicate with each other to transport HTML documents. It is also the first entity in a URL (uniform resource locator) address, indicating the type of protocol.

MODULE 9 Definitions I

ICA International Communication Association.

icon A graphic image or symbol used to represent software and hardware entities such as application programs, files, disk drives, documents, network connections, etc., on a computer screen.

identification The process of verifying a user's identity.

identity theft The crime of stealing someone's personal data and using it for financial gain.

idle The duration of time when an operating system is ready but not being used.

IEEE Institute of Electrical and Electronics Engineers, a professional organization formed by the merger of the American Institute of Electrical Engineers (AIEE) and the Institute of Radio Engineers (IRE). The IEEE works for the advancement of electrical and electronics engineering and the related disciplines of science and engineering. [http://www.ieee.org]

IEEE 394 *See* **FireWire**.

IEEE-488 A protocol for interconnecting digital programmable instruments (DPI).

IEEE 802 A set of standards developed by Institute of Electrical and Electronics Engineers (IEEE) for local area networks. (*See* Module 0.)

impedance (Z) The opposition that a circuit offers to the flow of alternating current. Table 9-1 lists impedance relations for different circuits.

TABLE 9-1 Impedance

CIRCUIT	IMPEDANCE (Z)
f, R	$Z = R\ \Omega$ WHERE: R = RESISTANCE
f, R, X_L, L	$Z = \sqrt{R^2 + X_L{}^2}\ \Omega$ WHERE R = RESISTANCE $X_L = 2\pi f L$ = INDUCTIVE REACTANCE f = FREQUENCY OF AC SIGNAL

(continues)

TABLE 9-1 Impedance (continued)

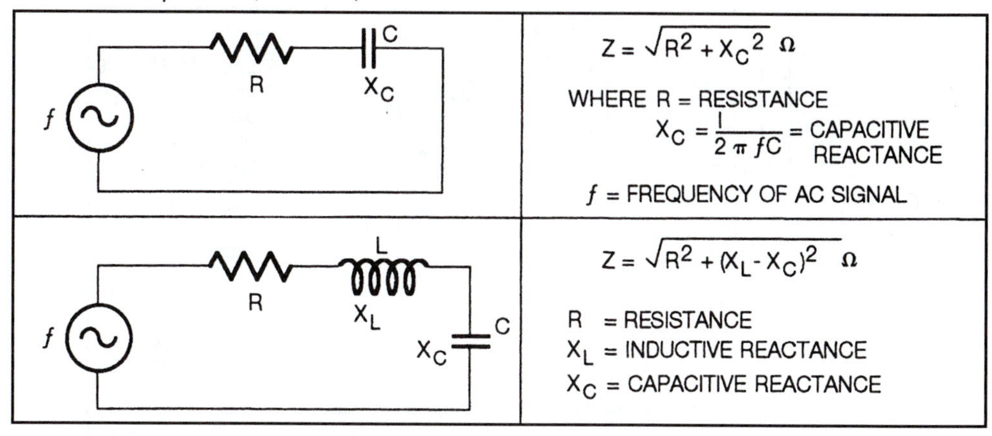

f, R, C, X_C	$Z = \sqrt{R^2 + X_C^2}\ \Omega$ WHERE R = RESISTANCE $X_C = \frac{1}{2\pi fC}$ = CAPACITIVE REACTANCE f = FREQUENCY OF AC SIGNAL
f, R, L, X_L, C, X_C	$Z = \sqrt{R^2 + (X_L - X_C)^2}\ \Omega$ R = RESISTANCE X_L = INDUCTIVE REACTANCE X_C = CAPACITIVE REACTANCE

impedance matching network A combination of passive electrical components used to provide impedance matching between two points having different impedances in an electrical circuit. Table 9-2 illustrates various impedance matching techniques.

TABLE 9-2 Impedance matching techniques

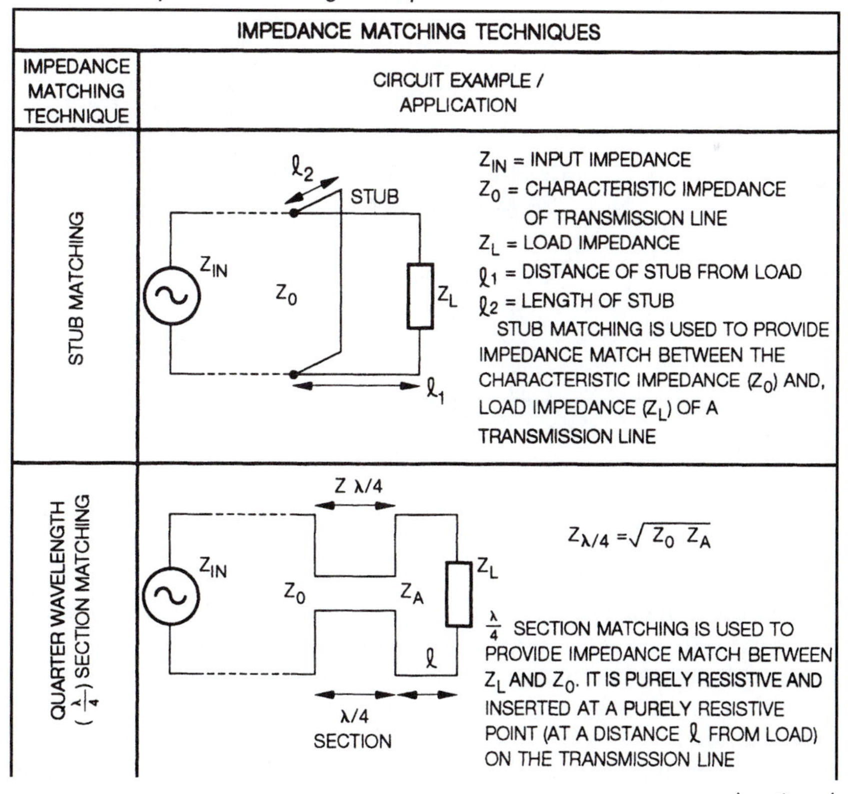

IMPEDANCE MATCHING TECHNIQUES	
IMPEDANCE MATCHING TECHNIQUE	CIRCUIT EXAMPLE / APPLICATION
STUB MATCHING	Z_{IN} = INPUT IMPEDANCE Z_0 = CHARACTERISTIC IMPEDANCE OF TRANSMISSION LINE Z_L = LOAD IMPEDANCE ℓ_1 = DISTANCE OF STUB FROM LOAD ℓ_2 = LENGTH OF STUB STUB MATCHING IS USED TO PROVIDE IMPEDANCE MATCH BETWEEN THE CHARACTERISTIC IMPEDANCE (Z_0) AND, LOAD IMPEDANCE (Z_L) OF A TRANSMISSION LINE
QUARTER WAVELENGTH ($\frac{\lambda}{4}$) SECTION MATCHING	$Z_{\lambda/4} = \sqrt{Z_0 Z_A}$ $\frac{\lambda}{4}$ SECTION MATCHING IS USED TO PROVIDE IMPEDANCE MATCH BETWEEN Z_L AND Z_0. IT IS PURELY RESISTIVE AND INSERTED AT A PURELY RESISTIVE POINT (AT A DISTANCE ℓ FROM LOAD) ON THE TRANSMISSION LINE

(continues)

TABLE 9-2 Impedance matching techniques (continued)

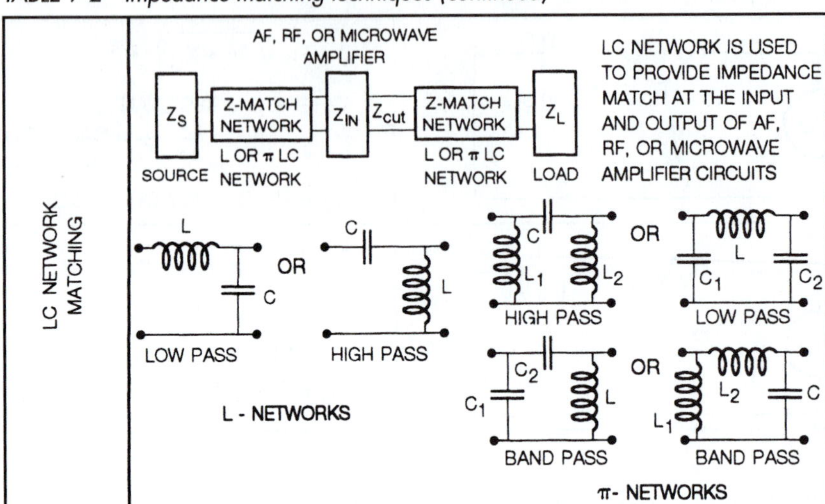

impersonation Pretending to be someone else.

incidence angle The angle between a light ray incident on a surface and the perpendicular to the surface, as shown in Figure 9-1.

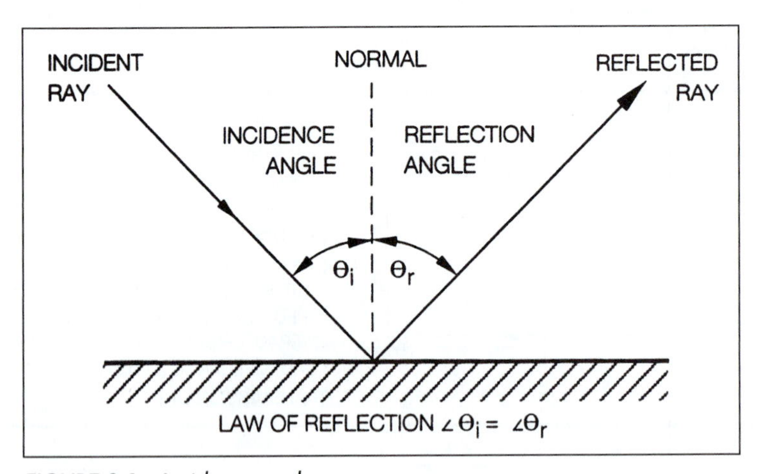

FIGURE 9-1 Incidence angle

independent common carrier A non-Bell company providing common carrier telephone service to customers other than *AT&T*. *See also* **AT&T**.

independent telephone company A non-Bell company providing telephone service in an area of franchise by villages, towns, or cities.

index of refraction The ratio of the speed of light in a vacuum to the speed of light in a given medium. It is represented by the letter n. Table 9-3 lists indices of refraction for different materials.

TABLE 9-3 Index of refraction

MATERIAL	INDEX OF REFRACTION (n=c/v)
Water	1.33
Air	1.00
Diamond	2.42
Germanium	4.1
Silicon	3.5
Fused Silica	1.46
Plexiglass	1.49

index profile A plot that shows how the refractive index of an optical fiber changes with respect to radial distance. *See also* **optical fiber**.

individual service A telephone line arranged to serve one customer. Additional stations may be added on the line as extensions for that customer.

inductance The property of a coil or any electrical component to oppose a change in the flow of current through it. Inductance is given by:

$$L = \frac{N\phi}{I}$$

Where: N = number of turns in coil
ϕ = magnetic field
I = current

inductive reactance The measure of an inductor's opposition to the flow of AC.

$$X_L = 2\pi fL$$

Where: X_L = inductive reactance (ohms)
f = frequency (Hz)
L = inductance (Henry)
π = 3.14159 . . .

industrial, scientific, and medical (ISM) bands Radio frequency bands assigned by the International Telecommunication Union (ITU) for industrial, scientific, and medical applications. These bands were intended for equipment that radiate microwaves as a by-product of their function, but these bands have been employed in cordless phones and wireless LANs (Table 9-4).

TABLE 9-4 ISM bands

ISM BAND	UNITED STATES (FCC) FREQUENCY ALLOCATIONS	BANDWIDTH (MHz)	EUROPE (ETSI) FREQUENTLY ALLOCATIONS	BANDWIDTH (MHz)
Industrial (I-band)	902–928 MHz	26	890–906 MHz	16
Scientific (S-band)	2.4–2.4835 GHz	83.5	2.4–2.5 GHz	100
Medical (M-band)	5.725–5.850 GHz	125	5.725–5.875 GHz	150

information rate *See* Module 33.

infrared (IR) Electromagnetic radiation having wavelengths in the range of 0.7 to 1000 micrometers (above the visible and below the microwave spectrum). Infrared radiation is invisible to the human eye.

in-house system A communication system within an establishment that can be accessed by users from different departments on a time-sharing basis.

injection laser diode (ILD) A semiconductor PN junction that emits monochromatic coherent light by stimulated emission when the current through the junction is increased above a certain threshold. If the junction current is below the threshold, the ILD behaves like a light-emitting diode. For fiber optic communications, an ILD couples higher optical power into the fiber and allows higher modulation frequencies.

input/output (I/O) device A device used to input data to or retrieve it from a computer.

insertion loss Loss in power at the load due to insertion of a device at some point on the transmission line.

integrated circuit (IC) A circuit containing interconnected active and passive circuit components formed by thin film deposits on a semiconductor substrate.

integrated services digital network (ISDN) An end-to-end digital network that supports voice and nonvoice services. (*See* Module 34.)

intelligent terminal A data terminal that has processing capabilities in addition to its function as a data input/output device. Also called *smart terminal.* Contrast to *dumb terminal.*

INTELSAT International Telecommunications Satellite Consortium. Owns and operates a global system of geosynchronous satellites that provides international telecommunication services all over the world. (*See* Module 46.) [http://www.intelsat.com]

interactive system A real-time communication system in which a user interacts with a computer program using a terminal and a communication line. Normally, the user enters data or initiates an inquiry and waits for a response by the computer program before continuing on.

interconnect company A company that supplies telecommunications equipment to be connected with the telephone company lines. Interconnect companies were born as a result of the Carterphone Decision (1968). The Federal Communications Commission (FCC) decided that telecommunication equipment supplied by companies other than the telephone companies could be connected with the telephone company lines if registered with the FCC.

inter-exchange carrier (IXC) A long-distance telephone carrier that provides connections between central offices (exchanges).

interface Circuitry that provides compatibility between devices, transmission media, etc.

intermodulation noise In a multiple-channel transmission system using frequency division multiplexing, noise introduced in a channel of interest due to intermodulation frequency products.

international access code A code used in conjunction with the country code, city code, and subscriber's number to make an international telephone call. For making an international call from the United States to other countries, the international access code is 011.

International Organization for Standardization (ISO) An organization consisting of representatives from national standards bodies from different countries, responsible for development of standards to facilitate international trade and services. The American National Standards Institute (ANSI) represents the United States in the ISO. The ISO has developed a standard for computer network architecture known as the *open system interconnect (OSI) reference model.* (*See* Module 35.)

International Telecommunication Union (ITU) The telecommunications agency of the United Nations, formed to provide standardized telecommunications practices and procedures on a worldwide basis. Figure 9-2 illustrates the past and present organizational structure. [http://www.itu.int]

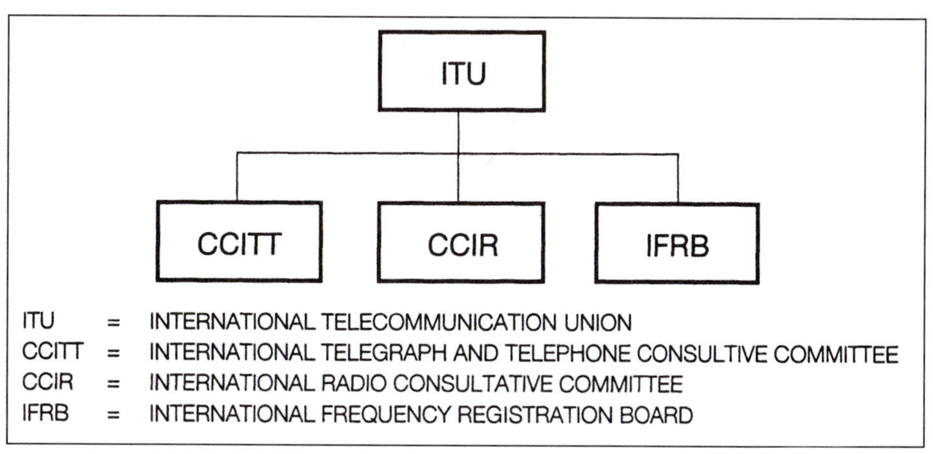

FIGURE 9-2 (a) International Telecommunication Union: Organizational structure (pre–1993)

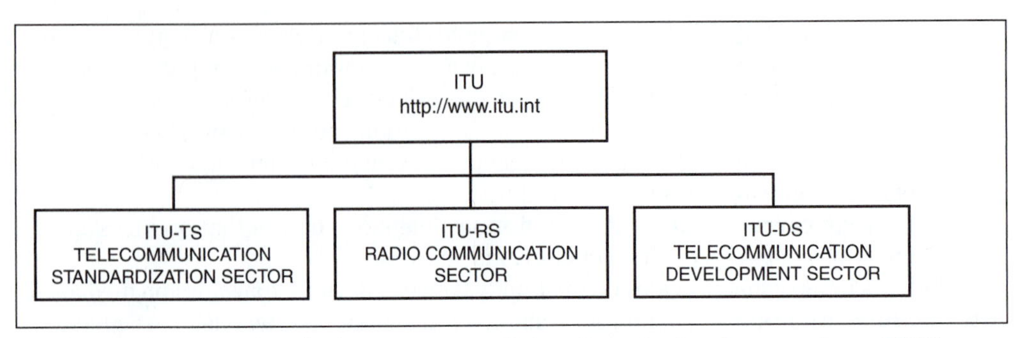

FIGURE 9-2 (b) International Telecommunication Union: Organizational structure (post–1993)

Internet A global network in which thousands of computer networks linked to millions of nodes allow millions of people to exchange information throughout the world. *Also see* Module 49 *for world Internet usage and growth of host computers.*

Internet Control Access Protocol (ICMP) A protocol used to transmit control and error data about IP traffic on a network. The most commonly used ping utility is based on ICMP.

Internet Message Access Protocol (IMAP) A protocol that allows e-mail client-server interaction by permitting the client to access and manage e-mail stored at the remote server.

Internet Protocol (IP) The protocol (set of rules) used to transport data through the Internet.

internetworking The use of software and hardware technologies to interconnect different types of computer networks, e.g., LAN to LAN, LAN to MAN, and LAN to MAN. *See also* **local area network, metropolitan area network,** *and* **wide area network**.

intranet An internal network belonging to an organization based on Internet technology, with the purpose of improving communication and productivity.

intrusion detection system (IDS) A system that alerts of attempts to breach the security of a network.

inverse multiplexer A type of multiplexer that accepts data from one high-speed line and transmits it on multiple low-speed lines combined at the other end into a high-speed link (Figure 9-3).

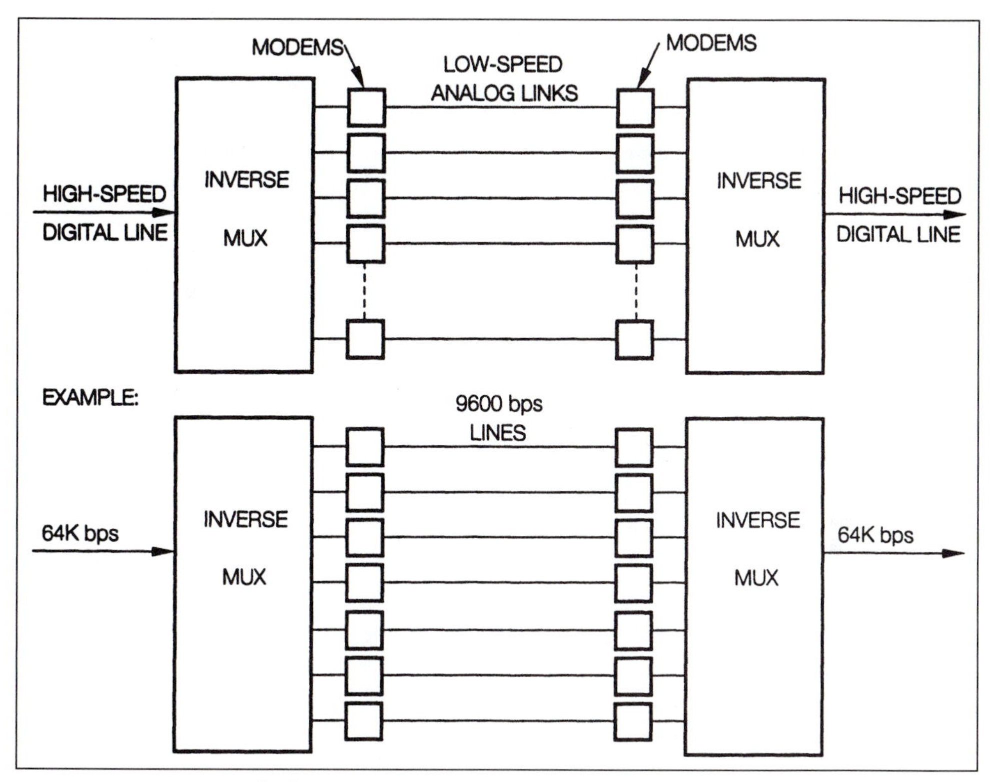

FIGURE 9-3 Inverse multiplexers

IP address A unique 32-bit address used by TCP/IP protocol to identify a device connected to the Internet. *See* Module 37.

IP aliasing A networking option that allows a single node to have more than one IP address assigned to it.

IP multicast A technique that allows simultaneous transmission of IP packets to multiple recipients.

IPv4 Internet Protocol version 4. *See* Module 37.

IPv6 Internet Protocol version 6. *See* Module 37.

irradiance The amount of power incident on a surface area. Irradiance (H) is given by:

$$H = \frac{P}{A} \text{ watts/cm}^2$$

Where: P = incident power (watts)
A = area (cm^2)

isotropic radiator An ideal source that uniformly radiates electromagnetic energy in all directions. The power density at any point on a sphere around an isotropic source is given by:

$$P = \frac{E^2}{120\pi} \text{ watts/meter}^2$$

Where: E = electric field strength (volts/meter)

ISP Internet service provider. A company that provides a connection to the Internet through its servers for a monthly or hourly fee.

ITU-RS International Telecommunication Union-Radiocommunication Sector. Formerly known as CCIR. [http://www.itu.int/ITU-R/]

ITU-TS International Telecommunication Union-Telecommunication Standardization Sector. Formerly known as CCITT. *See* Module 38. [http://www.itu.int/ITU-T/]

MODULE

10 Definitions J

J band A U.S. military microwave band that occupies a frequency range of 10–20 GHz. *See* Module 45.

jack A connecting device into which wires can be terminated by inserting a plug. *See* **RJ-11** *and* **RJ-45**.

jamming

1. In radio communication, an intentional interference of open-air radio frequency (RF) transmission to disrupt communication between two parties.
2. In a local area network (LAN), an intentional interference signal to alert other nodes that a collision has occurred.

jitter A type of signal impairment caused by slight variations in the time slots of the bits, if the signal bit is advanced or delayed with respect to the time slot allocated.

Johnson noise *See thermal noise in* Table 14-1.

Josephson junction A junction between two superconductors that has very high-speed switching capability at temperatures approaching zero degrees Kelvin.

jumbo group Six master groups combined to transmit 3,600 voice channels over a wideband channel [Figures 10-1 (a) and 10-1 (b)]. *See also* **group, master group,** *and* **super group**.

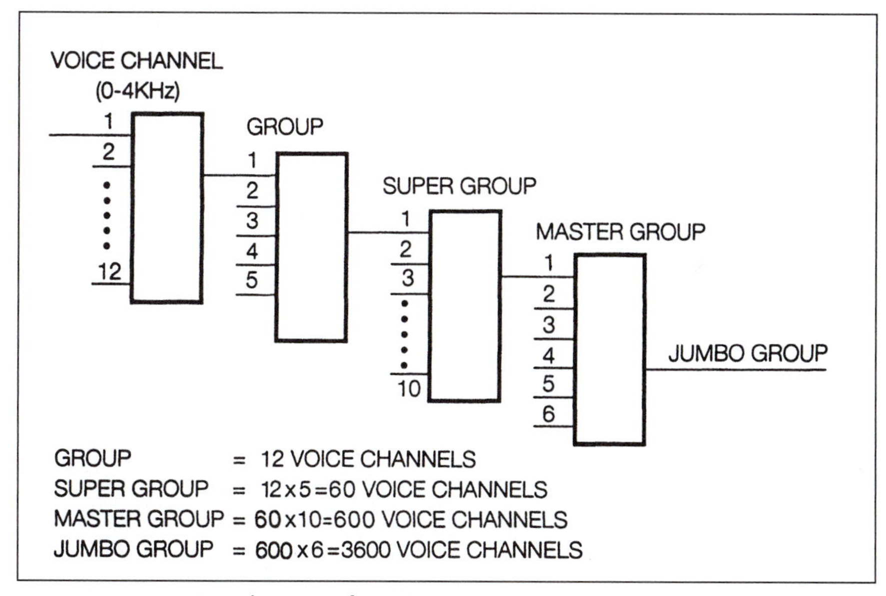

FIGURE 10-1 (a) Jumbo group formation

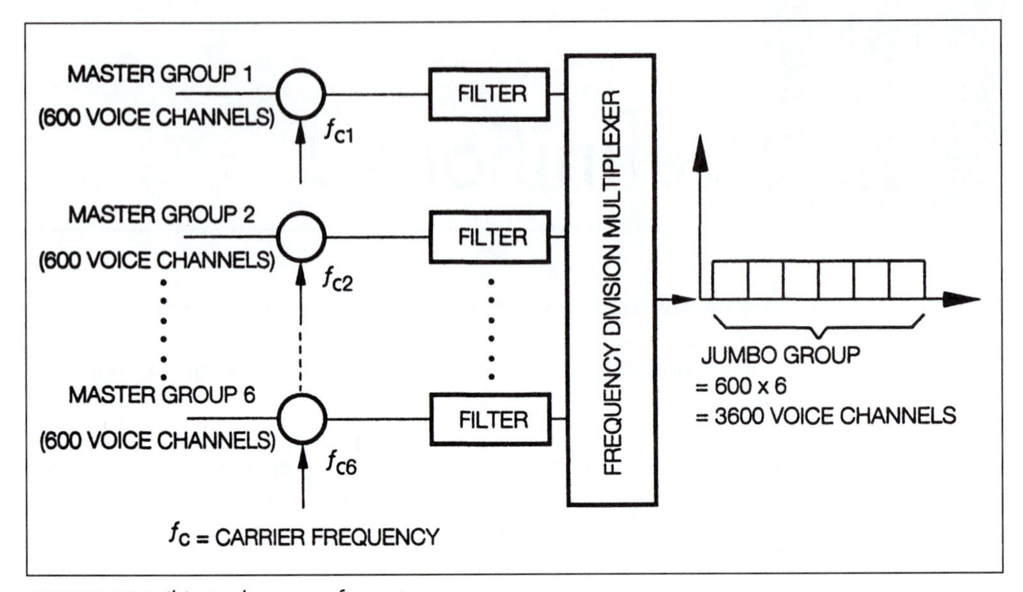

FIGURE 10-1 (b) Jumbo group formation

jumper A short length of wire that is used to provide a temporary connection between points in a circuit.

MODULE

11 Definitions K

k (kilo)

1. In binary notation

$$k = 2^{10} = 1024$$

2. In decimal notation

$$k = 10^{3} = 1000$$

K band Electromagnetic frequencies in the range of 10.9–36 GHz (2.75 to 0.834 centimeter wavelengths).

Ka band Electromagnetic frequencies in the range of 33–40 GHz (0.909 to 0.750 centimeter wavelengths). Part of the K band.

kbps (kilo bits per second) One thousand bits per second.

Kerberos A security protocol that allows faster authentication for logging into a system.

kernel The core processing and memory of an operating system.

Kerr-Cell modulator An electro-optic device consisting of a light polarizer, electrodes, and the Kerr-effect medium, used for the microwave modulation of laser light.

key An alphanumeric code used to encrypt or decrypt a message.

keyboard A device consisting of alphanumeric, and special function and control keys, used to enter data into and to interact with a terminal.

keyboard send and receive (KSR) A teleprinter set consisting of printer and keyboard (with no storage capability) that allows messages to be transmitted as they are keyed in and printed as they are received.

Key Service Unit (KSU)

1. The central part of a Key Telephone System. It consists of power supply, interrupter, feature block, and Key Telephone Unit (KTU) line slots. *See also* **Key Telephone System**.
2. In electronic telephone systems, the main control box.

Key Telephone System A telephone system consisting of telephone units with several buttons or keys, power supply, cables, and switching components offering special features like call holding, multiline pickup, intercom, line status display, etc. It is a cost-effective alternative to PBX and PABX systems for small businesses.

kHz (kiloHertz) One thousand cycles per second.

killer channel In a digital carrier, a channel that fails to maintain its assigned time slot. Its pulses encroach on other channels. Usually caused by CODEC failure or clock synchronization problems.

klystron An electron tube used to generate or amplify electromagnetic signals of very high frequencies (microwave) by velocity modulation of the electron beam passing through a series of resonant cavities constructed around its axis.

Ku band Electromagnetic frequencies in the range of 13–18 GHz. Used for satellite communications.

kW (kilowatt) One thousand watts.

kWh (kilowatt-hour) One thousand watt hours.

MODULE 12 Definitions L

L band The microwave frequency band extending from 390 MHz to 1.55 GHz.

lambert A unit of luminance. It is equal to the uniform luminance of an emitting or reflecting light at a rate of one lumen per square centimeter.

lambertian source An optical source that radiates light normally in all directions. The irradiance, radiance, and intensity of a lambertian source is proportional to the cosine of the angle from the normal that is parallel to the beam direction.

LAN *See* **local area network**.

large-scale integration (LSI) A method of fabricating integrated circuits in which a large number of circuits are integrated in a single chip. Typically, an LSI chip contains more than 100 logic gates. (See Table 22-1.)

laser Acronym for light amplification by stimulated emission of radiation. A laser device converts electrical energy into optical energy by emitting a monochromatic and coherent beam whose spectral width is very narrow. Semiconductor laser diodes are used as optical emitters in fiber optics communication systems.

laser diode A semiconductor PN junction diode in which the process of stimulated emission is used to emit monochromatic coherent light (Figure 12-1). Laser diodes are used as optical emitters in fiber optics communication systems.

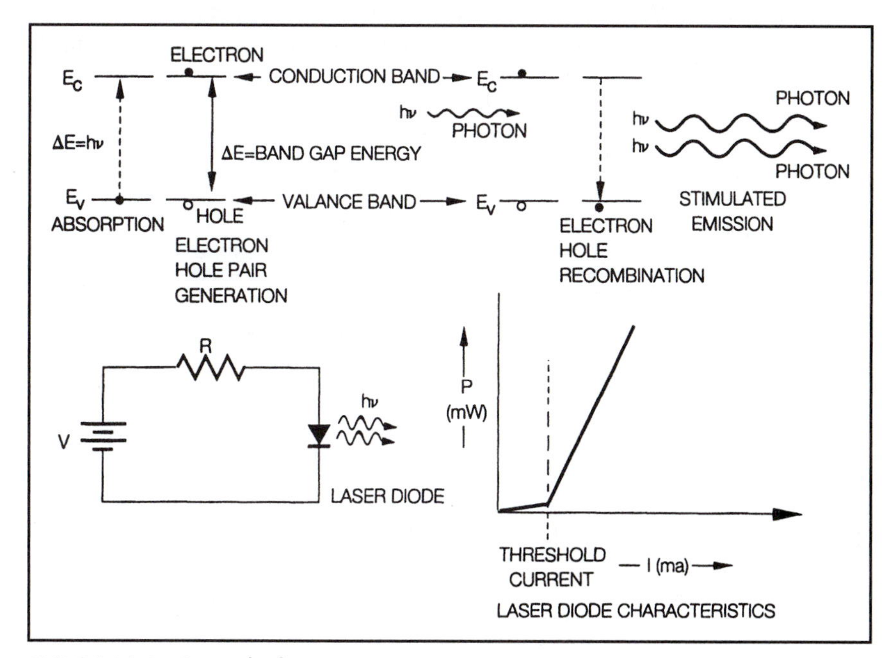

FIGURE 12-1 Laser diode

laser line width The range of frequencies emitted by a laser. Also known as *spectral width. See also* **spectral width.**

lasing threshold The minimum amount of current required to initiate stimulated emission in a laser.

launch angle The angle between the light ray incident on an optical fiber and the central axis of the optical fiber (Figure 12-2).

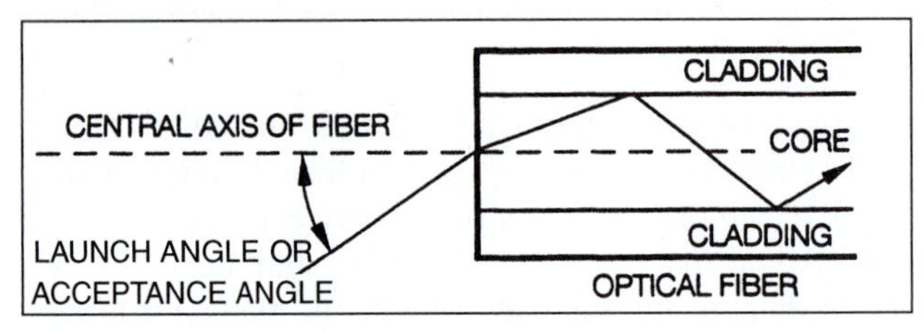

FIGURE 12-2 Launch angle

leased line A telephone line leased from a telephone company by a subscriber for exclusive use. Also called *private line* or *nonswitched line.*

light

1. Refers to the visible part of the electromagnetic spectrum, i.e., those electromagnetic waves (0.4 to 0.7 µm) that are visible to the human eye.
2. In lightwave or fiber optic communications, it refers to infrared (IR), visible, and ultraviolet (UV) frequencies that can be used to transmit information through an optical fiber.

light-emitting diode (LED) A semiconductor PN junction device that emits incoherent monochromatic light by electron hole recombination (spontaneous emission) (Figure 12-3). It is used as a display and as an optical emitter in fiber optics communication systems. *See also* **spontaneous emission** *and* **stimulated emission.**

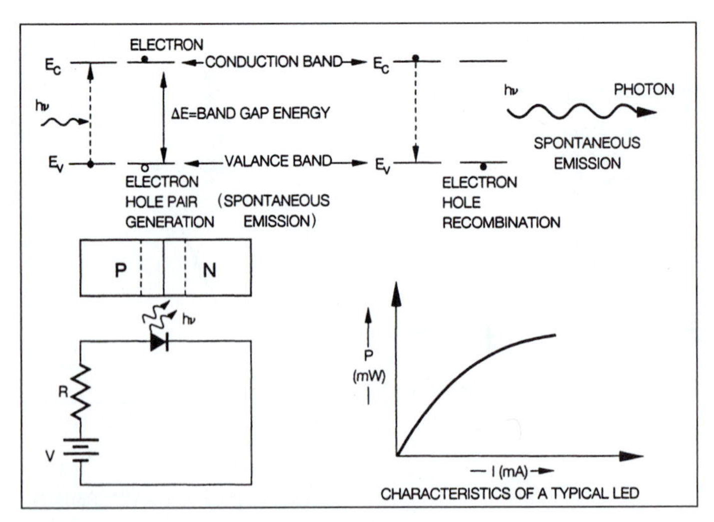

FIGURE 12-3 Light-emitting diode

light wave communications Transmission technology in which information is transmitted over a dielectric waveguide using light (IR, visible, and UV) frequencies. Also known as fiber optics communications. *See also* **fiber optics** *and* Module 27.

limited-distance modem A device that modulates and demodulates a digital signal for its transmission over a limited-distance analog link.

line conditioning The addition of components or equipment to a line to keep its characteristics within specified limits. Also called *conditioning. See also* **C conditioning** *and* **D conditioning**.

line driver A device consisting of a transmitter and receiver, used to extend the range of transmission between two communication devices that are directly connected. Line drivers are usually level converters.

line-of-sight communications A transmission system in which the radiated signal reaches the receiver directly from the transmitter through the air or free space without getting reflected by the earth or the ionosphere. *See also* **direct wave** *and* **free space optics (FSO)**.

link A physical path that connects two points.

link layer The second layer of the International Organization for Standardization's OSI model. This layer is responsible for the physical transmission of data between two network nodes. (*See* Module 35.)

liquid crystal display A passive graphic display device used in calculators, laptop computers, etc. LCDs use ambient light for their operation. They contain a fluid that partially blocks light when polarized. LCDs consume less power and are slower than LED display devices.

LISTSERV An e-mailing list server that allows an automated way of sending e-mail messages to all users belonging to a group.

load In terms of electrical circuits, it refers to: (1) a device that absorbs power, or (2) a device that receives power from a generator. In terms of computers, it refers to the input of data or programs into a computer's memory.

loading The addition of inductance to a transmission line to improve frequency response and to make it distortionless (Figure 12-4).

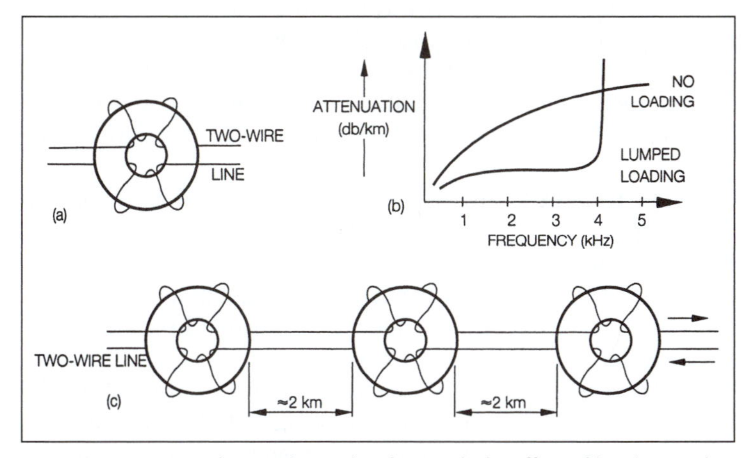

FIGURE 12-4 Loading coils: (a) loading coil; (b) effect of loading coils on attenuation; (c) in a telephone system, 88-mH loading coils are inserted approximately every two kilometers

local access and transport areas (LATA) A regional area in the United States within which a telephone company offers a local exchange and exchange services. For interLATA telephone service, long-distance telephone companies are employed.

local area network (LAN) A network that connects various telecommunication devices within a limited geographic area. *See also* **topology**.

local broadcast IP address The special Internet Protocol (IP) address (2555.255.255.255) used to broadcast packets to all hosts on the local network segment.

local call A telephone call for which the caller and the party being called belong to the same local exchange.

local exchange A telephone company's switching center to which all subscribers within a geographic area are connected by local loops. The local exchange is connected with other exchanges and trunks. Also called *central office* or *end office.*

local loop A line that connects the subscriber's telephone equipment with the local exchange.

local messages The messages related to local telephone call charges on a telephone bill.

local service area An area for which the telephone company charges local rates for telephone service.

logic circuit A circuit consisting of logic gates used to perform logical operations in digital devices responding to binary states. (*See* Module 30.)

logical unit (LU) In IBM's systems network architecture (SNA), a type of network addressable unit (NAU) consisting of software instructions that provide network access. A logical unit acts as a bridge between the user's application program and the network.

login spoofing A type of access control attack that replaces an authentic login screen with one supplied by the attacker.

long haul A long-distance telecommunications link. Contrast with *short haul.*

longitudinal redundancy check (LRC) An error-checking technique used in data communications in which a check character—known as a block check character (BCC)—is accumulated at the transmitter and receiver during the transmission of a data block. After the transmission, the transmitted BCC is compared with the received BCC. An equal condition indicates an error-free transmission of the previous data block. The BCC is usually obtained by exclusive ORing each bit with the previous bit of the data block. *See also* **CRC** *and* **vertical redundancy check**.

loop or line termination (LT) In integrated services digital network (ISDN) technology, it represents the termination of a line at the carrier's central switching office. *See also* Module 34.

loop start A method of signaling between a telephone set and a local exchange. When the handset of an analog telephone is picked up, current flows in the loop indicating a request for service. *See also* **ground start**.

loopback A diagnostic procedure used in analog and digital transmission devices. A test message is transmitted to the device being tested, which receives it and transmits it back for comparison with the original message transmission when the loopback is activated by approximately a 2700 Hz signal.

loss A decrease in a signal's power, voltage, or current during transmission between two points.

low earth orbit (LEO) satellite *See* **satellite communication** *and* Module 46.

lumen The unit of luminous flux. At a wavelength of 0.555 μm (peak sensitivity of the human eye), one watt equals 680 lumens.

luminance The luminous intensity of a surface in a given direction per unit-projected-area viewed from that direction. It is measured in candles per square meter.

lux The SI (Standard International) unit of illuminance. It is equal to one lumen per square meter.

MODULE

13 Definitions M

MAN *See* **metropolitan area network**.

mark A signaling state that represents binary 1. In asynchronous transmission, when no information is transmitted, the marking condition exits on the line, i.e., the line signals binary "1." When information is transmitted, a "0" level (called a *start bit*) is transmitted before the data bits, parity bits, and stop bits for each character. When the transmission is completed, the line returns to the marking condition. (*See* **asynchronous transmission**.)

master group Formed by combining five super groups to transmit 300 voice channels over a wideband channel (Figure 13-1). *See also* **group, jumbo group,** *and* **super group**.

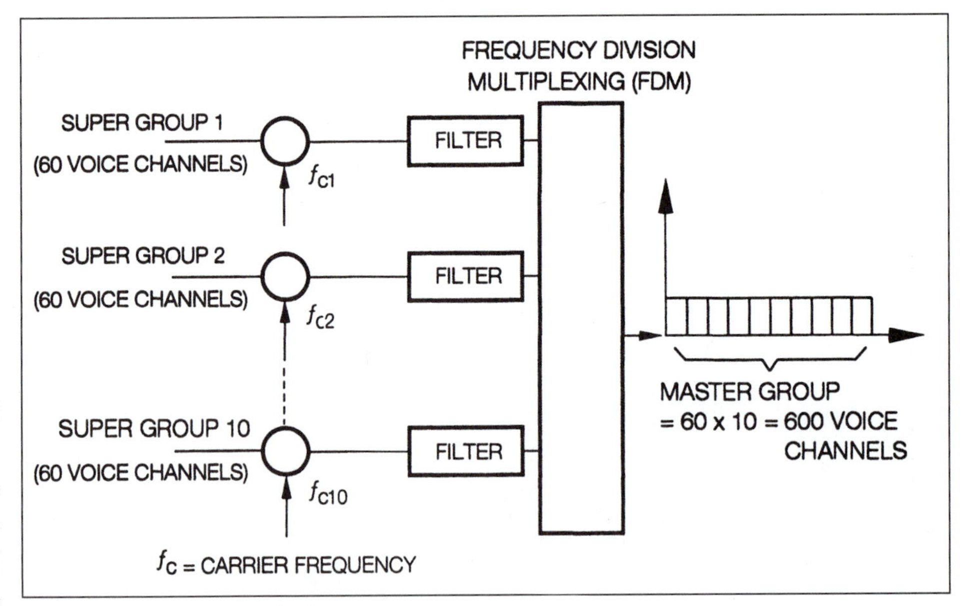

FIGURE 13-1 Master group formation

material dispersion In fiber optics communication systems, the dispersion of a signal caused by different wavelengths of light traveling at different speeds due to the frequency-dependent refractive index of the fiber. Material dispersion in a fiber is proportional to the spectral width of an optical source. *See also* **dispersion**.

Mbps Megabits per second. Also represented as Mb/s.

mean time before failure (MTBF) A measure of a component's reliability, it represents the average time duration until a component fails. (*See* Module 33.)

mean time to repair (MTTR) A measure of a component's reliability, it represents the average time required to repair a failed component. (*See* Module 33.)

media access unit (MAU) A type of transceiver that allows node connections to the backbone in a network.

medium

1. A material such as magnetic tape, magnetic disk, etc., on which data is recorded.
2. A material that is used as a channel for the transmission of information in the form of electromagnetic or acoustic signals between a transmitter and a receiver. Examples are copper cables, optical fibers, air, water, free space, etc.

medium earth orbit (MEO) satellite *See* **satellite communication** *and* Module 46.

megahertz (MHz) One million cycles per second.

$$1 \text{ MHz} = 1{,}000{,}000 \text{ Hz or } 1{,}000 \text{ kHz}$$

meridional ray A ray that propagates through a fiber by total internal reflection and passing through the optical axis of the fiber. *See also* **axial ray**, **optical fiber**, *and* **skew ray**.

message switching A transmission technique in which messages are transferred between two points that are not connected. This is done by store and forward, i.e., a switching center stores the message and forwards it to its destination when a line becomes available. Contrast to *circuit switching.*

metropolitan area network (MAN) A network that serves a metropolitan area. Its geographical size is smaller than a WAN (wide area network) but larger than a LAN (local area network). IEEE 802.6 standard addresses MANs. *See also* **local area network, topology,** *and* **wide area network**.

microcomputer A small computer consisting of a central processing unit (microprocessor), memory, and input and output devices.

microprocessor The central processing unit of a microcomputer. Contains circuitry to perform logical and arithmetic operations on data. Tables 13-1 and 13-2 compare different types of microprocessors.

TABLE 13-1 Microprocessors (1980s)

MICROPROCESSOR	# OF BITS	SPEED (MHz)	MANUFACTURER
8051/8052	8	12, 16	Intel
6804	8	2, 4	Motorola
Z80	8	6, 8, 18, 20	Zilog
8086/8088	8/16	5, 8	Intel
80186/80188	16	8, 10, 12.5, 16	Intel
80286	16	8, 10, 12.5, 16	Intel
68000 family	32	12, 16, 25, 33, 50	Motorola
80386	32	16, 20, 25, 33	Intel
80486	32	25, 33	Intel
88000	32	16, 20, 25, 33	Motorola
i860	32/64	33, 40	Intel

TABLE 13-2 Microprocessor (1990s–2005+)

MICROPROCESSOR	PROCESSOR SPEED (MHz)	BUS SPEED (MHz)	MANUFACTURER
Pentium MMX	133, 150, 166, 200 233, 266	66	Intel
Pentium II	233, 266, 300, 333, 350, 366, 400, 450	60, 66	Intel
Celeron	266 MHz–1.2 GHz	66, 100	Intel
Pentium III	450 MHz–1.2 GHz	100, 133	Intel
Pentium 4	1.3 GHz–2 GHz	100 (400 for data transfer)	Intel
Cyrix III	433, 466, 500, 533	66, 100, 133	Cyrix
AMD-K6-III	350, 366, 380, 400, 433, 450	100	AMD
AMD Athlon	1.9 GHz	200	AMD

microsecond (µs) One-millionth of a second.

microwave

1. Electromagnetic waves having wavelengths in the micron range. They cover a frequency band extending from 1 GHz to 1 THz.
2. A type of indirectly coupled or radiated telecommunications system in which microwave frequencies are used to establish a line-of-sight communications link between a transmitter and a receiver.

millisecond (ms) One-thousandth of a second.

$$1\text{ms} = 0.001 \text{ second} = 10^{-3} \text{ seconds}$$

minutes of use (MOU) A unit of traffic intensity measurement. One MOU refers to one fully occupied traffic path for one minute. Sixty MOU equal 1 erlang or 36 CCs. *Also see* **erlang, CCS,** *and traffic formulas in* Module 33.

mobile earth station A mobile unit capable of transmitting to and/or receiving information from communication satellites.

mobile telephone exchange A telephone exchange that provides service to mobile telephone customers.

mobile telephone service A telephone service in which mobile stations have access to a public switching telephone system by a radio link. *See also* **cellular radio/telephone**.

mode For electromagnetic waves, a specific relationship between the wave's electric field component, magnetic field component, and the direction of propagation. Figure 13-2 illustrates various modes of propagation.

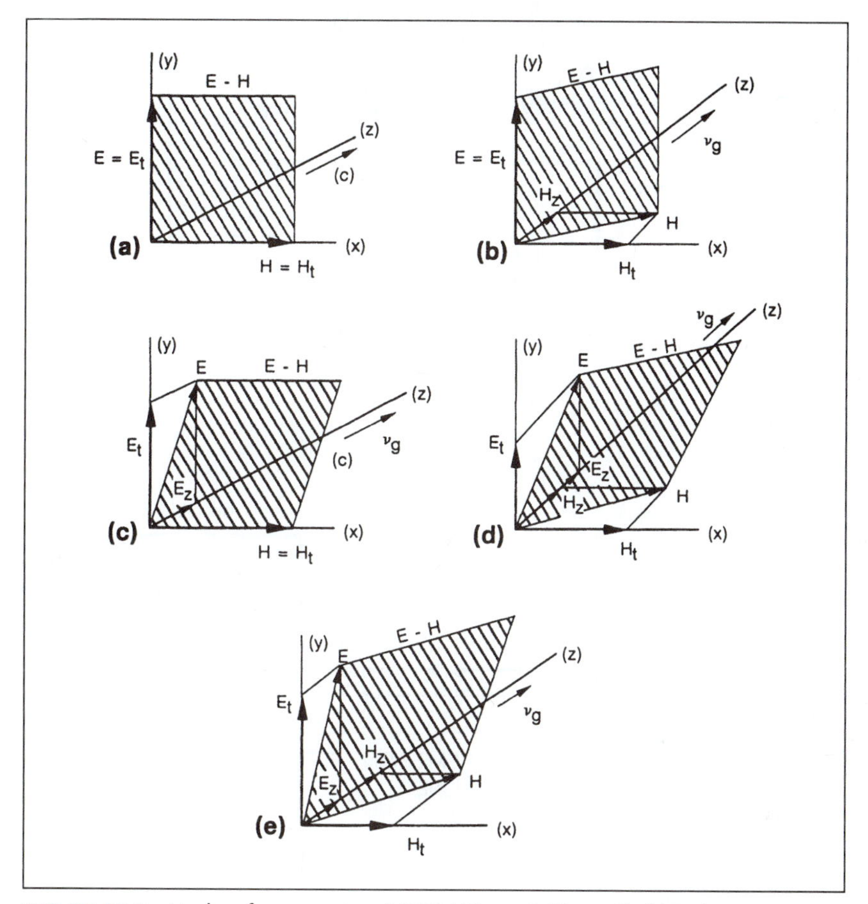

*FIGURE 13-2 Modes of propagation: (a) TEM ($E_z = 0, H_z = 0$); (b) TE ($E_z = 0, H_z$ finite); (c) TM (E_z finite, $H_z = 0$); (d) HE (E_z, H_z finite, $H_t > E_t$); (e) EH (E_z, H_z finite, $E_t > H_t$) (**Note:** Reprinted by permission of Prentice Hall, Englewood Cliffs, New Jersey, from Dennis Roddy and John Coolen,* Electronic Communications*, 3d ed., p. 717. ©1984.)*

modem Contraction for *mod*ulator-*dem*odulator. A device, consisting of a modulator/demodulator, that is used to convert a digital signal to analog tones for transmission over an analog line, and then convert the analog signal back to digital form at the other end of the link (Figure 13-3). *See also* **CODEC**.

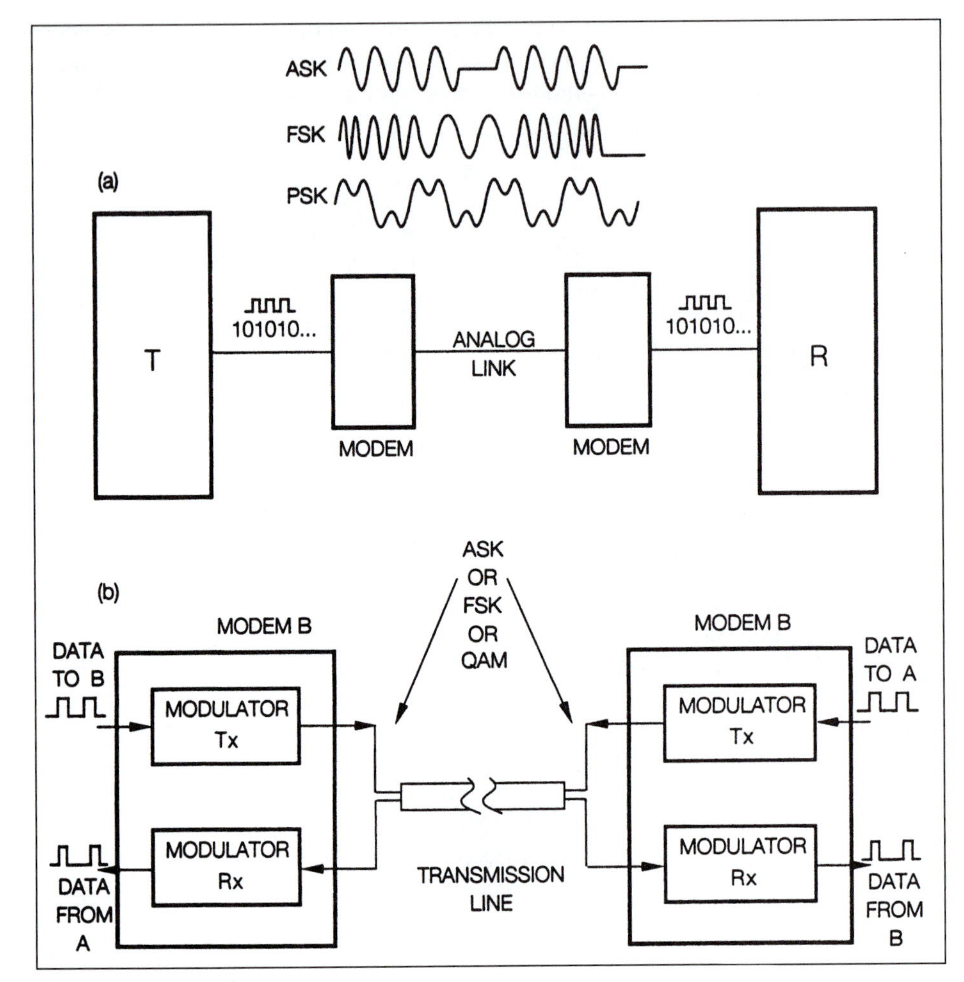

FIGURE 13-3 Modem

modem eliminator A device that permits the connection of two data-terminal-equipment devices without using modems. It is also called a *null modem* and can be obtained by simply reversing the transmit and receive pins 2 and 3 of the RS-232 interface cable. (*See* Module 39.)

modulation The technique of varying one of the characteristics of a carrier (high-frequency) signal, such as amplitude or frequency and phase, in accordance with the instantaneous changes in the information signal (low-frequency) (Figure 13-4). Tables 13-3, 13-4, and 13-5 list various types of analog, digital, and RF modulation techniques.

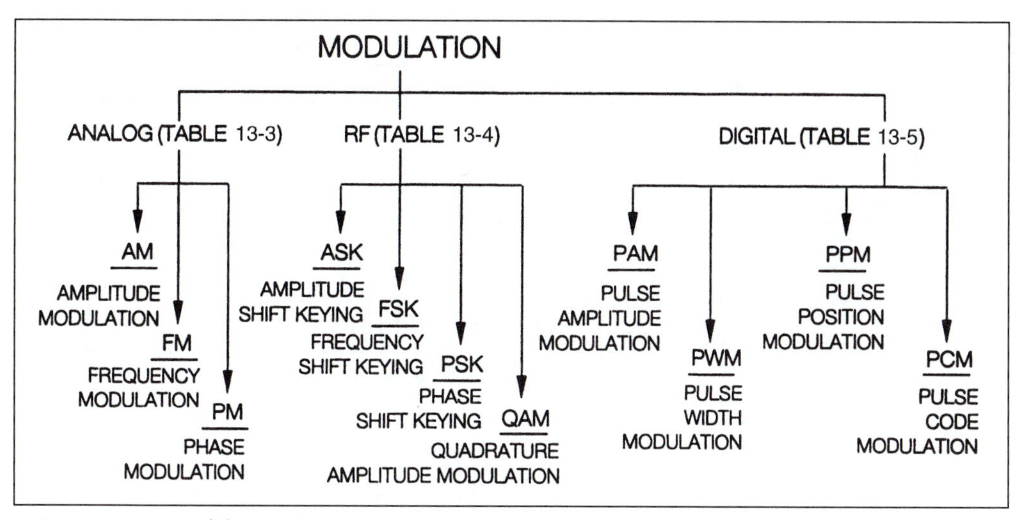

FIGURE 13-4 Modulation

TABLE 13-3 Analog modulation

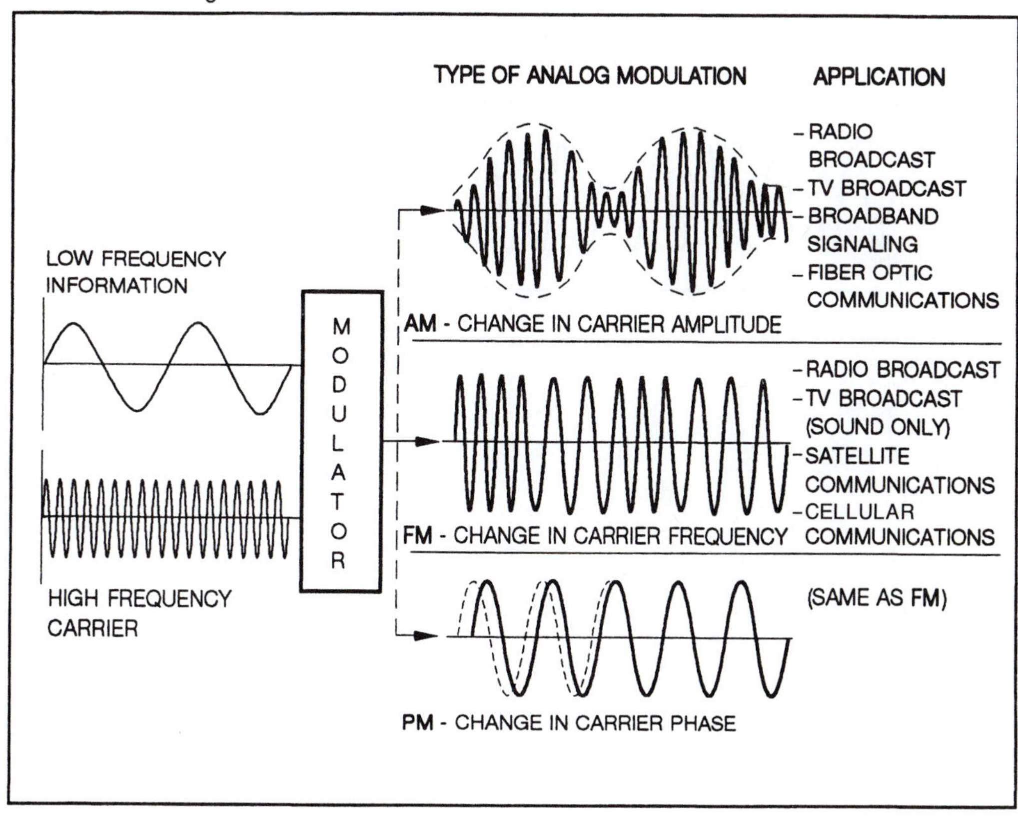

TABLE 13-4 RF modulation

GRAPHIC REPRESENTATION	APPLICATION
MARK / SPACE: 0 1 0 0 1 0; DATA SIGNAL; ANALOG CARRIER	
ASK	LOW-SPEED MODEMS
f_s f_m f_s f_s f_m f_s; FSK; f_s = SPACE FREQUENCY f_m = MARK FREQUENCY	MEDIUM-SPEED MODEMS
0° 180° 0° 0° 180° 0°; PSK	HIGH-SPEED MODEMS

(continues)

TABLE 13-4 RF modulation (continued)

GRAPHIC REPRESENTATION	APPLICATION
QAM PHASOR DIAGRAM 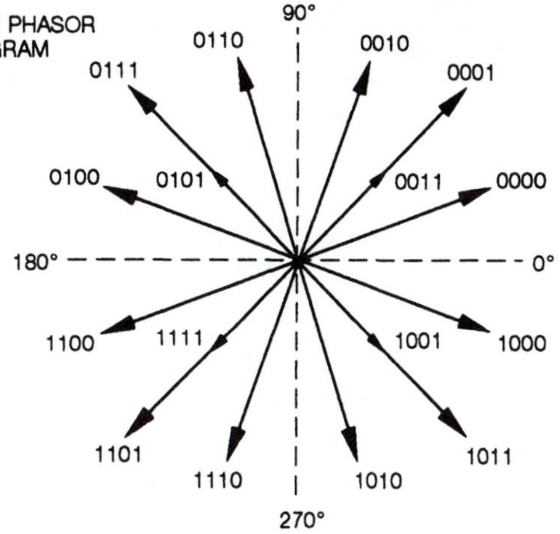 90° · 0110 · 0010 · 0111 · 0001 · 0100 · 0101 · 0011 · 0000 · 180° · 0° · 1100 · 1111 · 1001 · 1000 · 1101 · 1011 · 1110 · 1010 · 270°	4-BITS ALLOW SIXTEEN POSSIBLE STATES PER SIGNALING PULSE. FOR EXAMPLE, FOR 2400 BAUD SIGNAL, 4 X 2400 = 9600 BPS, INFORMATION RATE. USED IN HIGH-SPEED MODEMS.
16-POINT QAM SIGNAL CONSTELLATION (3 AMPLITUDES, 12 PHASES) 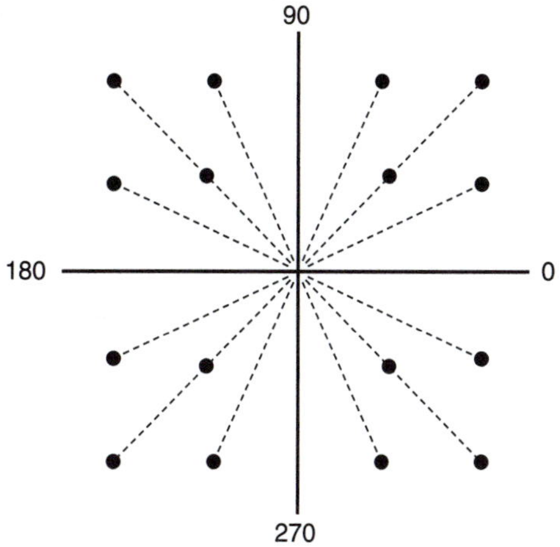 90 · 180 · 0 · 270	

TABLE 13-5 Digital modulation or pulse modulation

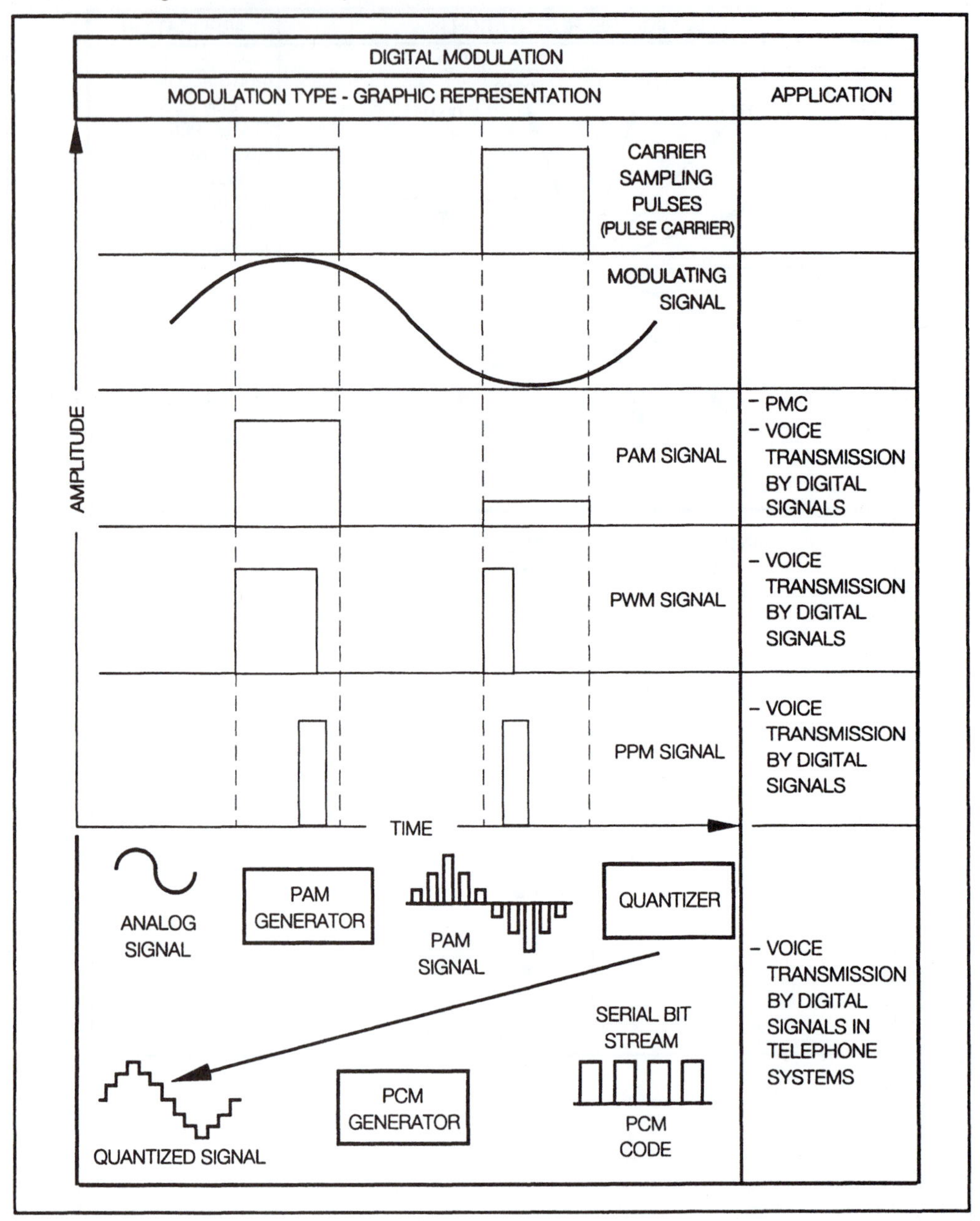

DIGITAL MODULATION	
MODULATION TYPE - GRAPHIC REPRESENTATION	APPLICATION
CARRIER SAMPLING PULSES (PULSE CARRIER)	
MODULATING SIGNAL	
PAM SIGNAL	- PMC - VOICE TRANSMISSION BY DIGITAL SIGNALS
PWM SIGNAL	- VOICE TRANSMISSION BY DIGITAL SIGNALS
PPM SIGNAL	- VOICE TRANSMISSION BY DIGITAL SIGNALS
ANALOG SIGNAL; PAM GENERATOR; PAM SIGNAL; QUANTIZER; QUANTIZED SIGNAL; PCM GENERATOR; SERIAL BIT STREAM; PCM CODE	- VOICE TRANSMISSION BY DIGITAL SIGNALS IN TELEPHONE SYSTEMS

monitor A device used to display the output of a computer.
monochromatic radiation Radiation consisting of a single wavelength or color.
Moore's Law The number of transistors in integrated circuits doubles every 18 months. Table 13-6 illustrates the transistor count trend for CPUs.

TABLE 13-6 Evolution of transistor trend for CPUs

DECADE	CPU	NUMBER OF TRANSISTORS/CPU
1970s	4004	
	8008	1,000–10,000
	8080	
	8086	10,000+
1980s	286	100,000+
	386 Processor	
	486 DX Processor	1,000,000
1990s	Pentium Processor	1,000,000+
	Pentium II Processor	
	Pentium III Processor	10,000,000+
200X		1,000,000,000

Morse code Code originally developed by Samuel Morse to transmit information using the telegraph. Radiotelephone Morse characters are represented by groups of *dots* and *dashes*, which are based on the duration of electrical impulses. A short-duration electrical impulse represents a dot and a long-duration impulse is a dash. (*See* Module 40.)

motherboard The main circuit board in a computer. The CPU, ROM chips, SIMMs, DIMMs, RIMMs, and interface cards are plugged into the motherboard of a computer.

mouse A pointing and input device that allows the movement of a cursor to select a program on a screen with the click of a button.

μ-law A North American version of companding (*com*pressing/ex*panding*) algorithm used for telephone systems; it allows analog channels to maintain acceptable signal-to-distortion levels over a wide dynamic range of input signals. *Also see* **A-law, companding**, *and* Module 33 *for formulas.*

multidrop line A single line that terminates several devices at different locations. Also called *multipoint line.*

multifiber cable A cable that contains two or more fibers.

multimode dispersion In a multimode optical fiber, the dispersion of a signal caused by a difference in path lengths for different modes of propagation. Also known as *modal* or *intermodal dispersion. See also* **dispersion, material dispersion,** *and* **wave guide dispersion** (Table 4–2).

multimode fiber A type of optical fiber that supports multiple-mode propagation. *See also* **optical fiber**.

multiplexer A device that performs the multiplexing of a number of information channels in order to transmit them over a common channel. Figure 13-5 illustrates a digital multiplexer. *See also* **frequency-division multiplexing (FDM), multiplexing,** *and* **time-division multiplexing (TDM).**

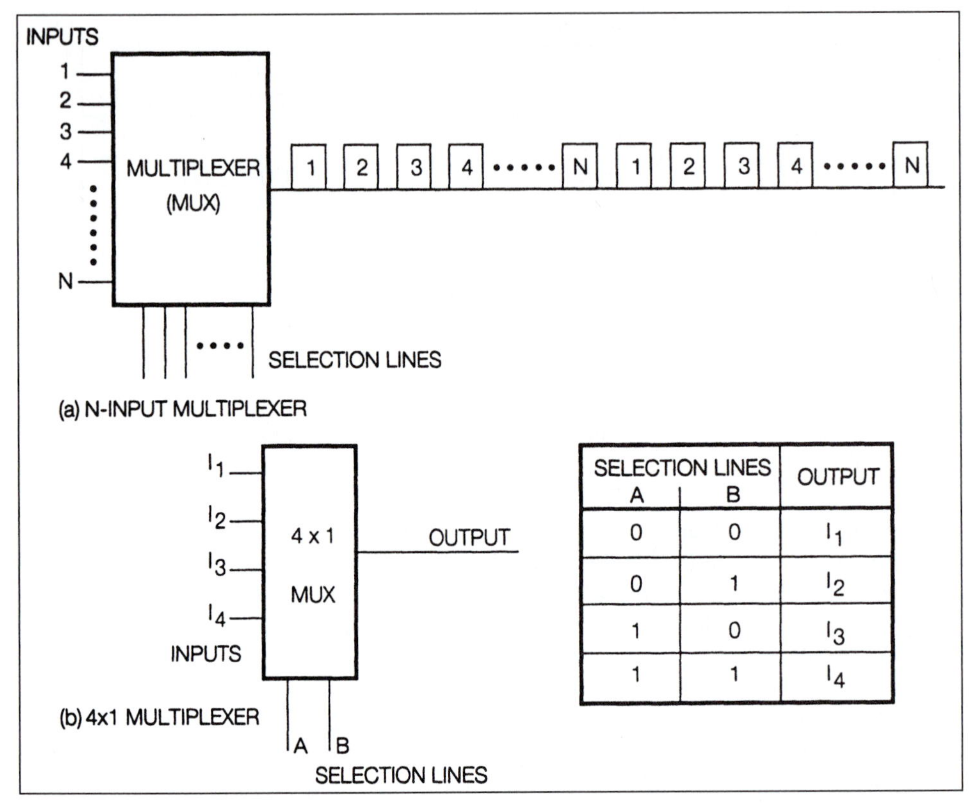

SELECTION LINES A	B	OUTPUT
0	0	I_1
0	1	I_2
1	0	I_3
1	1	I_4

FIGURE 13-5 Multiplexer

multiplexing A technique of transmitting two or more narrow band channels over a broadband by: (1) dividing the bandwidth of a broadband channel into narrow bands whereby a number of information channels can be transmitted simultaneously (frequency-division multiplexing [FDM]), or (2) using the entire bandwidth of a broadband channel to transmit a number of information channels not simultaneously but one at a time (time-division multiplexing [TDM]). *See also* **frequency-division multiplexing (FDM)** *and* **time-division multiplexing (TDM)**.

multipoint network A network in which the communication channel is shared by two or more terminals.

multiprocessing Use of more than one processor in a multiprocessor system to execute simultaneously two or more programs or sets of instructions.

multiprogramming A technique that allows linked or resident execution of two or more programs by one computer, not necessarily simultaneously.

multi-tasking The simultaneous execution of two or more tasks or the simultaneous use of a single program that can carry out many functions.

MODULE 14 Definitions N

NAK *See* **negative acknowledgement.**

narrow band channel Subvoice-grade channel capable of transmitting data at 100 to 200 bps. Usually applied to telegraph, metering, or alarm circuits, allowing more than one per channel.

negative acknowledgement (NAK) A control character used in data communication to indicate that the previous data block received by the receiver was erroneous and that the receiver is ready to receive the retransmission of the previous data block.

neper A logarithmic unit used to express attenuation of electromagnetic signals on a transmission line or in free space.

$$1 \text{ db} = 8.686 \text{ nepers}$$

network A number of points connected by communications channels.

network architecture An overall plan that defines the design of a communication network in terms of software, hardware, protocols, and topologies. *See also* **open systems interconnection, systems network architecture (SNA)**, *and* Module 35.

network layer The third protocol layer in the OSI model, responsible for transmitting data from source to destination.

network termination 1 In integrated services digital network (ISDN) technology, a device responsible for the carrier's side of a connection. It performs signal conversion and maintenance of the loop's electrical characteristics. *See also* Module 34.

network termination 2 In integrated services digital network (ISDN) technology, an intelligent device responsible for the user's side of the connection. It performs switching, multiplexing, and concentration of a number of TE1 terminals. *See also* Module 34.

network topology The physical and logical arrangement of nodes in a network. *See also* **topology**.

network virtual terminal An abstract network in which incompatible terminals can be linked to each other with the help of protocol software.

NNX A code format used in the older central office code. N represented any digit from 2 to 9 and X represented any digit from 0 to 9. NNX presently means any first three digits of a seven-digit telephone number, not an area code. *See also* **numbering plan area**.

node A point in a network where two or more communications channels terminate. A node generally refers to intelligent devices such as computers, processors, controllers, etc.

noise Random electrical disturbances that interfere with the reception and reproduction of electric signals and cause the deterioration of signal fidelity. Noise causes errors in data transmission, and it is one of the basic factors that sets a limit on the rate of transmission. There are many sources that produce noise. They can generally be classified as internal (Table 14-1) and external (Table 14-2).

TABLE 14-1 Internal Noise Sources

NOISE TYPE	DESCRIPTION / DEFINITION
Thermal	Also known as Johnson noise, white noise, and Gaussian noise, thermal noise is generated by the random motion of free electrons in conductors. The root mean square noise voltage generated within a conductor equals to $V_n = \sqrt{4\,kTRB}$ where *Vn = RMS noise voltage* *T = temperature in degrees Kelvin* *k = Boltzmann's Constant = 1.38 × 102^{-23} J/K* *R = resistance of conductor in ohms* *B = bandwidth in hertz* The average noise power generated is given by $P = kTB$ watts The average power is independent of resistance value.
Shot	In vacuum tubes, shot noise occurs due to random fluctuations in the emission of electrons from the cathode. The root mean square value of shot noise current is given by $I_n = \sqrt{2\,I_{dc}qB}$ where *I_{dc} = dc current (rms amperes)* *q = electron charge (1.6 × 10^{-19} Coulombs)* *B = bandwidth (hertz)* In semiconductors, shot noise occurs due to random fluctuation in current due to the discrete-particle nature of charge carriers. The charge carrier flows in a random path of motion. The magnitude of root mean square value of shot current in a pn-junction diode is given by

(continues)

TABLE 14-1 (continued)

	$$I_n = \sqrt{2(2 + 2I_o)qB}$$ where *I = dc current through the junction (amperes)* *I_o = reverse saturation current (amperes)* *q = electron charge (1.6 × 10^{-19} Coulombs)* *B = bandwidth (hertz)*
Transit Time	In a semiconductor device, noise is generated by diffusion of charge carriers back to the source or emitter because the transit time for a junction is comparable to the time period of the input signal.
Generation-Recombination	In semiconductor devices, noise is generated by random generation of charge carriers due to random ionization of impurity atoms by thermal excitation.
Flicker	In semiconductor devices, noise is generated due to variations in the conductivity because of fluctuation in the charge carrier density.
Partition	In semiconductor devices, noise is generated by random fluctuation in the division of current among two or more paths. Partition noise is more significant in transistors than in diodes.

TABLE 14-2 External Noise Sources

NOISE TYPE	DESCRIPTION / DEFINITION
Man-made or Industrial	It is generated by spark-producing devices such as engine ignition systems, fluorescent lights, etc. Also present on AC power lines in form of voltage surges produced by switching on and off of motors. Man-made noise covers a wide spectrum of frequencies ranging from low frequencies up to 500 MHz.
Atmospheric	It is caused by disturbances such as lightning in the atmosphere. Atmospheric noise covers the entire radio spectrum. The intensity of atmospheric noise is inversely proportional to frequency. Therefore its intensity is significant at low frequencies, and at frequencies above 20 MHz it is insignificant.

(continues)

TABLE 14-2 (continued)

Extraterrestrial	This type of noise covers noise coming from the sun and other stars. Solar radiation follows an 11-year cycle known as the solar cycle. Solar storms following this cycle cause significant disturbances in the earth's atmosphere and affect ionospheric radio communication. Noise coming from other stars is known as *cosmic* noise and is significant in the frequency range 8 to 1500 MHz.

Figure 14-1(a) and Figure 14-1(b) illustrate noise and a classification tree of noise. *See also* Module 33 *for noise formulas.*

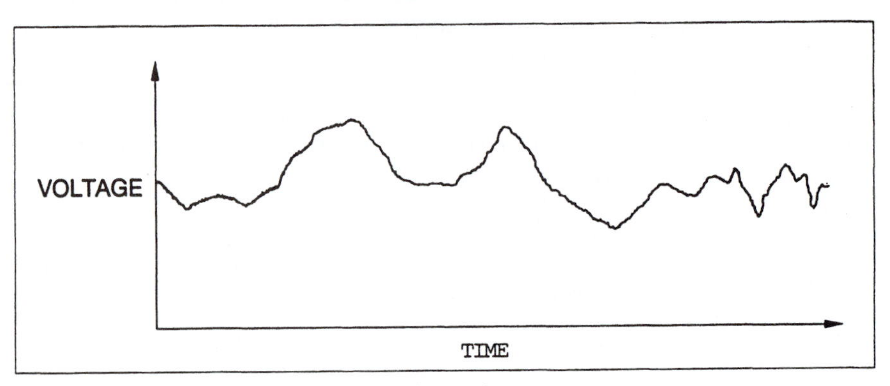

FIGURE 14-1(a) Noise: random electrical signal

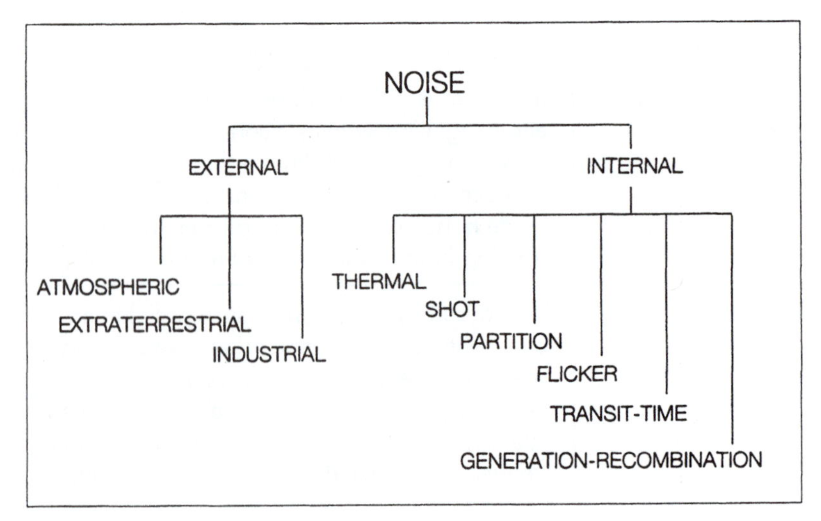

FIGURE 14-1(b) Classification tree of noise

noise equivalent power (NEP) A figure of merit used to measure the performance of a radiation detector. It is defined as the amount of incident radiant power required to produce an rms signal-to-noise ratio (SNR) equal to one at the detector output, i.e., the output rms voltage or current equals rms noise voltage or current.

$$NEP = \frac{\text{incident radiant power}}{V_s/V_n}$$

$$= \frac{HA}{SNR} = \frac{HA}{V_s/V_n} \text{ in watts}$$

Where: H = irradiance (W/cm^2)
A = detecting area of detector (cm^2)
V_s = signal voltage at detector output (volts)
V_n = noise voltage
SNR = signal-to-noise ratio

noise factor (F) For any electrical circuit, the ratio of the signal-to-noise power ratio at the input to the signal-to-noise power ratio at the output.

$$F = \frac{\text{signal-to-noise ratio at the input of circuit}}{\text{signal-to-noise ratio at the output of circuit}}$$

$$= \frac{P_{si}/P_{ni}}{P_{so}/P_{no}}$$

Where: P_{si} = signal power at input
P_{so} = signal power at output
P_{ni} = noise power at input
P_{no} = noise power at output

noninteractive system A type of input/output control system in which there is no interaction between the user and the computer during the execution of the program. Often referred to as *batch processing.*

non-return-to-zero (NRZ) A digital data encoding technique in which the duty cycle of the binary pulse representing a bit is 100 percent, i.e., the voltage level remains the same for the time period of the bit. Figure 14-2 illustrates unipolar and bipolar NRZ signals. *See also* **digital signal**.

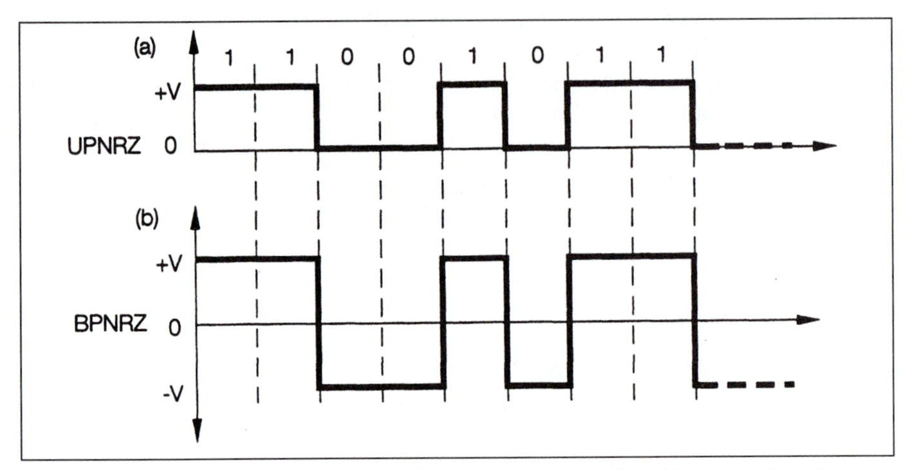

FIGURE 14-2 Non-return-to-zero signals: (a) UPNRZ, unipolar; (b) BPNRZ, bipolar

nonswitched line *See* **leased line**.

ns (nanosecond) One-billionth of a second.

NTSC A television broadcast standard developed by the National Television Systems Committee. (*See* Module 41.)

null modem *See* **modem eliminator**.

numbering plan area (NPA) A geographic subdivision (represented by a 3-digit code) that is a part of the national numbering plan for direct distance dialing.

numerical aperture (NA) The light-gathering power of an optical fiber (Figure 14-3). Mathematically, it is given by:

$$NA = \frac{\sqrt{n_1^2 - n_2^2}}{n_0}$$

$$= \sin^{-1} \Theta_{0max}$$

Where: n_1 = refractive index of fiber core
n_2 = refractive index of fiber
n_0 = refractive index of launching medium (usually air)
$\Theta_{0\,max}$ = maximum acceptance angle

NXX A format used to represent a code for the central office. N represents any digit from 2 to 9 and X represents any digit from 0 to 9.

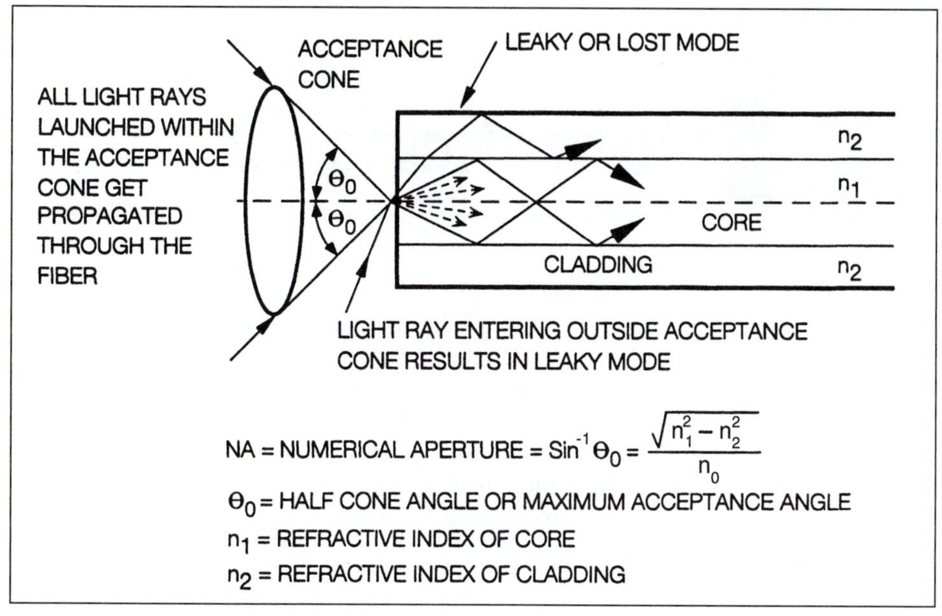

FIGURE 14-3 Numerical aperture

Nyquist sampling theorem A principle stating that each cycle of a sampled analog signal must be sampled at least twice to accurately reproduce that signal at the receiver. The sampling frequency is given by:

$$f_s \geqslant 2f_a$$

Where: f_s = sampling frequency
f_a = the highest frequency of the analog signal

The sampling of an analog signal is required to convert it into digital form. The higher the sampling frequency at the transmitter, the better the reproduced signal at the receiver.

Nyquist theorem The maximum amount of binary information that can be transmitted over a noiseless channel is equal to twice the bandwidth of the channel.

$$C = 2W \text{ bps}$$

Where: C = transmission speed of the channel
W = bandwidth of the noiseless channel

MODULE 15 Definitions O

octet A group of eight bits. Also called a *byte*.

odd parity A noninformation bit added to a group of information bits to make the total number of ones odd. It is used for error detection. *See also* **parity check**.

off-hook A condition that causes current (called loop current) to flow from the local exchange when the telephone's handset is removed from its cradle. Electronic phones generate a digital code when off-hook.

off-line system A system in which there is no connection between the data input/output devices and the central processing unit. The data is stored on a storage medium, such as magnetic tape or diskette, and then processed later.

on-hook The opposite condition of *off-hook*. By replacing the telephone handset on its cradle, current flow from the local exchange is stopped.

on-line system A system in which there is a direct connection between the central processing unit and the data input/output devices (terminals).

open systems interconnection (OSI) A network architecture standard developed by the International Organization for Standardization (ISO). Also called OSI reference model, it is a seven-layer network architecture that defines the protocol standards for information exchange between two OSI-compatible devices. (*See* Module 35.)

open-wire A transmission line consisting of two conductors that are separated by air (Figure 15-1). The major disadvantage of open-wire is radiation loss, which puts a limit on its bandwidth. The radiation loss is proportional to the signal frequency. That is why open-wire is used for low-frequency transmission only. For transmitting high frequencies, tuned open-wires are used. *See also* **two-wire line**.

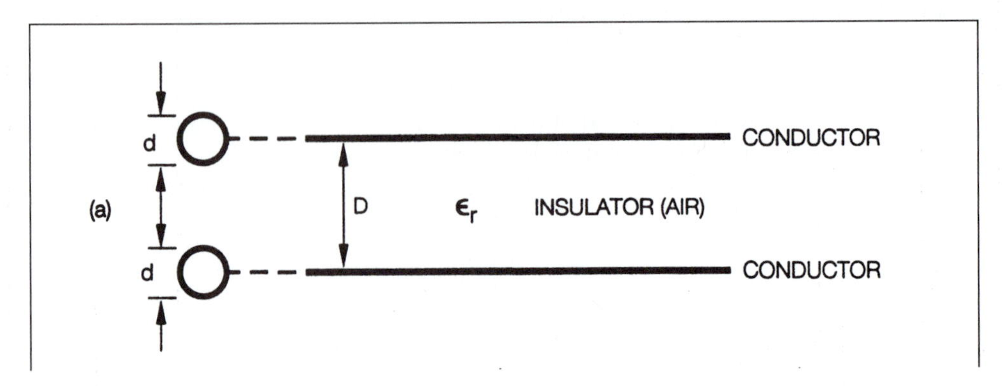

FIGURE 15-1 Open-wire line: (a) two-wire open line; (b) transverse view of electric and magnetic flux between the two conductors of the line (continues)

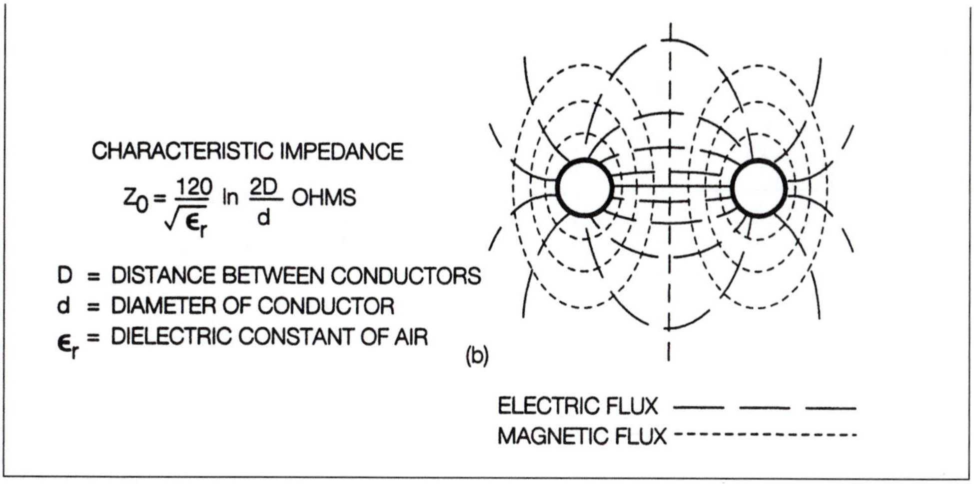

CHARACTERISTIC IMPEDANCE

$$Z_0 = \frac{120}{\sqrt{\epsilon_r}} \ln \frac{2D}{d} \text{ OHMS}$$

D = DISTANCE BETWEEN CONDUCTORS
d = DIAMETER OF CONDUCTOR
ϵ_r = DIELECTRIC CONSTANT OF AIR

(b)

ELECTRIC FLUX — — —
MAGNETIC FLUX - - - - - - - -

FIGURE 15-1 (continued)

operating system (OS) Software that controls the internal operation of a computer. It provides scheduling, memory assignment, data management, debugging, input and output control, compilation, etc. Also called *master control program.*

optical attenuator A device used to reduce the intensity of light in fiber optics communication systems.

optical character recognition (OCR) A technique in which a light-sensitive scanning device reads printed characters and converts them to digital signals.

optical disk A data storage device in which an optical beam (laser) is used to record information. It has a greater storage capacity than the magnetic disk. *See* **CD-ROM, DVD.**

optical fiber A thin strand of dielectric material (e.g., glass, plastic) that consists of core and cladding and is used as a dielectric waveguide for transmission of light signals in fiber optics communications. Table 15-1 illustrates different types of optical fibers and their characteristics. *See also* **fiber cladding, fiber core,** *and* **fiber optics.**

TABLE 15-1 Types of optical fiber

TYPES OF OPTICAL FIBER						
TYPE OF FIBER	FIBER DIAGRAM / TYPES OF RAY PATHS d = CORE DIAMETER λ = WAVELENGTH I = AYJAL RAY 2 = MERIDIONAL 3 - SKEW RAY	REFRACTIVE INDEX PROFILE n= REFRACTIVE INDEX r = DISTANCE	FIBER CHARACTERISTICS			APPLICATION
			SOURCE	BW	DISPERSION	

(continues)

TABLE 15-1 Types of optical fiber (continued)

MONOMODE STEP-INDEX FIBER	CENTRAL RAY; n_2; CORE; n_1; 1; n_2 CLADDING; $d \geq \lambda$	n_2; n_1; n_2; r; n	LASER OR LED	10–100 GHz-Km	SMALL	LANS, MANS, WANS, SONET, FDDI, Gigabit Ethernet, FTTH, FTTC, SUBMARINE CABLE, e.g., TAT-8, TAT-9, etc.
MULTIMODE STEP-INDEX FIBER	n_2; 1; CORE; n_1; 2; n_2 CLADDING; $d >> \lambda$	n_2; n_1; n_2; r; n	LASER OR LED	200 + MHz-Km	HIGH	LANS, MANS, DATA LINKS, etc.
MULTIMODE GRADED-INDEX FIBER	CLADDING; 1; CORE; 3; 2; $d >> \lambda$	n_2; n_1; n_2; r; n	LASER OR LED	(200 MHz-Km) – (4 + GHz-Km)	HIGH	LANS, MANS, WANS, TRUNKS CONNECTING CENTRAL OFFICES, etc.

optical repeater A circuit used to amplify an optical signal in fiber optics communication links.

optical time domain reflectometry (OTDR) A technique used for characterizing and testing an optical fiber waveguide. A high-intensity optical pulse is launched into the fiber and the resulting reflections and backscatter are measured and analyzed for attenuation with respect to distance, location of defects, breaks, faults, and splices.

optical transmitter An optical source (e.g., LED, laser diode) that converts an electric signal into an optical signal, so that the signal can be transmitted through a dielectric medium such as an optical fiber. (*See* Module 27.)

optics The branch of physics that deals with the nature and properties of light. *See also* **geometric optics**.

optoelectronic device A device that converts an electronic signal into an optical signal or vice versa. These devices are used as optical transmitters and optical receivers in fiber optics communication systems.

oscillator An electronic device that generates AC signals by converting DC power to AC power. The frequency of AC oscillations is determined by the time constant of the circuit. An oscillator can also be defined as an amplifier with a positive feedback (Figure 15-2). Table 15-2 lists four types of electronic oscillators.

TANK CIRCUIT

AMPLIFIER

OUTPUT

FEEDBACK NETWORK

FIGURE 15-2 Oscillator (block diagram)

TABLE 15-2 Types of Oscillators

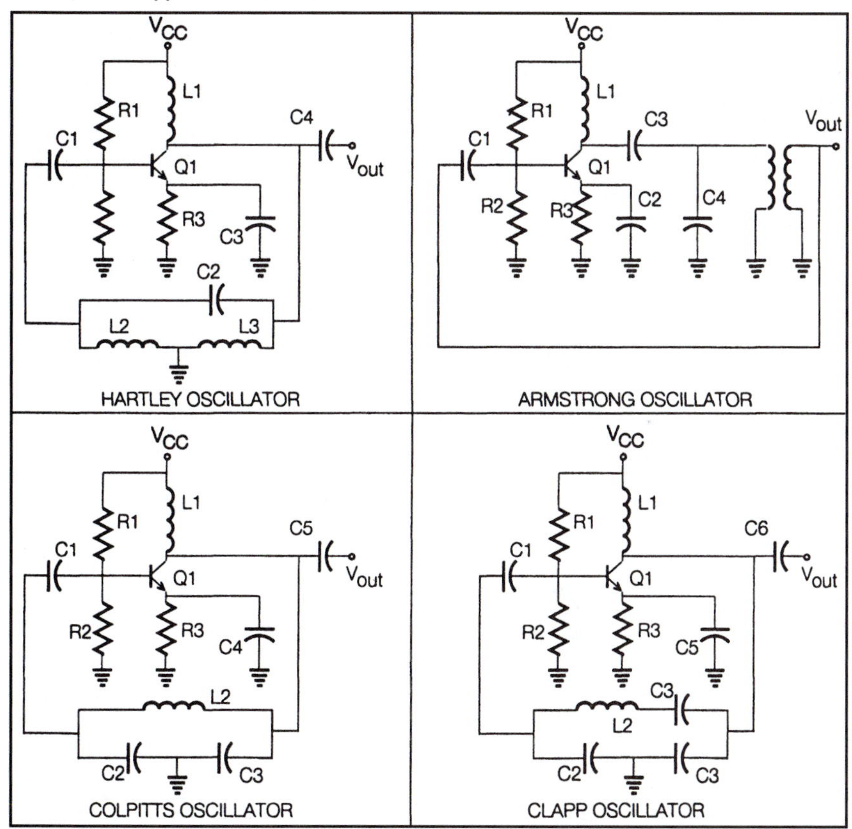

other common carrier (OCC) A common carrier that did not belong to the Bell System.

overflow A telephone call that cannot be completed due to excess traffic on a route. It is diverted over an alternate route or the result is a blockage and network busy signal (480 Hz + 620 Hz) tone that is on for 0.5 second and off for 0.5 second.

overhead bit A bit appended with a message in order to provide error checking, framing, etc. Also called *control bit.*

MODULE 16 Definitions P

PA Public address system, such as a loudspeaker, paging, etc.

PABX *See* **exchange, private branch.**

packet A group of bits, consisting of data and control signals, that is switched and transmitted as a composite whole. Data, address, control, and error detection information is arranged in a defined format.

packet switched network A network that carries data in the form of packets. *See also* **packet, packet switching**.

packet switching A data communications technique in which a message is divided into defined length units (packets) that are transmitted to a destination via a single path or multiple paths. The receiver arranges the received packets in proper order to retrieve the transmitted message.

PAD

1. A resistor network used as an attenuator or impedance matching device (Figure 16-1).
2. Packet assembler/disassembler: protocol conversion hardware or software that allows user terminals or devices to access a packet-switched network.

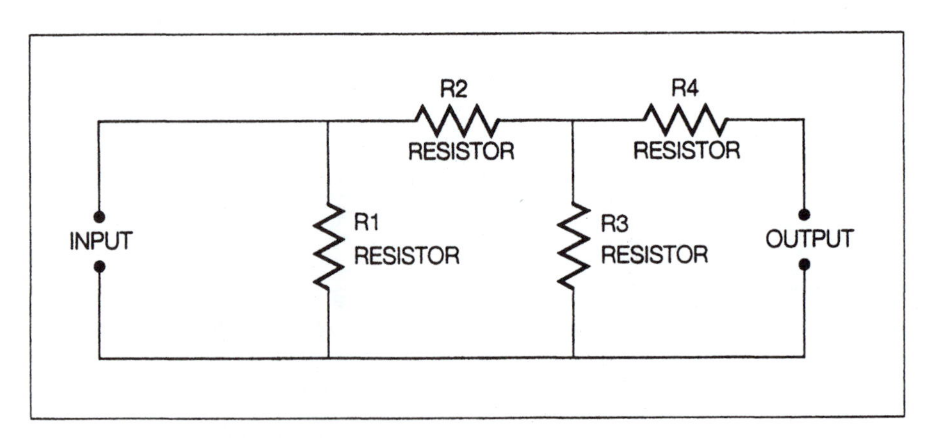

FIGURE 16-1 PAD network

PAL *See* **phase alternate line.** (*Also see* Module 41).

PAM *See* **pulse amplitude modulation**.

parallel port A female 25-pin port on a computer that allows transmission of data in parallel mode, usually to a printer. Parallel ports are designated as LPT1 and LPT2 (Figure 16-2).

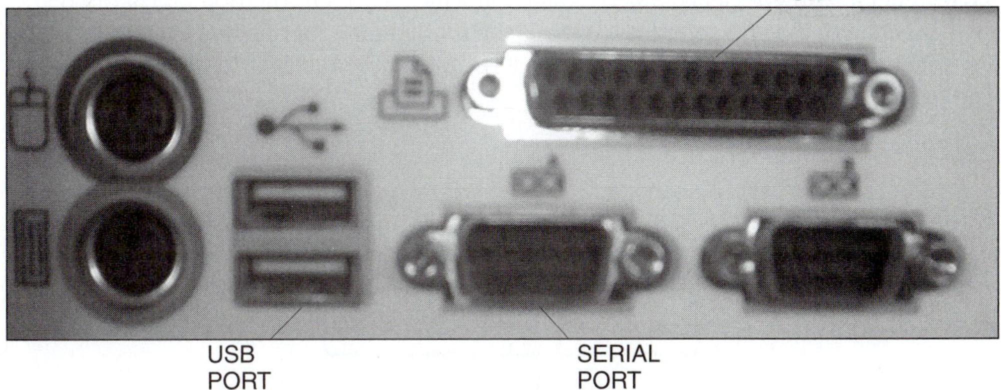

FIGURE 16-2 Computer ports (parallel, serial, and USB)

parallel transmission A data transfer technique in which all bits of a character or bytes are transmitted simultaneously to a desired destination over separate communication lines or on different carrier frequencies on one communication line (Figure 16-3).

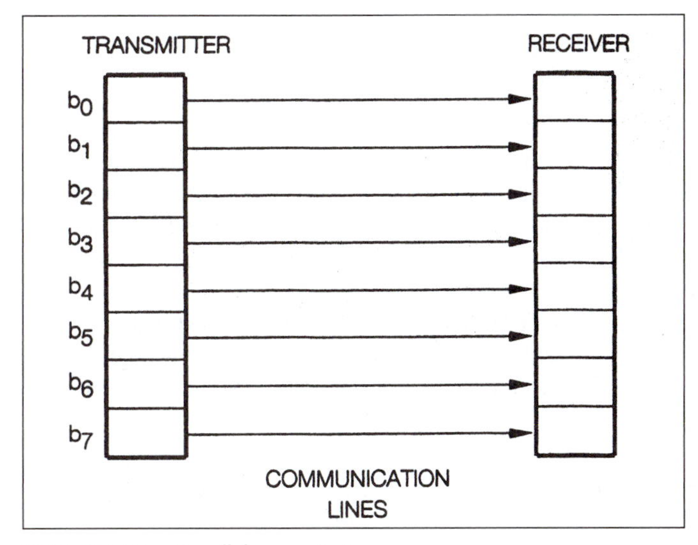

FIGURE 16-3 Parallel transmission

parity bit *See* **parity check.**

parity check An error-checking technique in which a noninformation bit (parity bit) is added to the character bits to make the total number of one bits even or odd (Figure 16-4). The transmitted parity is verified at the receiver for error-free data transmission.

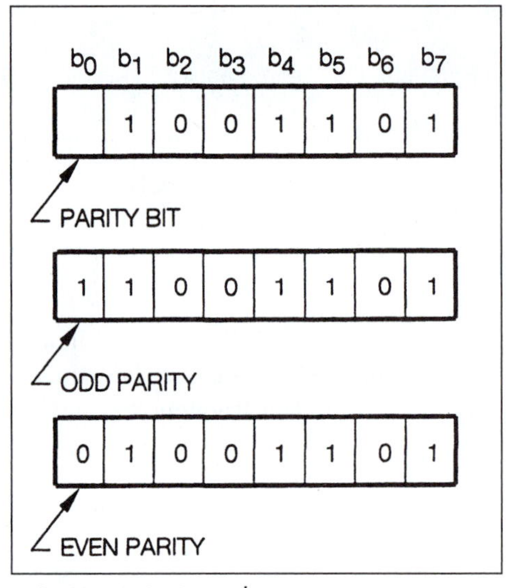

FIGURE 16-4 Parity bit

party line A subscriber's line to which multiple stations are connected.

passive attack An attempt by an unauthorized user such as a hacker to eavesdrop network traffic in order to obtain information that could be used to launch an active attack. A passive attack is more difficult to detect than an active attack.

PBX *See* **exchange, private branch**.

PDM Pulse duration modulation. *See* **pulse time-modulation**.

peer-to-peer A networking configuration in which computers are connected to each other, thus allowing each computer to act as client as well as server to other computers.

phase

1. In a periodic signal (a) a fraction of the period that has elapsed (measured with respect to the start of the signal), expressed in degrees or radians; or (b) an angular relationship between voltage and current (Figure 16-5).
2. One of the circuits of the polyphase system.

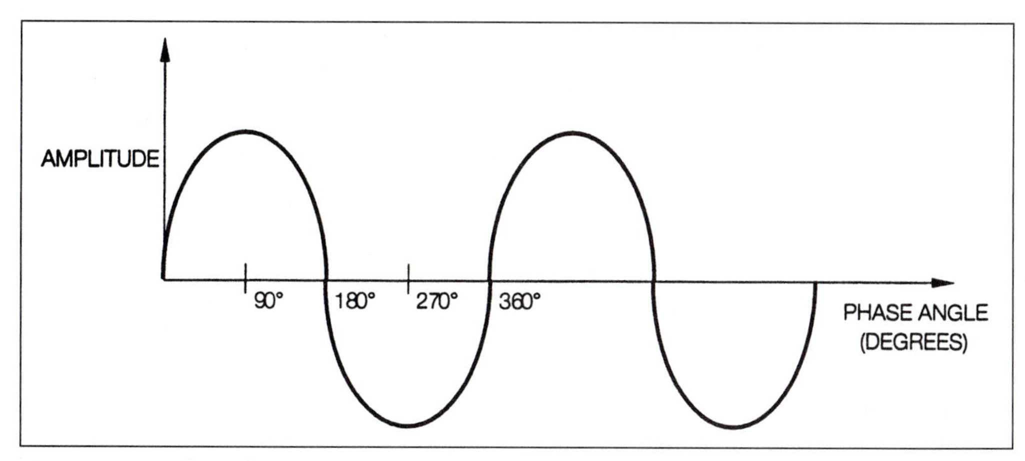

FIGURE 16-5 Phase diagram

phase alternate line (PAL) A color television broadcasting standard developed by West Germany and England. It uses 625 lines per frame, 25 frames per second, and a 50 Hz field frequency. *See also* **NTSC, SECAM** *and* Module 41.

phase angle The angle that indicates the phase relationship between two periodic signals (Figure 16-6).

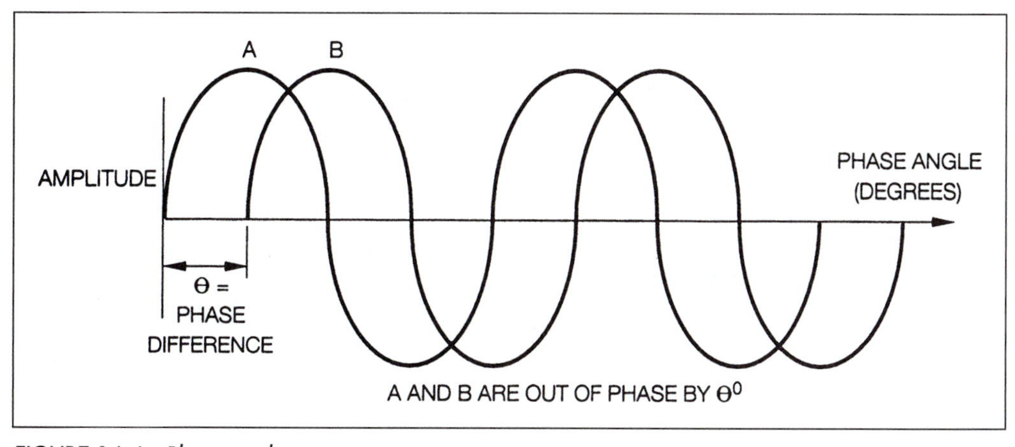

FIGURE 16-6 Phase angle

phase jitter A random deviation in the phase of a transmitted signal caused by fluctuations in the frequency. It interferes with the synchronization of the signal.

phase modulation (PM) An analog modulation technique in which the phase of the carrier signal is varied in accordance with the changes in the information signal. (See Table 13-3.)

phase shift keying (PSK) A type of RF modulation technique in which the phase of an analog carrier signal is varied in accordance with the variation in the digital information signal. (See Table 13-4.)

photodiode A semiconductor PN junction diode that converts quantum wave (IR, visible, UV) radiation into electrical signals. Photodiodes are reverse biased to reduce charge-carrier transit time and junction capacitance for high-speed operation. Photodiodes are used in fiber optics communication systems. (*See* Figures 1-29, 4-4, and 4-9.)

physical layer (See **the OSI Reference Model** *in* Module 35.)

PIN diode A type of photodiode in which an intrinsic semiconductor region is sandwiched between p-type and n-type semiconductor regions (Figure 16-7). The width of the depletion region is wider than an ordinary pn junction diode. Therefore, more photons can be absorbed. The capacitance of a PIN diode is much less than a pn junction diode. This reduces the response time. *See* also Module 27.

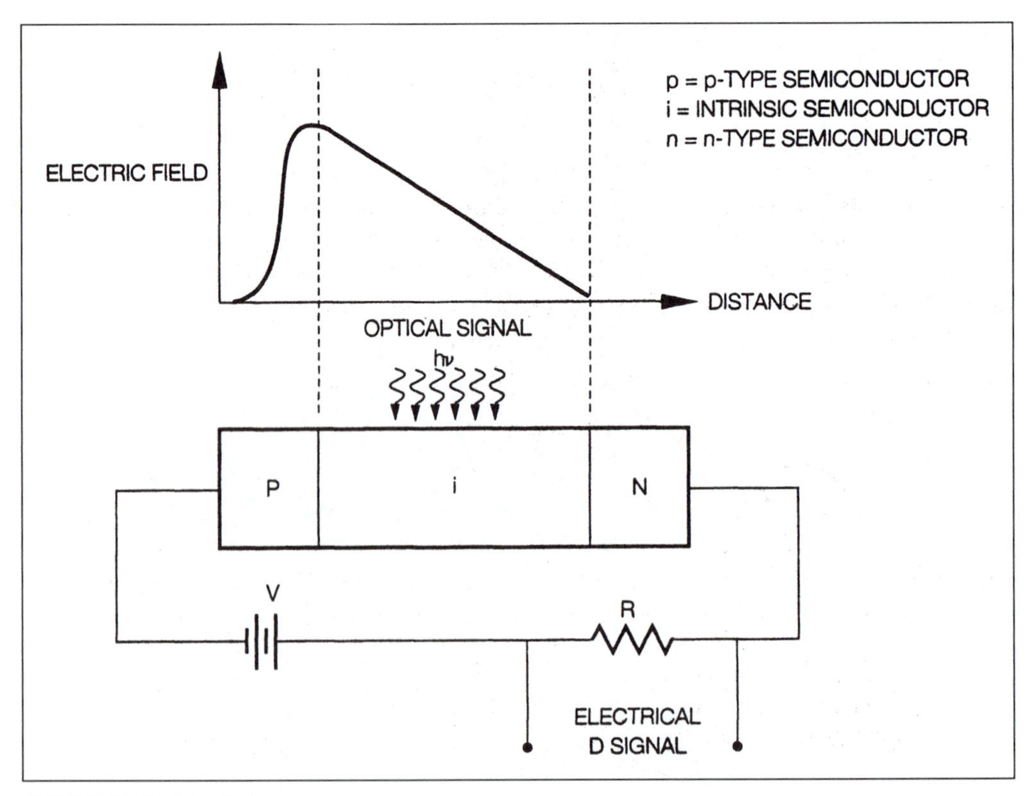

FIGURE 16-7 Pin diode

ping A utility that sends a packet to an IP address and waits for a response to check if the destination node is connected to the Internet or not. It verifies the connection and also measures the response time.

plastic-clad silica fiber A type of optical fiber in which a glass core is surrounded by a plastic layer or clad.

PM *See* **phase modulation**.

point-of-presence (POP) Since the AT&T divestiture, the physical-access location within the local access and transport area (LATA) of a long-distance and/or interLATA common carrier. The point to which the telephone company terminates a subscriber's circuit for long-distance dial-up or leased-line communications.

point-to-point connection A network or circuit configuration that connects two points without using any intermediate processing node. There may be switching facilities in the connection.

Point-to-Point Protocol (PPP) A protocol that allows a host to be connected directly to a single computer. It is used with modems that provide a connection between two computers.

polarization

1. The formation of dipoles in molecules.
2. A property of electromagnetic waves that defines the relationship between the electric-field vector and the direction-of-propagation vector of an electromagnetic wave (Figure 16-8). *See also* Module 29.

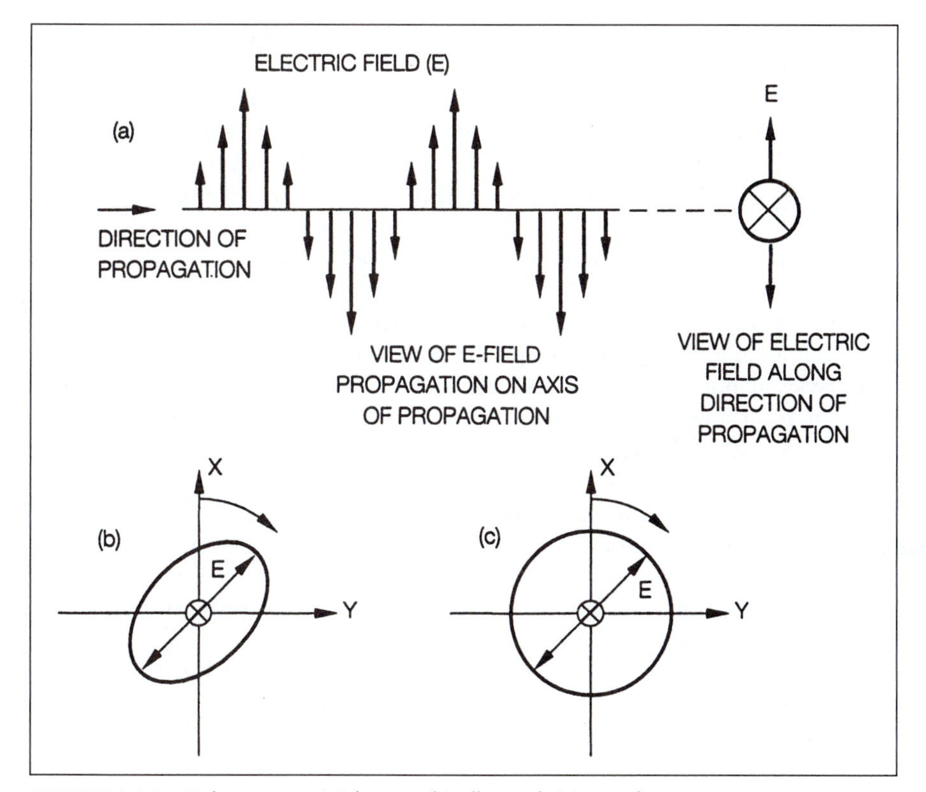

FIGURE 16-8 Polarization: (a) linear; (b) elliptical; (c) circular

polling A technique used in data communications systems whereby communications devices connected to a multipoint line are periodically invited by the controller unit for the transmission of data.

port

1. A point of access to a communication circuit, switch, network, computer, electronic device, etc. A port acts as a physical or electrical interface through which one gains access to a system via input or output devices.
2. A physical connector that allows an interface cable from a peripheral device (e.g., monitor, floppy drive, printer, mouse, etc.) to be attached to a computer (Figure 16-2).

POTS *P*lain *O*ld *T*elephone *S*ervice.

PPM Pulse position modulation. *See also* **modulation** *and* **pulse-time modulation**. (See Table 13-5.)

PPS Pulses per second.

pre-emphasis A technique used in an FM transmitter to boost the amplitude of higher modulating frequencies prior to modulation in order to maintain a constant signal-to-noise ratio (SNR) for higher modulating frequencies. *Also see* **de-emphasis.**

presentation layer (See **the OSI Reference Model** in Module 35.)

primary access interface An ISDN level of service recommended by ITU-TS (CCITT). It has a bit rate of 1.536 Mbps that is divided into 23 64 kbps B-channels and one 64 kbps D-channel. *See also* **integrated services digital network (ISDN)**. (See Module 34.)

primary center

1. In the United States, a class 3 office that connects toll centers and can act as a toll center for its local central office.
2. Internationally, a switching office that (a) is connected to local central offices, (b) passes trunk traffic, and (c) is equivalent to the U.S. class 4 office.

primary group A group of information channels combined together at the first level of multiplexing (FDM). *See also* **group, jumbo group, master group,** *and* **super group**.

primary station A network node that is responsible for the transmission of information on a link.

private key An encryption key used in asymmetric encryption to decrypt the encrypted message. In contrast to a public key, it is not known to all users of the cryptosystem, but only to an individual user who can use it to encrypt or decrypt, not both, for a single transaction.

private line *See* **leased line.**

private network A network established, owned, and operated by a private organization. Contrast with **public switched telephone network**.

programmable read only memory (PROM) An information storage device that allows the alteration of stored information. Contrast to read only memory (ROM), in which stored information cannot be altered *See also* **read only memory**.

PROM *See* **programmable read only memory**.

propagation delay A measure of time for a signal to pass through a device or a system. Factors like device turn-on time, device turn-off time, reactance, etc., contribute to signal propagation delay.

protocol A set of rules or procedures used to exchange information between two devices in a communications network. *See* Modules 35 *and* 38.

protocol converter A device that performs protocol translation in order to enable two protocol-incompatible devices to communicate with each other.

proxy server A software application that can cache information on an Internet server to act as a go-between for a client and a server. It provides additional security to the server, and acts as an application, e.g., HTTP, FTP, etc., to facilitate the communication on a network.

PTT (Post, Telephone & Telegraph) A government agency that provides the services for mail, telephone, and telegraph in most countries of the world, except North American countries.

public data network (PDN) A communication network (switched circuit network or packet-switched network) that is used for the transmission of data in digital form.

public key An encryption key used in asymmetric encryption that can be used to confirm the signature of an incoming message or to encrypt a message so that the intended user of the message can decrypt the message with the help of a private key. It is known to all users of the cryptosystem, and is used in conjunction with a private key. *See also* **private key**.

public service commission (PSC) The state regulatory body that is responsible for communications regulations and other public utility regulations. Also called *Public Utilities Commission (PUC)*, *corporate commission*, and *railway commission.*

public switched telephone network (PSTN) A network established and operated by any common carrier that provides circuit switching between public users.

Public Utilities Commission (PUC) *See* **public service commission**.

pulse A momentary, sharp variation in the voltage or current level of a signal whose value is normally constant (Figure 16-9).

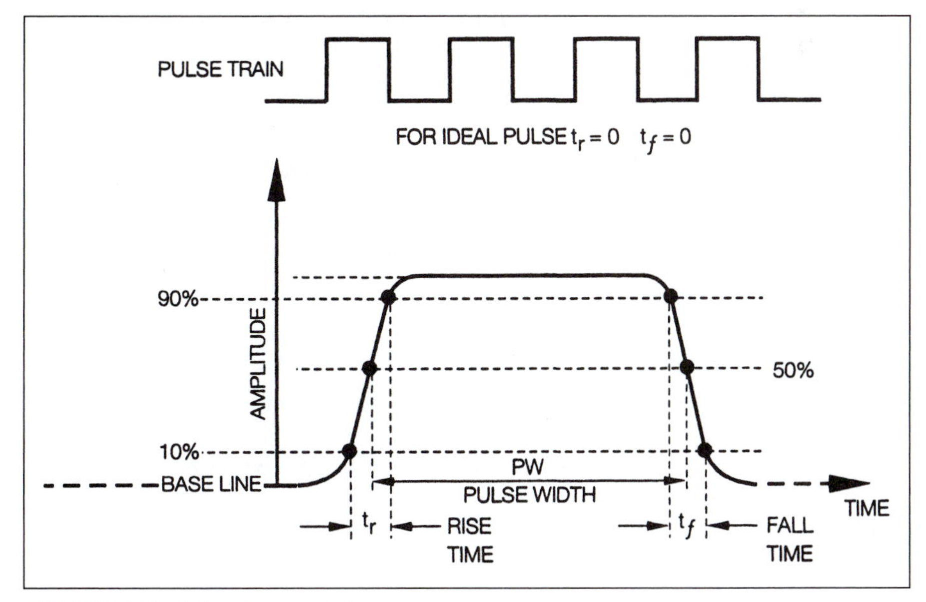

FIGURE 16-9 Pulse characteristics

pulse amplitude modulation (PAM) A form of pulse modulation in which the amplitude of a pulse carrier is varied in accordance with the changes in the successive samples of the information signal. (See Table 13-5.)

pulse carrier A high repetition rate signal used in pulse modulation. (See Table 13-5.)

pulse code modulation (PCM) A pulse modulation technique in which an analog signal is first sampled at regular intervals and converted into a PAM signal, i.e., into a pulse train where the amplitude of each pulse is proportional to the amplitude of the information signal at the sampling time. The PAM signal is then quantized and converted into a series of coded pulses whose bit combination reflects the quantized amplitude value of the PAM signal. Figure 16-10 illustrates the application of PCM in a telephone system (see Table 13-5). *See also* **CODEC**.

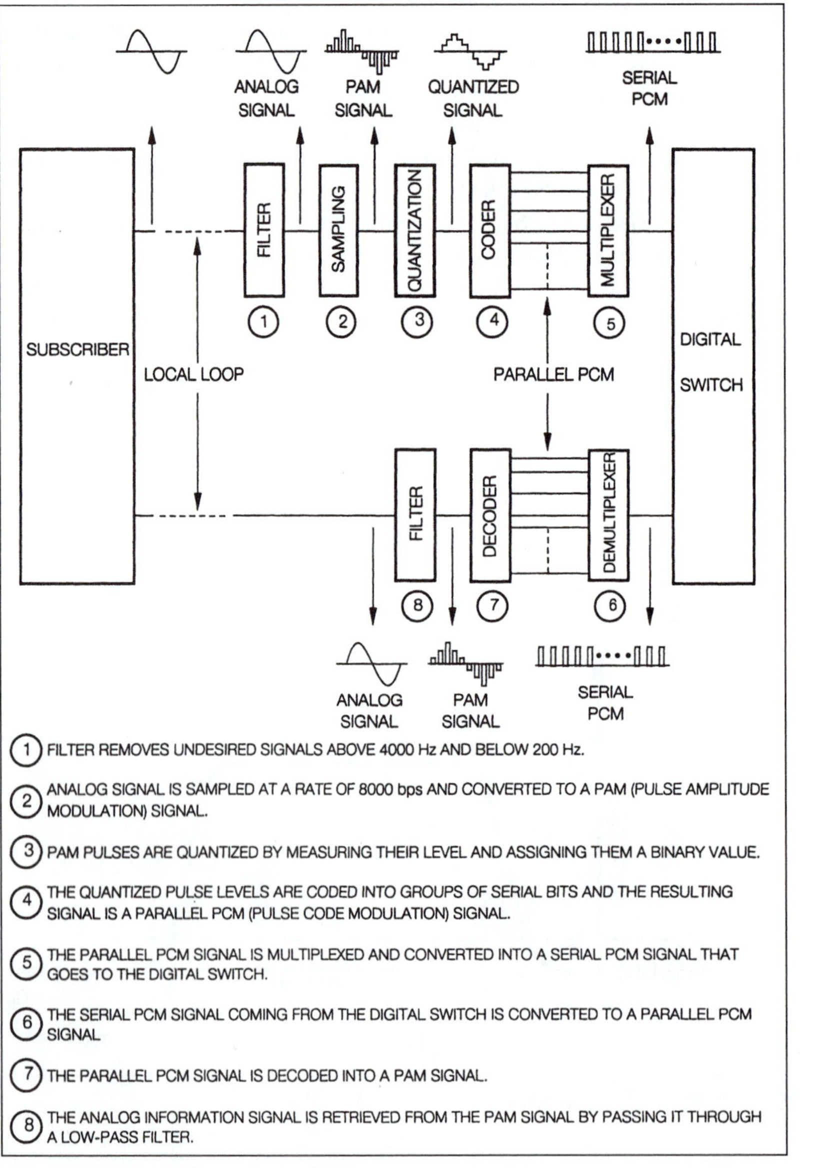

FIGURE 16-10 Pulse code modulation

pulse dispersion *See* **dispersion**.

pulse modulation A form of modulation in which one of the characteristics (such as amplitude or time) of a pulse carrier is varied in accordance with the changes in the amplitude of the information signal. It is used for the digital transmission of analog signals. Table 13-5 illustrates various forms of pulse modulation techniques. *See also* **pulse amplitude modulation, pulse code modulation,** *and* **pulse-time modulation.**

pulse-repetition frequency The rate at which pulses are generated or transmitted by a system. Usually expressed in cycles per second.

pulse-time modulation A form of pulse modulation in which time-dependent quantities (such as width, duration, or position of pulse) are varied in accordance with the changes in successive samples of the information signal. (See Table 13-5.)

push-button dialing The use of keys or buttons to generate a series of digits to establish a circuit connection. It uses dual tone multiple frequency (DTMF) signaling. The dialing time per digit is the same (100 milliseconds per digit) in contrast to rotary dialing in which the dialing time per digit varies from digit to digit. Also called *tone dialing* and *touch-tone*. *See also* **dual tone multifrequency (DTMF)**.

PWM Pulse width modulation. *See also* **pulse modulation** and **pulse-time modulation.** (See Table 13-5.)

MODULE

17 Definitions Q

quad A cable consisting of two twisted pairs of conductors, each separately insulated.

quadbit A group of four bits that allows representation of 16 unique signal levels (Table 17-1).

TABLE 17-1 Quadbit

SIGNAL LEVEL	b_3	b_2	b_1	b_0
0	0	0	0	0
1	0	0	0	1
2	0	0	1	0
3	0	0	1	1
4	0	1	0	0
5	0	1	0	1
6	0	1	1	0
7	0	1	1	1
8	1	0	0	0
9	1	0	0	1
10	1	0	1	0
11	1	0	1	1
12	1	1	0	0
13	1	1	0	1
14	1	1	1	0
15	1	1	1	1

quadrature amplitude modulation (QAM) A modulation technique that uses both amplitude and phase modulation. (See Table 13-5.)

quadrature phase shift keying (QPSK) A phase shift key technique that uses four phases (Figure 17-1).

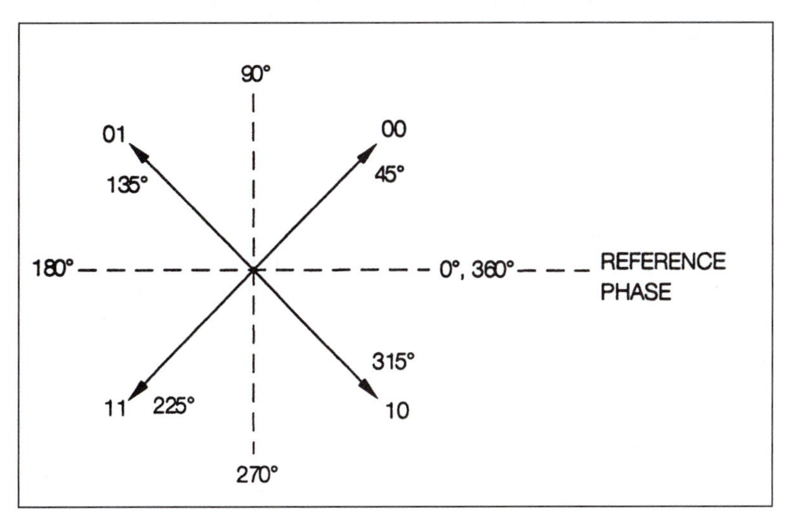

FIGURE 17-1 Quadrature phase shift keying phaser diagram

quality of service (QoS) A network's ability to provide a specific level of service availability.

query A request for information, initiated while the computer system is processing data.

queue A series of items arranged in a sequence, such as telephone calls, computer jobs, or print jobs waiting for service or processing.

MODULE

18 Definitions R

radiant power The rate of transfer of radiant energy.

radiation The emission and propagation of electromagnetic energy in wave or particle form through space or any other medium.

radio The use of electromagnetic waves to transmit and receive information signals without any wire or waveguides between transmitter and receiver.

radio channel A band of frequencies used in radio communications.

radio common carrier (RCC) A common carrier licensed by the FCC to receive and transmit signals from mobile transmitters within a specific geographical area in order to provide public land-mobile radio service.

radio communications Communications by radio waves that are not guided by transmission lines or waveguides between transmitter and receiver.

radio detection and ranging (RADAR) The technique of using radio waves to detect and locate a remote object or a target. The transmitted microwave pulses are reflected by the target, the time for a round-trip is measured, and thus the distance to the target is determined. Figure 18-1 illustrates the block diagram of a radar system.

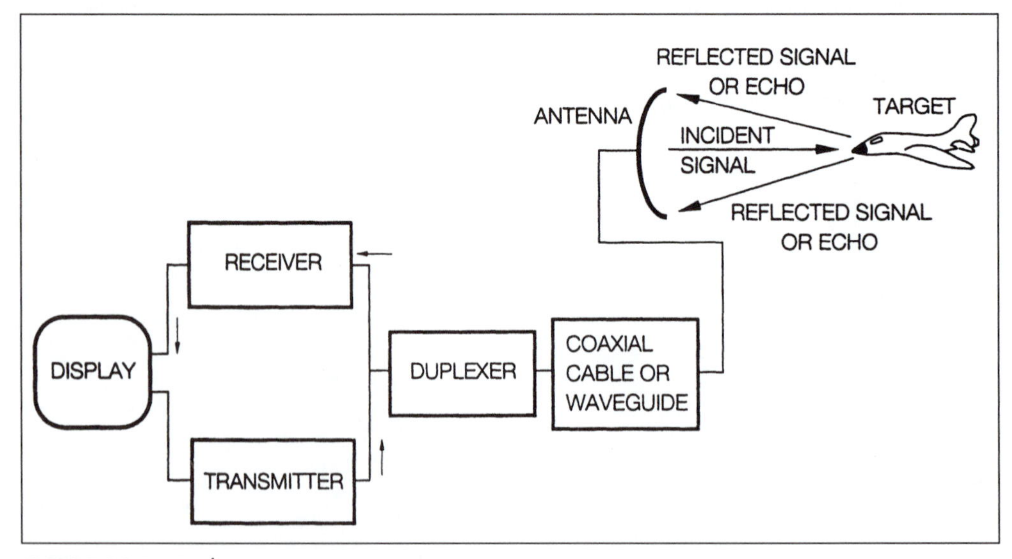

FIGURE 18-1 Radar

radio determination The use of radio signals to locate the position of, to determine the speed of, or to collect information about a remote object.

radio determination satellite service (RDSS) A service that uses a single or multiple satellites for radio determination.

radio frequency An electromagnetic wave used in radio communications. Figure 18-2 illustrates various modes of radio wave propagation. Table 18-1 lists various bands of radio frequencies. *See also* Module 29 *and* Module 45.

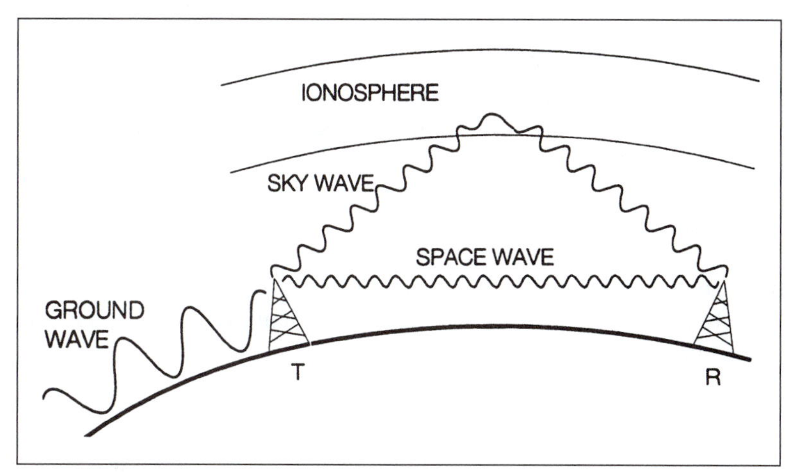

FIGURE 18-2 Radio wave propagation

TABLE 18-1 Radio frequencies

RF BAND	FREQUENCY	WAVELENGTH
ELF (extremely low frequency)	30–300 Hz	10–1 Mm
VLF (very low frequency)	3–30 kHz	100–10 km
LF (low frequency)	30–300 kHz	10–1 km
MF (medium frequency)	300–3000 kHz or 0.3–3 MHz	1–0.1 km
HF (high frequency)	3–30 MHz	100–10 m
VHF (very high frequency)	30–300 MHz	10–1 m
UHF (ultra high frequency)	300–3000 MHz	100–10 cm
SHF (superhigh frequency)	3–30 GHz	10–1 cm
EHF (extremely high frequency)	30–300 GHz	10–1 mm

radio frequency identification (RFID) A method used to remotely store and retrieve data using small devices called RFID tags (active or passive), which contain antennas to enable them to receive and transmit information in response to radio-frequency queries from an RFID transceiver. A passive RFID tag does not have a power source, but the small electrical current induced in its antenna due to an incoming radio-frequency signal enables it to send a short response. An active RFID tag has a power source and therefore has a longer range and ability to store information.

random access memory (RAM) A storage device that allows stored information to be accessed and retrieved randomly with equal access time for all storage locations, i.e., access time is independent of the memory location. Table 18-2 compares static RAMs (SRAM) and dynamic RAMs (DRAM).

TABLE 18-2 Static and dynamic RAMs

Static RAM (SRAM)	A type of RAM in which information is retained as long as power is applied.
Dynamic RAM (DRAM)	A RAM in which the contents of the memory must be continually refreshed (usually every three milliseconds), even with power applied, in order to retain stored information.

raster A scanning line pattern used to generate, record, or reproduce television, facsimile, or graphics on a screen.

rate

1. A charge for a specific service or equipment usage.
2. The data transmission capacity of a communication device or a channel, expressed in bits per second.

rate base The total amount of investment on which a rate of return for a regulated company is calculated.

rate center A specific geographic point used by telephone companies for determining distance-dependent telephone rates.

rate of return The ratio of net profits to the total amount of investment. The maximum rate of return for telecommunications companies is determined by appropriate state or federal regulatory agencies.

raw data Unprocessed data.

ray A geometric representation of the path of propagation of any form of radiant energy from its origin to its destination.

Rayleigh scattering The scattering of light in an optical fiber due to variations in the refractive index (such as submiscroscopic density variation, frozen impurities) that are compatible in size to the wavelength of the optical signal. Rayleigh scattering causes attenuation of the optical signal. Rayleigh scattering loss in an optical fiber is inversely proportional to the fourth power of the wavelength (Figure 18-3). *See also* Module 27.

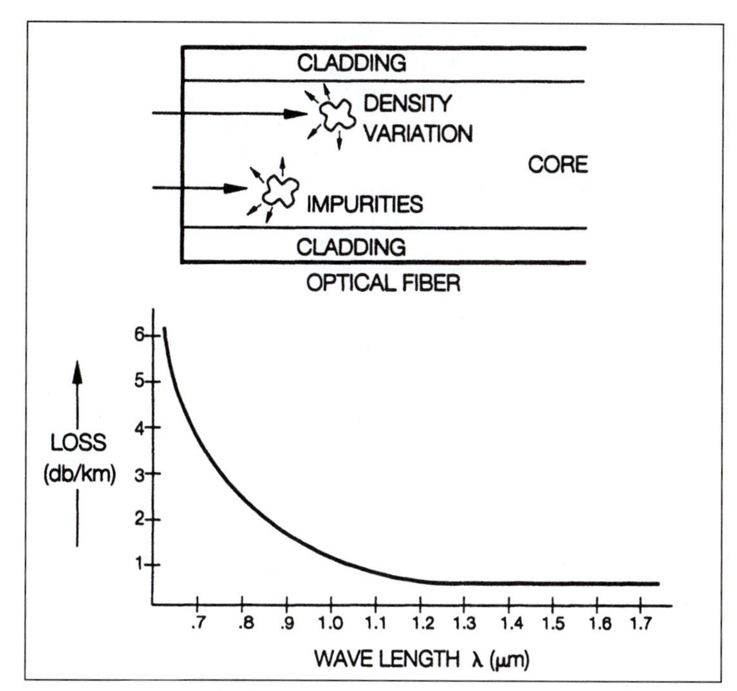

FIGURE 18-3 Rayleigh scattering

RBHC Regional Bell Holding Company. One of the seven companies formed by AT&T's divestiture (1984): Ameritech, Bell Atlantic, Bellsouth, NYNEX, Pacific Telesis, Southwestern Bell Corporation, and US West. Many RBHCs have merged with other companies to form new business partnerships.

RBOC Regional Bell Operating Company. *See also* AT&T.

reactance The opposition to the flow of an alternating current signal by a capacitor, an inductor, or a combination of two. It depends on frequency and is measured in ohms. The reactance offered by a capacitor is known as capacitive reactance and is given by:

$$X_c = \frac{1}{2\pi fC} \text{ ohms}$$

Where: f = frequency of AC signal (Hz)
C = capacitance (F)

The reactance offered by an inductor is known as inductive reactance and is given by:

$$X_L = 2\pi fL \text{ ohms}$$

Where: f = frequency of AC signal (Hz)
L = inductance (H)

read only memory (ROM) A type of data storage device in which information is stored permanently and cannot be altered, unless the device has a programmable option. Table 18-3 lists and compares different types of ROMs.

TABLE 18-3 Types of read only memory

MROM	Mask ROM—a type of ROM programmed at the time of manufacturing. The data stored is permanent and cannot be altered.
PROM	Programmable ROM—a type of ROM that can be programmed by the user. Once programmed, the contents of PROM are permanent.
EPROM	Erasable Programmable ROM (also called UV PROM)—a type of ROM that can be programmed and the contents later erased by shining UV light on the quartz window of the memory integrated circuit in order to reprogram it.
EEPROM	Electronically Erasable Programmable ROM—a type of ROM in which old data stored can be erased electronically by writing new data into it.

real time A data processing or transmission mode in which the response to an input in an interactive session is fast enough to control the subsequent input or events.

receive only (RO) A device or a system that can receive transmitted information but cannot transmit.

receiver

1. One of the three components (transmitter, channel, receiver) of a communications system.
2. A device that receives the transmitted signal and converts it to a visible or audible form.

redundancy

1. In data communications, the part of a message that can be eliminated without the loss of essential information.
2. In computer systems, the provision of duplicate, backup equipment to prevent system failure.
3. In data base systems, the storage of the same information at two or more locations.

redundancy check An automatic or programmed check based on the systematic insertion of components or characters used for checking purposes. *See also* **CRC, longitudinal redundancy check, vertical redundancy check**.

redundant array of independent drives (RAID) A system of sharing or replicating data by using multiple hard disk drives to increase system performance and fault tolerance in network servers.

reflectance The ratio of reflected power to incident power. In optics applications, it is generally expressed as a percent; in communications applications, it is expressed in decibels (db).

reflection The phenomenon in which a wave incident on an interface between two dissimilar media is reflected back into the medium from which it originated. The angles of incidence and reflection are equal and in the same plane. *Also see* **incidence angle**.

refraction The phenomenon in which a light wave changes direction as it strikes an interface between two dissimilar media (Figure 18-4). When a wave passes at an angle from a lower refractive index medium to a higher refractive index medium, it is bent toward the normal. When a wave passes at an angle from a higher refractive index medium to a lower refractive index medium, it is bent away from the normal.

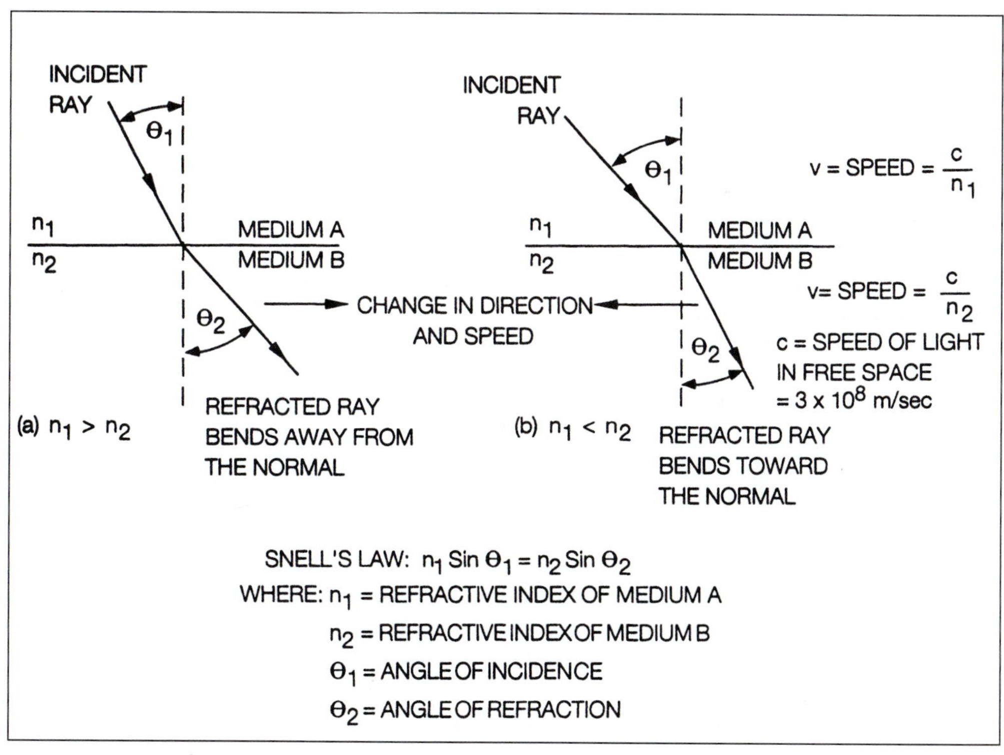

FIGURE 18-4 Refraction

refractive index *See* **index of refraction**.

refresh rate The rate per unit of time that a displayed image on a CRT display is refreshed in order to appear stable and flicker free. In Europe, the rate is 50 times per second; in the United States, 60 times per second.

regional center A class 1 telephone office that connects various sectional centers of the telephone system of a country together. In the United States, each regional center has a multiple circuit group running from one center to the other.

register A device that provides temporary storage of binary information. A register consists of cascaded flip-flops. Each flip-flop stores a single bit. Figure 18-5 illustrates various types of registers.

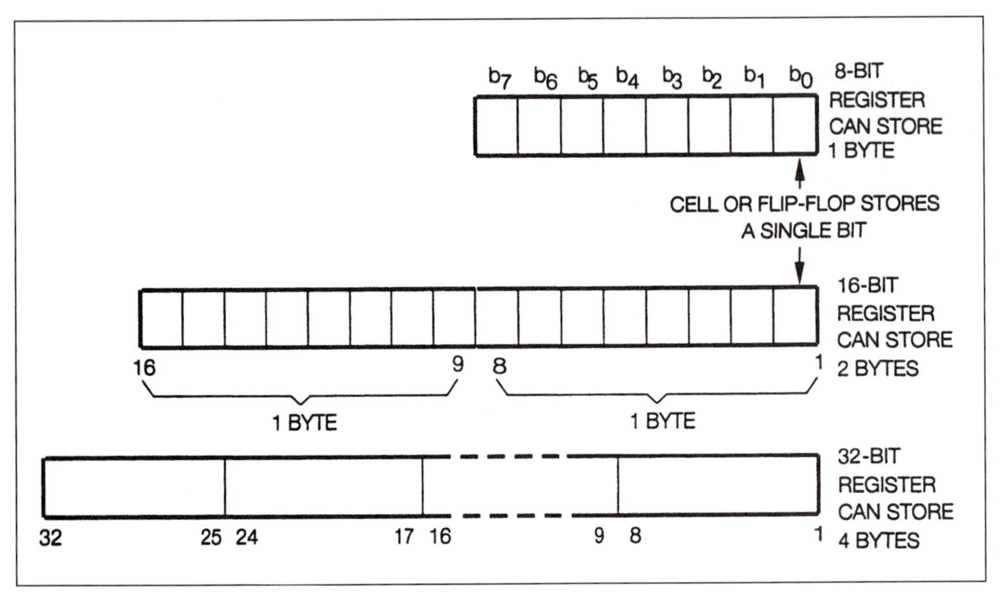

FIGURE 18-5 Registers

relational data base A type of data base in which each information file has some relationship with other information files.

relay An electrical device by means of which one circuit is indirectly controlled by a change in the same or another circuit. It normally consists of an electromagnet and armature to open and close contacts.

remote job entry (RJE) A mode of computer operation that allows submission of data processing jobs via a communication link from a remote site.

remote terminal A computer terminal used to enter data from some distance to the central processing unit where the data is processed.

repeater

1. In analog transmission systems, a device that amplifies attenuated signals to restore their original transmission levels for retransmission.
2. In digital transmission systems, a device that receives an attenuated and/or distorted pulse train and reconstructs it for signal retransmission.

3. In fiber optics communication systems, a device that detects a low-power optical signal, converts it to an electrical signal, amplifies it, and retransmits it in optical form using a light-emitting diode (LED) or laser diode.
4. In microwave transmission systems, a device that detects a low-power signal, amplifies it, and retransmits it to extend the range of propagation.

Figure 18-6 illustrates the various types of repeaters.

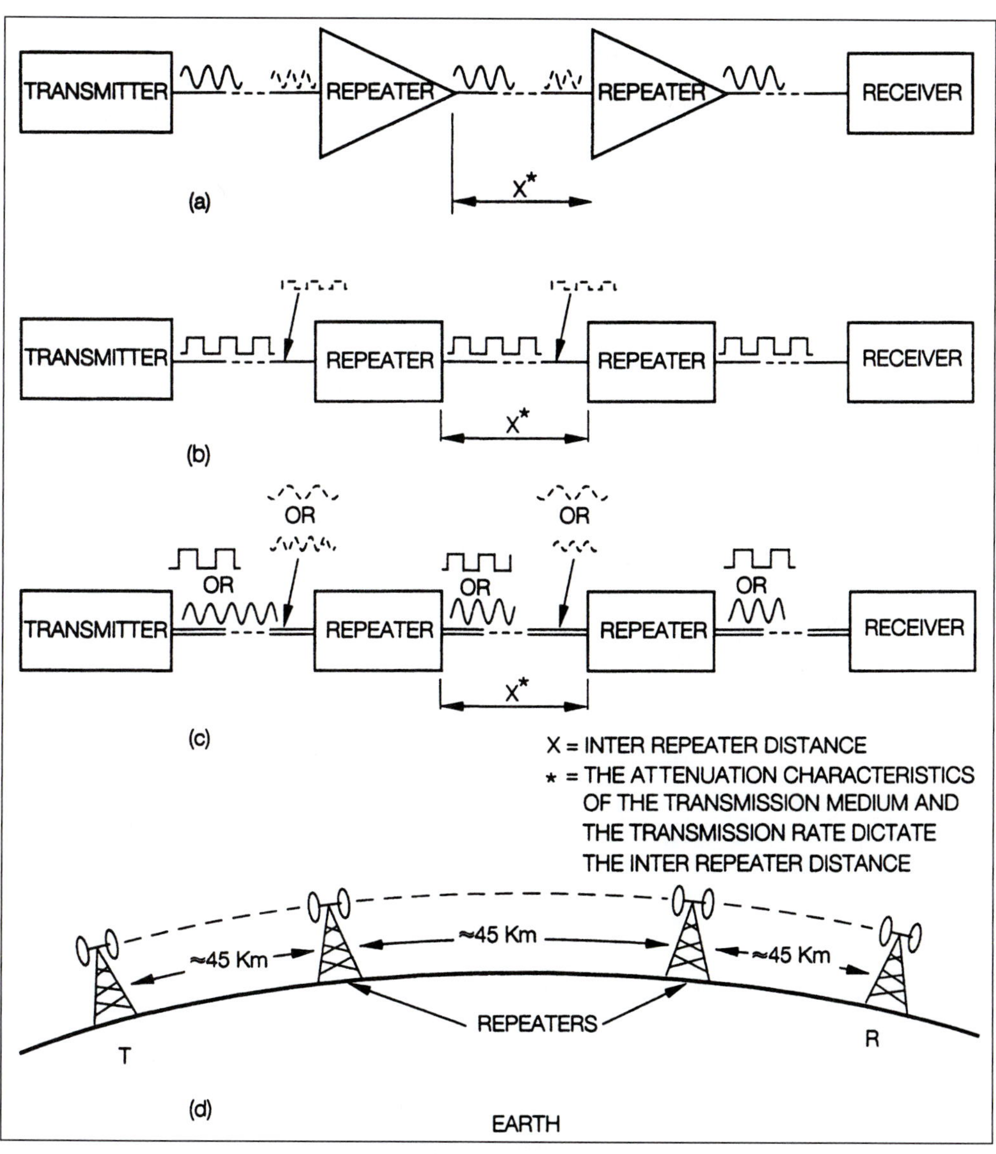

FIGURE 18-6 Repeater: (a) analog transmission system; (b) digital transmission system; (c) fiber optics communication system; (d) microwave transmission system

request for proposal (RFP) A follow-up to a request for information (RFI). Sent to an interested supplier to solicit a configuration proposal, with prices compatible to the user's requirements.

resistance The property of a material or a device to oppose the flow of current through it. It depends on the nature and the dimensions of the material and on the temperature. The unit of resistance is the ohm.

resistor A passive device that resists and controls the flow of current in a circuit.

resolution The measure of the ability of a visual system to reproduce fineness of detail.

response time

1. In an interactive system, the elapsed time between the end of an inquiry and the beginning of the response.
2. The time required by a device to reach a certain percentage (e.g., 63%, 98%, etc.) of its final value.

responsivity (R) A figure of merit used to measure the dependence of the signal output of a photodetector on the input radiant power. Generally expressed in volts/watts (voltage responsivity) or amperes/watts (current responsivity), the voltage responsivity is given by:

$$R = \frac{\text{output RMS signal voltage}}{\text{incident RMS radiant power}}$$

$$R = \frac{V}{P}$$

$$R = \frac{V}{HA} \text{ volts/watts}$$

Where: H = irradiance (watts/cm^2)
A = detection area (cm^2)

The current responsivity is given by:

$$R = \frac{\text{output RMS signal current}}{\text{incident RMS radiant power}}$$

$$R = \frac{I}{P}$$

$$R = \frac{\eta e}{E} = \frac{\eta e}{h\upsilon} \text{ amperes/watts}$$

Where: I = output rms signal current
P = incident rms radiant power
η = quantum efficiency of the detector
e = charge of an electron (1.6×10^{-19} C)
E = photon energy in electronvolts (eV)
h = Planck's constant (6.625×10^{-34} J-s)
υ = frequency of a photon (Hz)

retransmissive star In a fiber optics communication system, a passive component that allows the distribution of the optical signal from a single fiber to multiple fibers (Figure 18-7). *See also* Module 27.

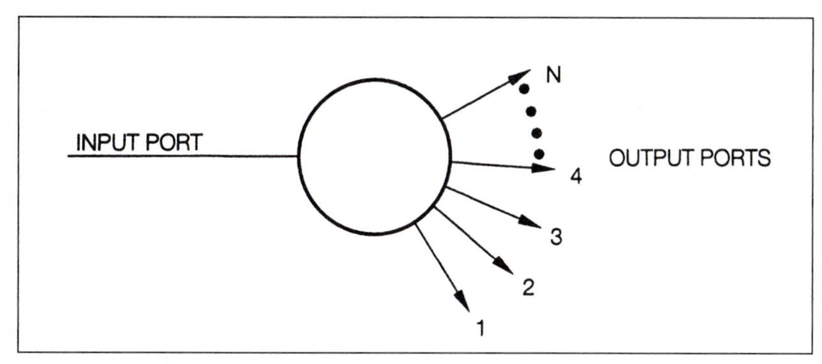

FIGURE 18-7 Retransmissive star

return loss The ratio in decibels of the incident power to the reflected power on a transmission line. It is a measure of the dissimilarity of the impedance of the transmission line and its load impedance.

$$\text{return loss} = 10 \log\left(\frac{\text{reflected power}}{\text{incident power}}\right)$$

return-to-zero (RZ) A technique of digital data transmission in which the voltage level returns to zero after each encoded bit. Figure 18-8 illustrates unipolar and bipolar return-to-zero transmission schemes. Contrast to *non-return-to-zero (NRZ)* transmission. *See also* **non-return-to-zero** *and* **digital signals**.

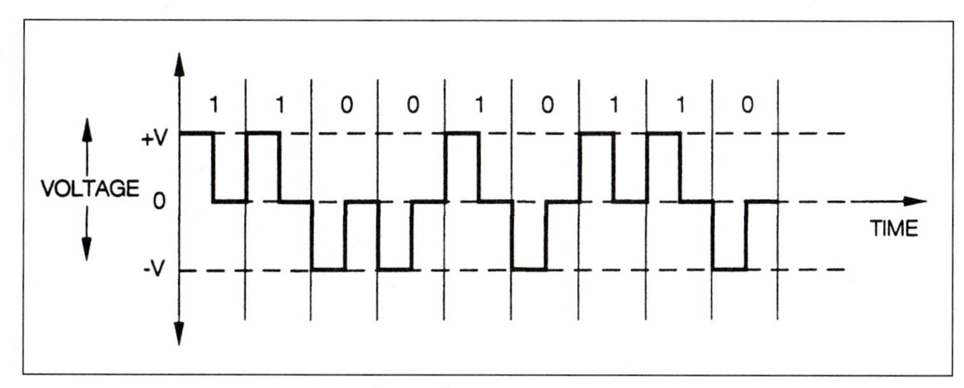

FIGURE 18-8 Return-to-zero signal waveform

reverse channel A channel that allows simultaneous data transmission in both directions on a half-duplex line. The reverse channel generally has a small bandwidth and a lower transmission rate than the forward channel. It is normally used for positive/negative acknowledgement of received data. Reverse channels reduce the need for modem turnaround time.

RFI

1. Request for information. A general notification of an intended purchase of telecommunication equipment sent to a potential supplier to determine interest and to seek product information.
2. Radio frequency interference.

RGB monitor Red, green, and blue monitor. A color display unit for computers.

ring network A network topology in which each node is directly connected to two adjacent nodes. Access to the network is passed sequentially from one node to another by means of polling from a master node or by passing an access token (a special bit pattern) from one node to another (Figure 18-9). The failure of one node can cause the network to be inactive unless bypass circuitry or redundancy is built into the system. *See also* **topology** *and* Module 48.

ringing signal An AC or a DC signal transmitted over a line to alert a party of an incoming call.

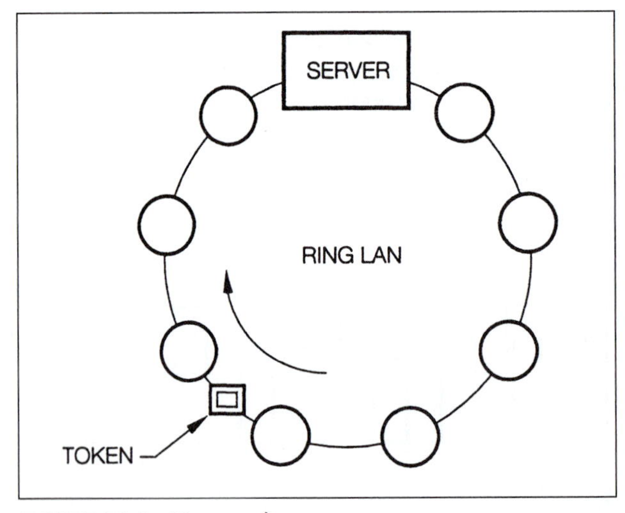

FIGURE 18-9 Ring topology

rise time The time required for the leading edge of a pulse to increase from 10 to 90 percent of its final value. *See also* **pulse**.

RJ-11 Registered Jack type 11. A standard connector used with unshielded twisted pair (UTP) cables for telephone lines (Figure 18-10).

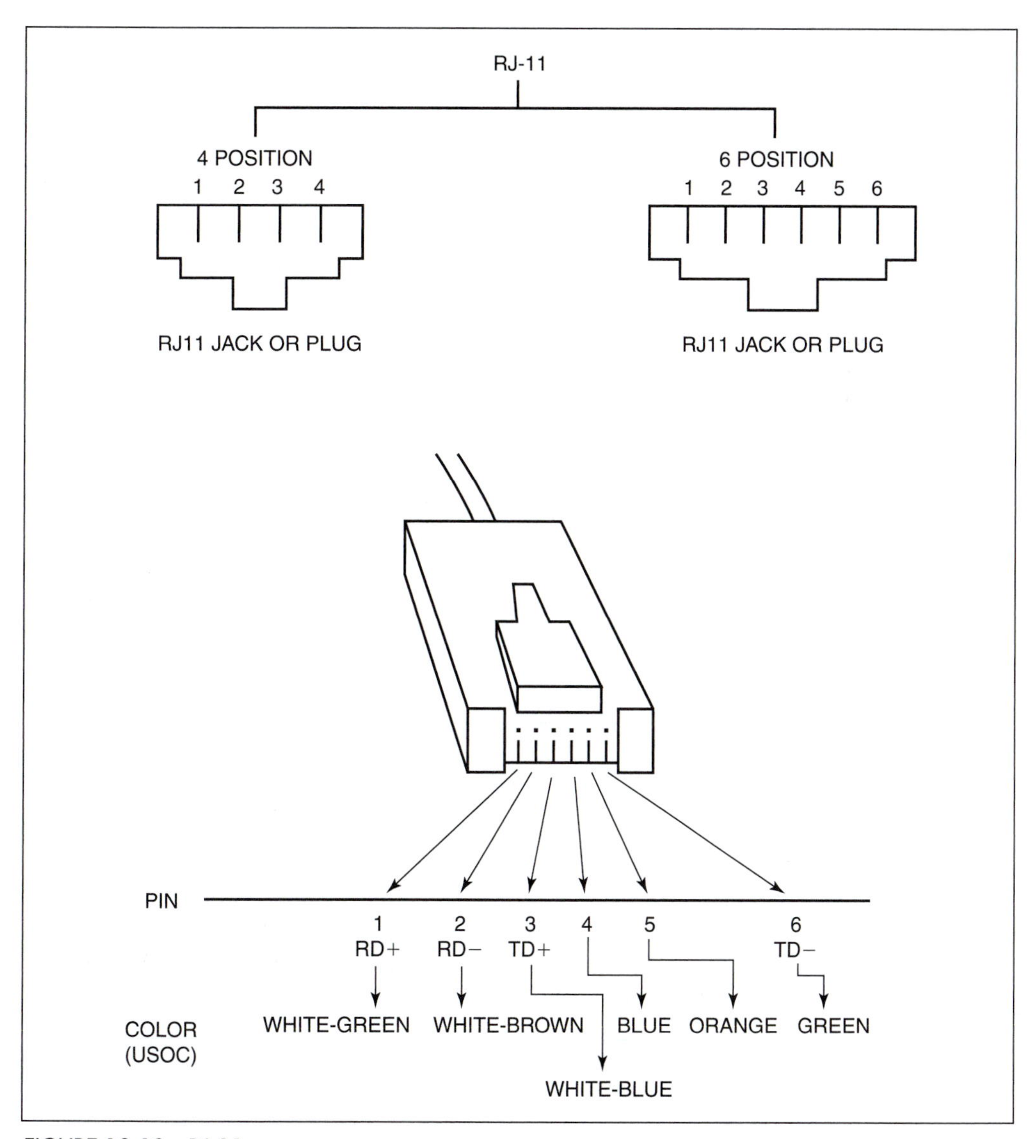

FIGURE 18-10 RJ-11

RJ-45 Registered Jack type 45. A standard connector used with shielded twisted pair (STP) and unshielded twisted pair (UTP) cables for computer networks (Figure 18-11).

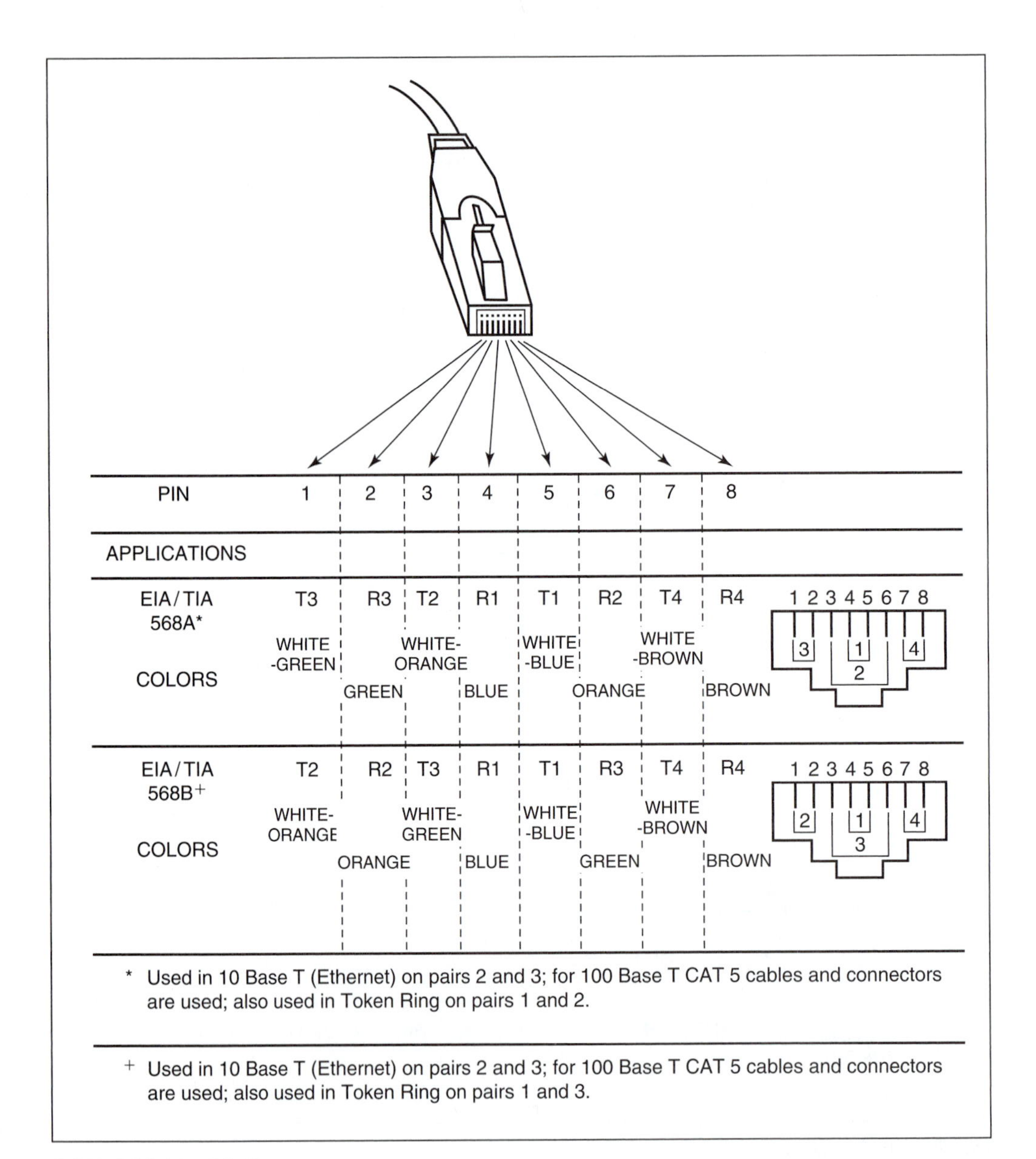

FIGURE 18-11 RJ-45

rotary dial The rotary telephone calling device. When wound up and released, it generates current pulses to identify the called party. It is slower than DTMF dialing (Figure 18-12). *Also see* **dual tone multifrequency (DTMF)**.

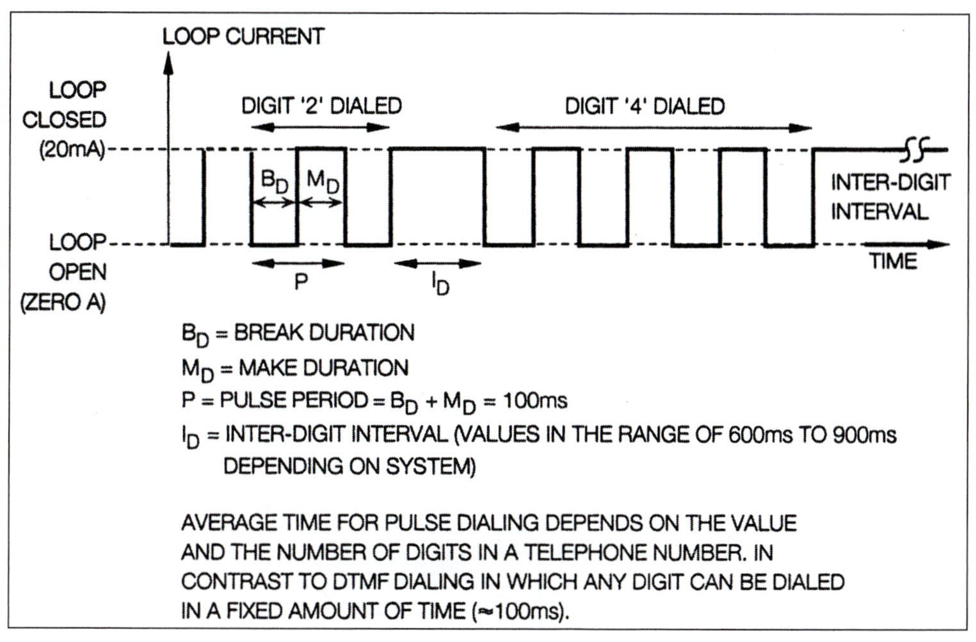

FIGURE 18-12 Rotary dial pulses for digit 2 and digit 4

route An assigned communication path for the transmission of information from source to destination.

routing The process of selecting a communication path to transmit information from source to destination.

RPQ Request for price quotation.

RSA A public key algorithm developed by Rivest, Shamir, and Adelman (RSA). It is one of the most widely used encryption and authentication methods for transmitting information over the Internet.

RS-232-C An Electronics Industries Association (EIA) recommended physical interface for asynchronous transmission between DTE (data termination equipment) and DCE (data communication equipment). It is a 25-pin interface and allows a transmission speed up to 20 kbps at a maximum distance of 50 feet. (*See* Module 39.)

RS-422-A An EIA recommended specification for asynchronous transmission between DCE and DTE. It improves on the capabilities of RS-232-C. (*See* Module 39.)

RS-423-A An EIA recommended specification for the electrical characteristics of an unbalanced voltage digital interface circuit.

RS-449 An EIA recommended specification that improves on the capabilities of RS-232-C. (*See* Module 39.)

RS-449-1 Addendum 1 to RS-449.

RTS Request to send. An RS-232-C control signal between DCE and DTE that initiates data transmission on the line.

MODULE

19 Definitions S

sampling A statistical method in which a signal level is sensed at specific time intervals in order to get a general representation of signal behavior. It is used in analog-to-digital conversion of signals. *See also* **Nyquist sampling theorem**.

satellite A man-made vehicle placed in an orbit around earth to receive information from one point on earth and retransmit it to another point on earth, thus providing a long-distance communication link. *See also* **geosynchronous satellite** *and* Module 46.

satellite communications A communication technique in which a geosynchronous satellite is used to relay information from one earth station to another (Figure 19-1). (*See* Module 46.)

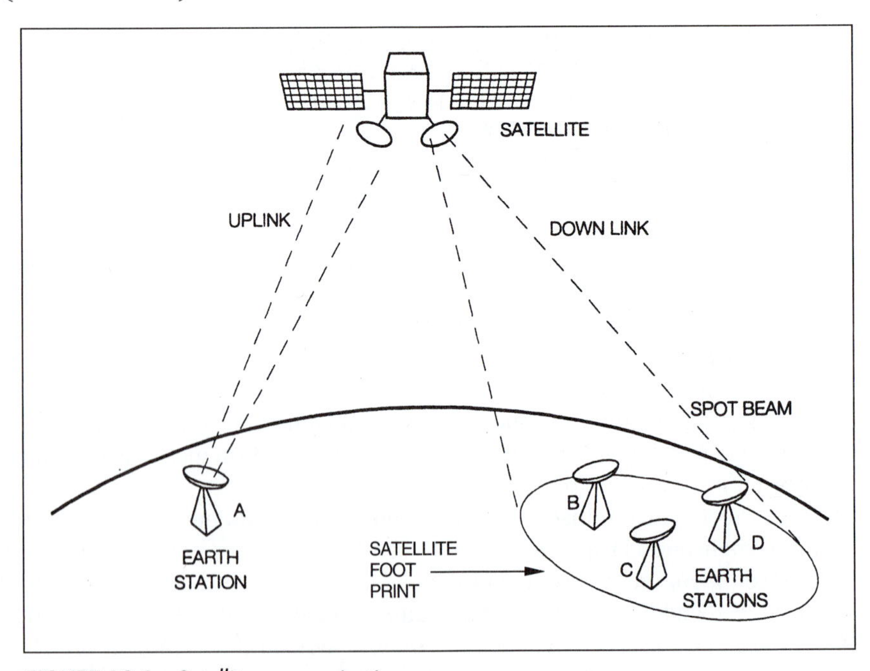

FIGURE 19-1 Satellite communications

satellite earth station *See* **earth station**.

satellite relay An active satellite that relays information between two earth stations.

scattering The change in the direction of optical waves or photons after they collide with a small particle or as they pass through inhomogeneous material.

scrolling The process of moving information either vertically or horizontally on the screen of a display terminal.

SECAM *S*equential *C*olor *a*nd *M*emory. A color television standard with 625 lines per frame, 25 frames per second, and a 50 Hz field frequency. (*See* Module 41.)

secondary channel An auxiliary or a control channel, in contrast to the *main channel.*

secondary station A station or a node that receives information from a primary station. It cannot initiate a transmission unless permission is granted by a primary station. A secondary station is usually a temporary designation during a session.

serial data Representation of data by a series of binary digits occurring in a sequence one after another.

Serial Line Internet Protocol (SLIP) A protocol that uses a computer's serial port to establish connection via a modem to an ISP for transmission of data.

serial port Male 9-pin or 25-pin ports on the computer allowing transmission of data in serial mode, i.e., bit by bit transmission. Serial ports are designated as COM1, COM2, COM3, and COM4. *See* Figure 16-2.

serial transmission A method of transmission in which data bits of a character or a message are transmitted sequentially one bit at a time on a transmission link. Serial transmission can be classified into two methods: asynchronous and synchronous (Figure 19-2). In asynchronous transmission, each character has its own synchronizing information in the form of start/stop bits. In synchronous transmission, the data are transmitted in blocks and each character does not have start/stop bits. Rather, the use of a common clock pulse at the transmitter and receiver achieves synchronization.

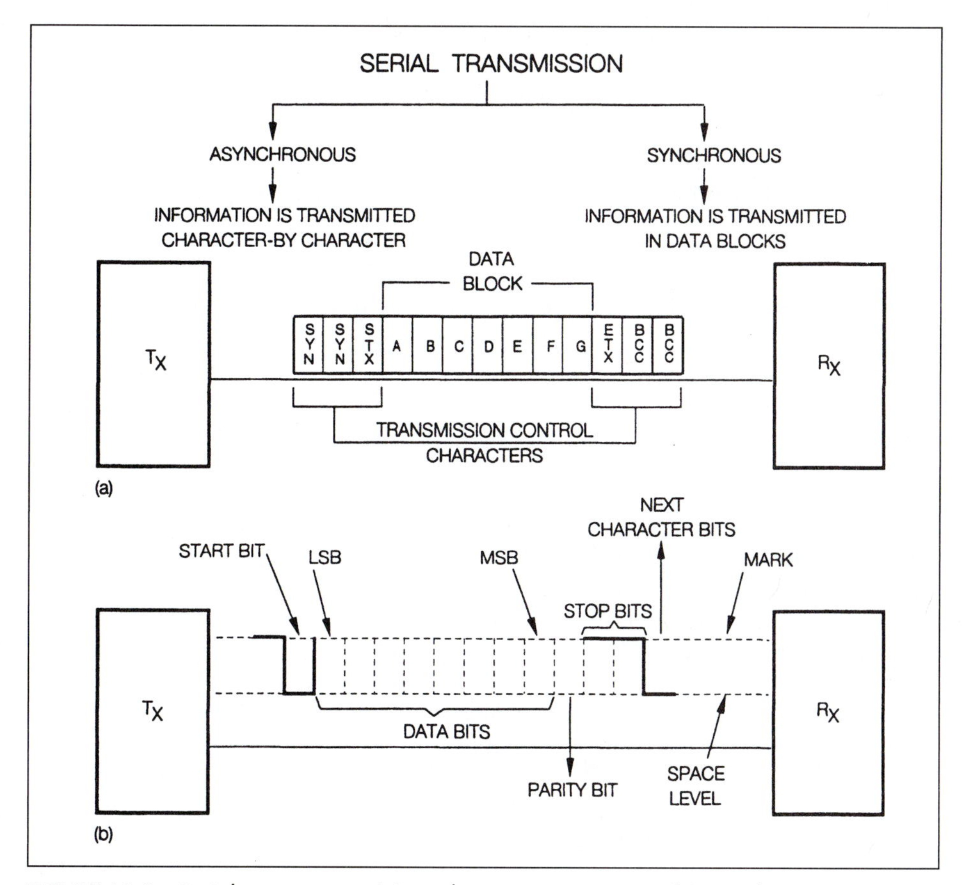

FIGURE 19-2 Serial transmission: (a) synchronous transmission; (b) asynchronous transmission

server A computer that runs specialized software to provide information or services to a client in a network.

session

1. A period during which an end user communicates with an interactive computer.
2. A connection between two stations for information exchange purposes.
3. In IBM's system network architecture (SNA), a connection between two network addressable units (NAUs).

session layer The fifth layer of the International Organization for Standardization's (ISO) open systems interconnection (OSI) reference model for network architecture. (*See* Module 35.)

Shannon's law A mathematical expression developed by Claude Shannon in 1948. It defines the theoretical capacity of a channel in terms of bits per second:

$$\text{channel capacity} = W \log_2(1 + SNR) \text{ bps}$$

Where: W = channel's bandwidth (Hz)
SNR = channel's signal-to-noise ratio

SHF Superhigh frequency. That part of the electromagnetic spectrum with frequencies in the range of 3 to 30 GHz.

shielding A protective enclosure that surrounds a transmission line to reduce radiation loss and interference.

sideband A frequency band on the upper or lower side of the carrier frequency that contains the frequencies generated by the process of modulation. Figure 19-3 illustrates sidebands for AM modulation schemes.

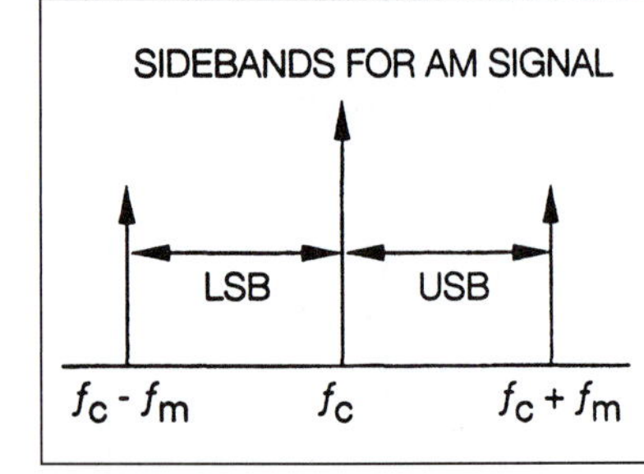

FIGURE 19-3 Sidebands

side tone The transmission and reproduction of an acoustic signal on a local path from the transmitter to the receiver of the same telephone set.

signal A physical, time-dependent energy value used for the transmission of information from one point to another through free space, air, wire, waveguide, etc.

signal-to-noise ratio (SNR) The relationship of the magnitude of the signal power and noise power at a given point in a communication system. It is generally expressed in decibels (db). *Also see* **noise** *and* Module 33.

$$SNR = \frac{\text{signal power}}{\text{noise power}} = \frac{S}{N}$$

$$(SNR)_{db} = 10 \log\left(\frac{S}{N}\right)$$

SILO A FIFO (first in first out)-type memory.

SIMM (single inline memory module) A small circuit board that contains memory chips in a computer. SIMMs can hold 8, 16, 32, and 64 megabytes (MB) of random access memory (RAM) on a single module.

Simple Mail Transfer Protocol (SMTP) The protocol used for transmitting e-mail on the Internet.

simplex (SPX) A circuit or a transmission link that allows the flow of information in only one direction (Figure 19-4). *See also* **half duplex** *and* **duplex**.

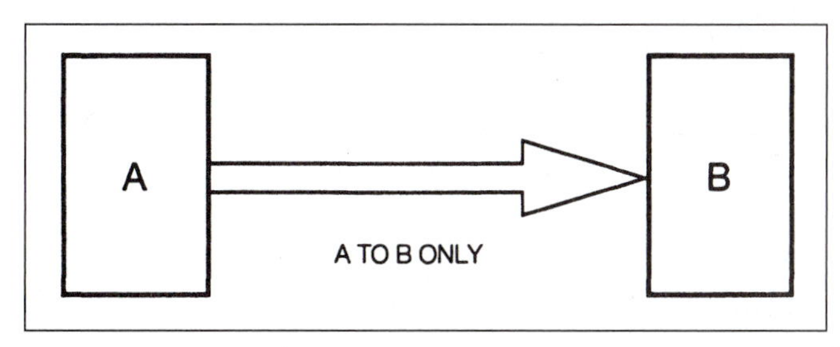

FIGURE 19-4 Simplex transmission

sine wave A wave whose displacement is the sine function of an angle proportional to time or space or both (Figure 19-5).

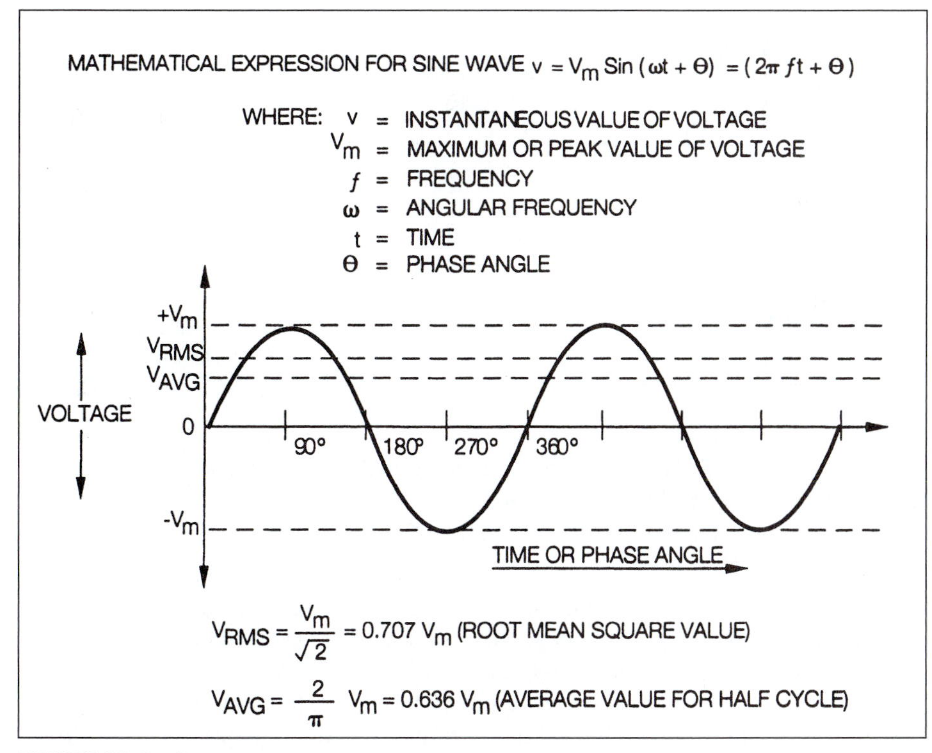

FIGURE 19-5 Sine wave

single sideband (SSB) An amplitude modulation scheme in which one of the two sidebands and the carrier frequency are filtered out to make efficient use of the bandwidth and transmitted power. Figure 19-6 illustrates the SSB frequency spectrum.

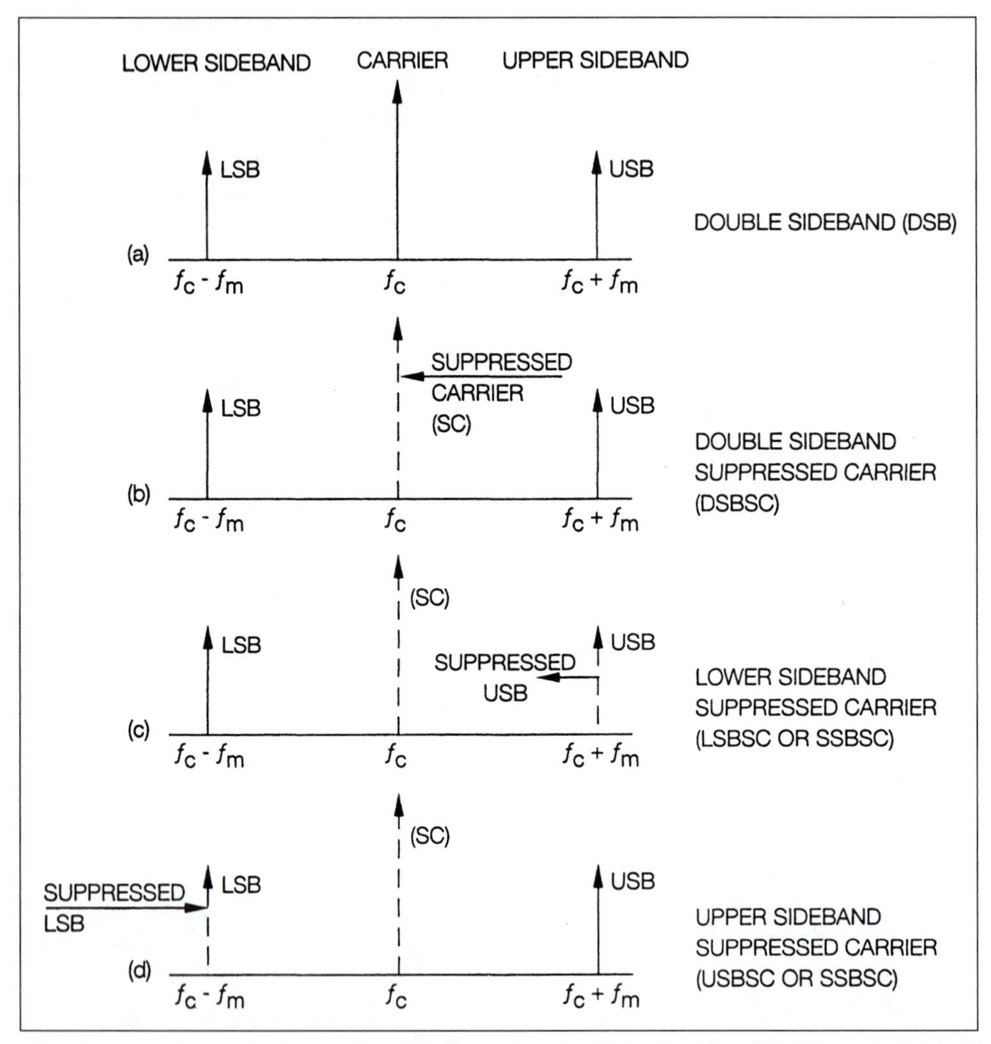

FIGURE 19-6 DSB and SSB signal spectra for an AM signal: (a) double sideband (DSB); (b) double sideband suppressed carrier (DSBSC); (c) lower sideband suppressed carrier (LSBSC or SSBSC); (d) upper sideband suppressed carrier (USBSC or SSBSC)

single-mode fiber An optical fiber that allows a single mode of optical signal to propagate through due to a very small core diameter. Also called *monomode fiber. See also* **optical fiber** *and* Module 27.

sink The part of a communication system that receives information from the source.

skew ray A ray that follows a helical path around the core of a graded index fiber. It never intersects the axis of the fiber. Contrast to *meridional ray. See* **graded index fiber**.

sky wave A radio wave that is reflected by the ionosphere. (See Figure 18-2.)

slave A device or a station that functions under the control of a master device or a station.

SLIC Abbreviation for *S*ubscriber *L*ine *I*nterface *C*ircuit. Performs network interface functions.

smart terminal A computer terminal that has data processing capabilities in addition to its function as a data input/output device.

SNA *See* **systems network architecture** *and* Module 35.

Snell's law A mathematical expression that describes the phenonmeneon of refraction by relating the angles of incidence and refraction and the refractive indices of the two media:

$$n_1 \sin \Theta_1 = n_2 \sin \Theta_2$$

(*See also* **refraction**.)

software A set of instructions that, when executed, performs specific task/s for a user. Examples of software are BIOS, operating systems (OS), and application programs (word-processing, spreadsheet, etc.).

source A device, circuit, or system that generates an electromagnetic signal in order to transmit information to another device, circuit, or system.

space In data communications, a signal level that represents a binary zero state.

space wave A type of electromagnetic wave that travels directly from transmitter to receiver. *See also* **radio frequency**.

specialized carrier A company that provides value-added services or communication facilities.

spectral width A measure of the range of wavelengths emitted by an optical source. Also called the *line width.* Figure 19-7 compares the spectral widths for various types of optical sources. *See also* **light emitting diode (LED)** *and* **laser diode (LD)**.

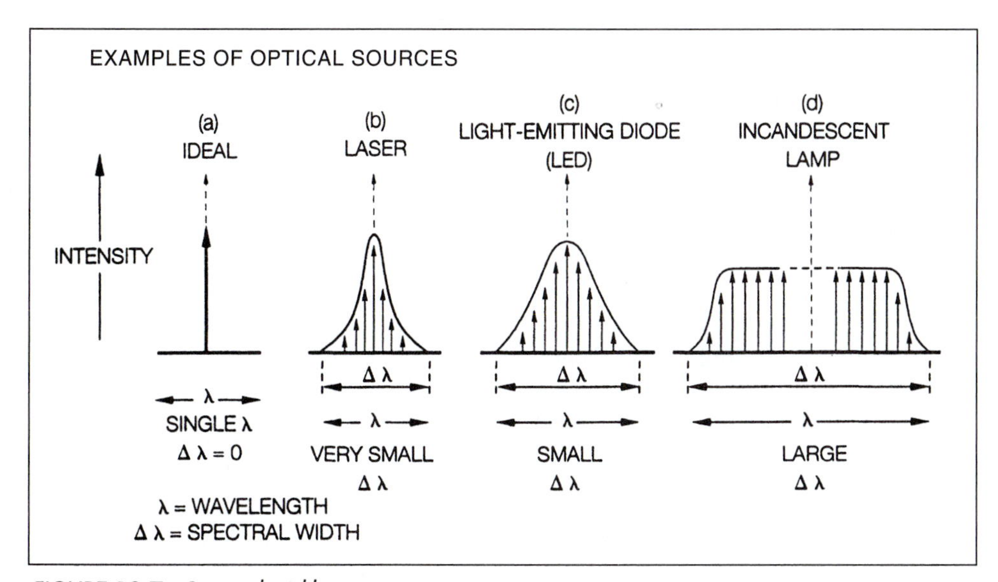

FIGURE 19-7 Spectral width

spectrum

1. A continuum of frequencies.
2. A graphical representation of the amplitude or phase components of a signal as function of frequency.

splice A permanent junction that connects two optical fibers.

splitter In a cable television distribution system, a device that allows a single input to be separated into two or more outputs without loss or impedance mismatch.

spontaneous emission The radiation emitted by an atom when it returns to a lower energy state from an excited state. In a light-emitting diode (LED), light is emitted by spontaneous emission. Figure 12-3 illustrates spontaneous emission.

spoofing A technique to change the origin of data, i.e., to make data appear to come from some place other than its original source.

spot beam In satellite communications, a down-link focused signal that covers a smaller geographical area. Figure 19-1 illustrates the satellite spot beam. *See also* Module 46 for **satellite communication**.

spread spectrum A wireless communications technology that modulates a signal over a wide range of frequencies. *See also* Module 47 for **wireless technologies and standards**.

square wave A periodic waveform that alternately assumes, for equal time intervals, one of two fixed levels. Figure 19-8 illustrates a square wave.

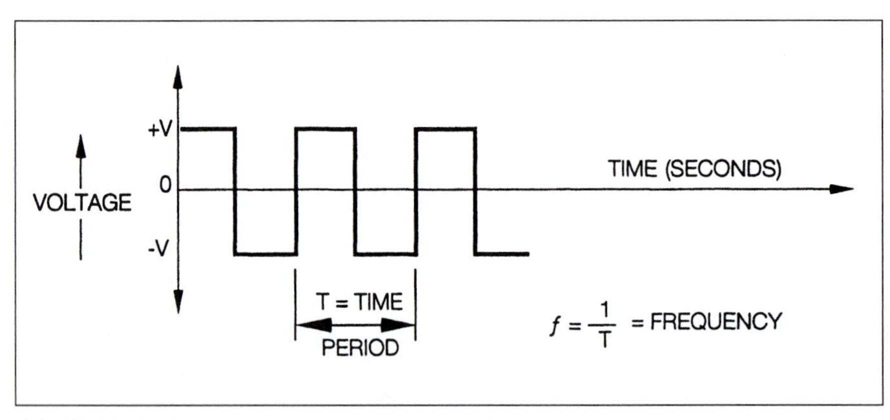

FIGURE 19-8 Square wave

stack An area of a memory for temporary storage of information on a LIFO (last in first out) basis.

standing wave A periodic wave generated due to the interference of incident and reflected signals on a transmission line. Figure 19-9 illustrates a standing wave. *See also* **impedance matching network**.

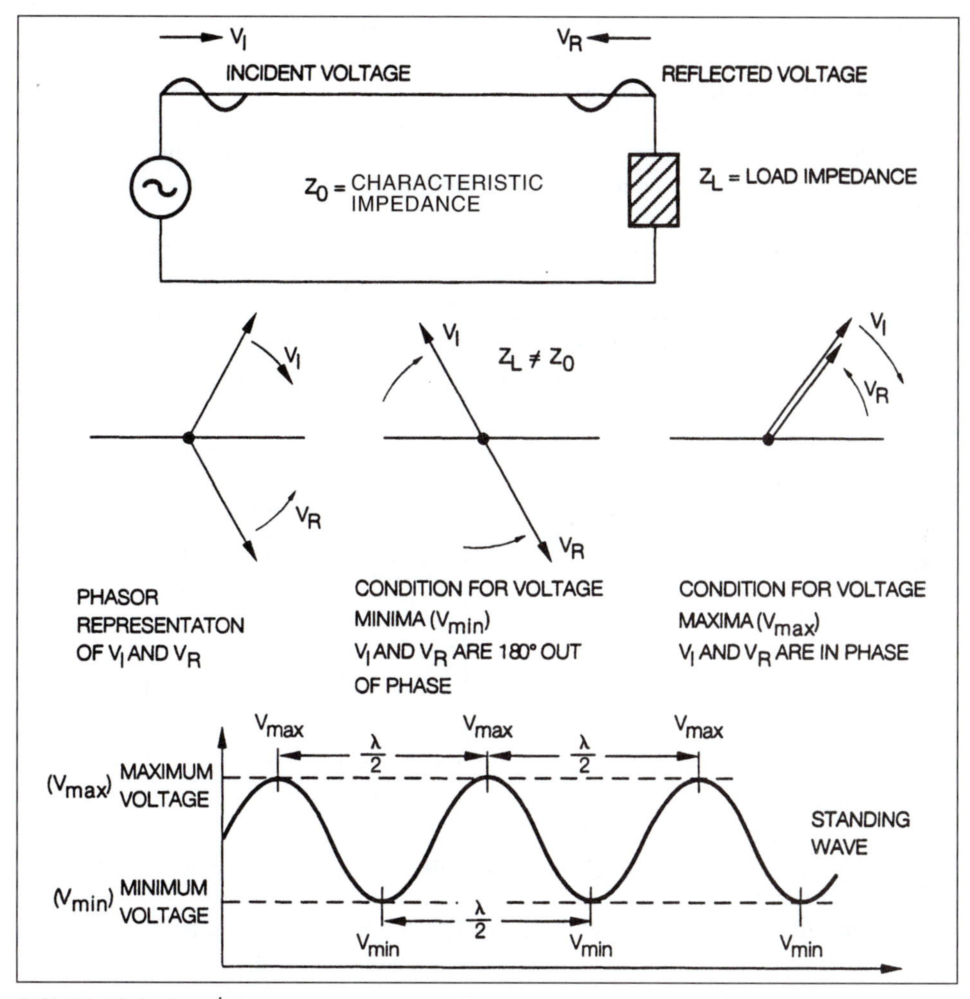

FIGURE 19-9 Standing wave

star A network topology in which all stations communicate with each other through a central station to which they are connected by a point-to-point link (Figure 19-10.) *Also see* **topology** *and* Module 48 for **computer networks**.

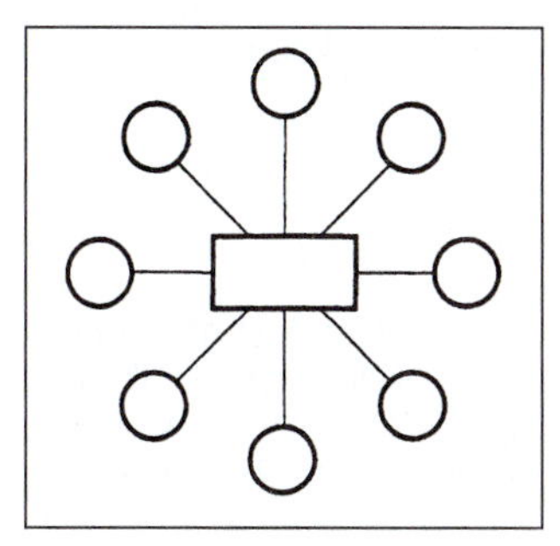

FIGURE 19-10 Star topology

star coupler A passive optical device in which the power from one or more inputs is distributed among a number of outputs.

start bit In asynchronous data transmission, a signal element that indicates the beginning of the transmission of a character. *See also* **asynchronous transmission**.

start of header (SOH) In data communications, a signal element that indicates the beginning of the message header. *See also* Module 40 for **EBCDIC** and **ASCII** codes for **SOH**.

start of text (STX) In data communications, a signal element that indicates the beginning of the message as well as the end of the message heading. *See also* Module 40 for **EBCDIC** and **ASCII** codes for **STX**.

state diagram A diagram that defines the relationship between the present state and the next state. It describes the overall operation of a synchronous system. Figure 19-11 illustrates a state diagram.

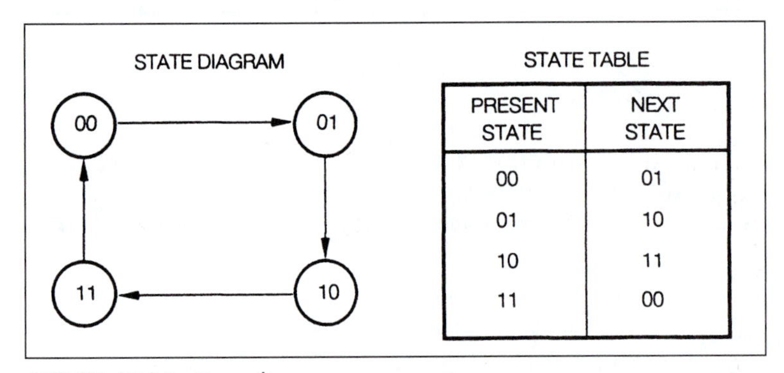

PRESENT STATE	NEXT STATE
00	01
01	10
10	11
11	00

FIGURE 19-11 State diagram

station An input or output point of a communication system, e.g., a telephone set, a computer terminal, etc.

statistical time-division multiplexing (STDM) A time-division multiplexing technique in which the efficiency of data transmission over a communication channel is increased by giving priority to certain channels depending on the expected transmission load. Figure 19-12 illustrates STDM. *Also see* **frequency-division multiplexing, time-division multiplexing,** *and* **multiplexing**.

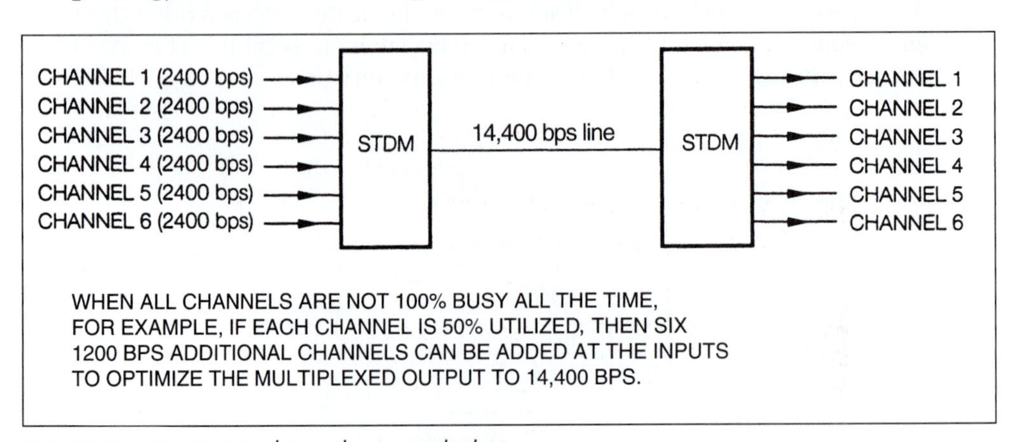

FIGURE 19-12 Statistical time-division multiplexing

step-by-step switching (SXS) An automatic telephone switching technique that uses successive step-by-step selector switches actuated by current pulses generated by a rotary telephone dial. The dialing of each digit causes the successive selector switches to carry the connection forward until the desired line is reached.

step index fiber A type of optical fiber in which the refractive index is uniform within the core but changes abruptly at the core cladding interface. *See also* **optical fiber** *and* Module 27.

steradian (Sr) The SI unit of a solid angle. One steradian is defined as a solid angle whose vertex is at the center of a sphere and which cuts off an area equal to r^2 (radius of the sphere) on the surface of the sphere.

stimulated emission Radiation emitted when a photon of energy equal to or greater than the energy difference between the excited state and the ground state of a quantum mechanical system impinges on an electron in the excited state stimulating it to return to ground state by emitting two photons. In lasers, light is emitted by stimulated emission. (Figure 12-1 illustrates stimulated emission.)

stop bit In asynchronous data transmission, a signal element that indicates the end of the transmission of a character. *See also* **asynchronous transmission**.

store and forward A method of switching by which complete or partial messages are buffered or stored and then transmitted toward their destination.

strap A metallic link used between alternate anodes of a multicavity magnetron in order to provide anode separation.

Strowger switch A step-by-step switch, named after its inventor, Almon B. Strowger.

subnet A small network that is a part of a larger network; it shares an IP address with other segments of the network and can be distinguished by its subnet address. In TCP/IP, subnets share the same prefix of the IP address. *Also see* **Transmission Control Protocol/Internet Protocol (TCP/IP),** *and* Module 37.

subscriber loop *See* **local loop**.

subscriber television (STV) A type of television broadcast in which the signal is scrambled and subscribers with decoders can receive it.

sub-voice-grade channel A channel that has a bandwidth smaller than the voice-grade channel (4 kHz).

super band The band of electromagnetic signals having a frequency range from 216 to 300 MHz. It is used for cable television.

super group An assembly of five twelve-channel groups that occupy adjacent bands in a frequency spectrum (Figure 19-13). *Also see* **group, master group,** *and* **jumbo group**.

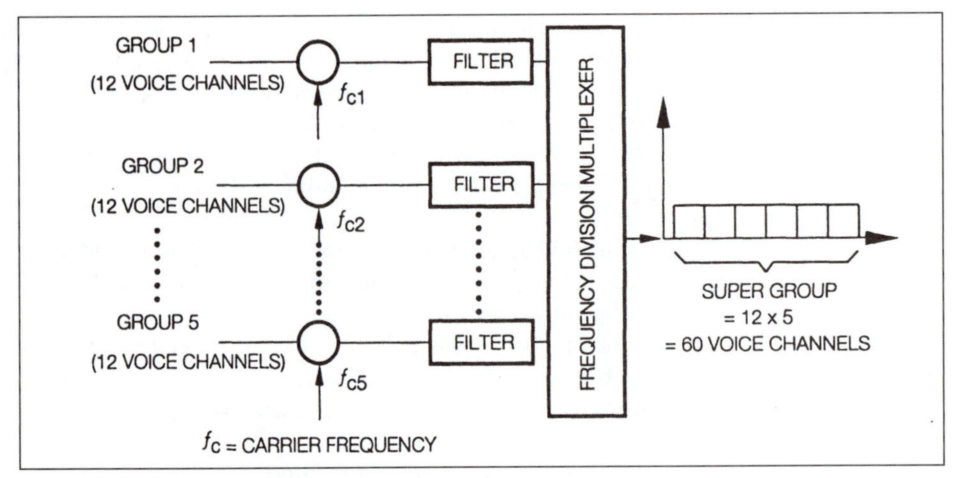

FIGURE 19-13 Super group formation

superheterodyne receiver A type of radio receiver developed by Major Edwin H. Armstrong during World War I. Since then, it has served as a model for receiver systems such as AM, FM, television, satellite systems, radar, etc. Figure 19-14 illustrates a block diagram of a superheterodyne receiver for AM reception. *See also* **heterodyning** *and* **beat frequency**.

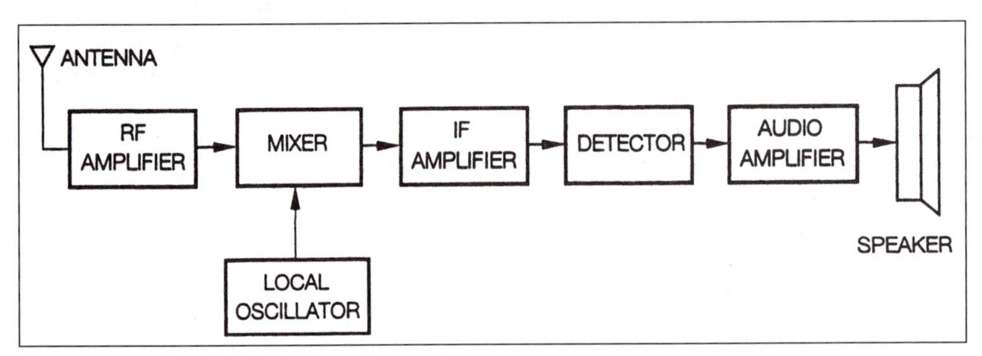

FIGURE 19-14 Superheterodyne receiver

superstation An independent television station that broadcasts its programs locally as well as to other areas via satellite.

supervision The process of monitoring a change of state between idle and busy conditions on a circuit.

SWIFT Abbreviation for Society for Worldwide Interbank Financial Telecommunications, a company that provides messaging services and interface software to financial institutions.

switched line A communication path between two points that can be established and maintained via a switching center for a communication session. Contrast to *leased line.*

switched message network A network service that provides interconnection of message devices.

Switched Multimegabit Data Service (SMDS) An earlier version of a broadband service specified by Bellcore.

switching center A location that (1) is capable of interconnecting circuits and (2) terminates multiple circuits.

switching center rank One of the six levels of the North American switching center hierarchy. Table 19-1 lists the ranking of the six levels:

TABLE 19-1 North American switching hierarchy

Regional Center	Class 1
Sectional Center	Class 2
Primary Center	Class 3
Toll Center	Class 4C
Toll Point	Class 4P
End Office	Class 5

sync character A transmission control character transmitted along with information to establish synchronization between the transmitter and the receiver.

synchronization The process of adjusting the receiver's clock to match the transmitter's clock in order to achieve an in-phase condition for the two clocks.

synchronous

1. A condition in which two events have a specific relationship under the control of a master clock.
2. A device by which events are synchronized and controlled by a master clock.
3. A condition in which the time interval between bits, characters, and events is the same.

synchronous communication A data transmission technique in which an entire message is transmitted in the form of data blocks; synchronization between transmitter and receiver is achieved by synch characters that are transmitted between successive data blocks. It is highly efficient compared to asynchronous communications because no synchronization (start/stop) bits per character are required. Only block synchronization is required. Figures 15-19(a) and (b) compare the two types of synchronous frame formats.

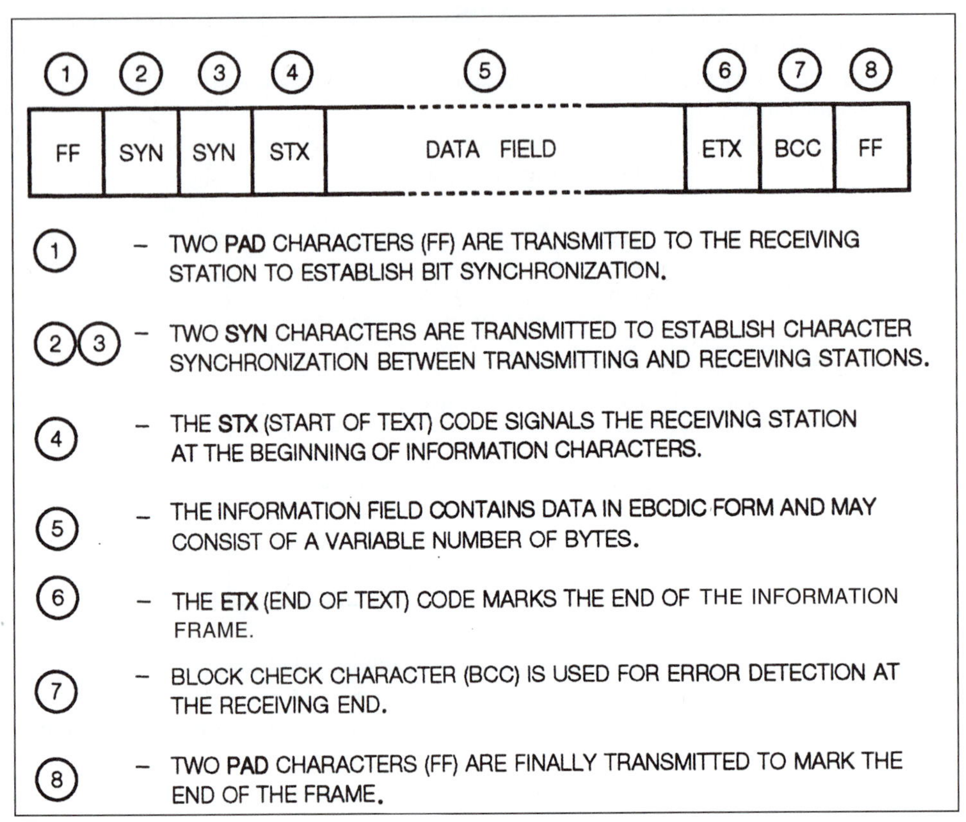

FIGURE 19-15(a) Binary synchronous frame format (character-oriented frame)

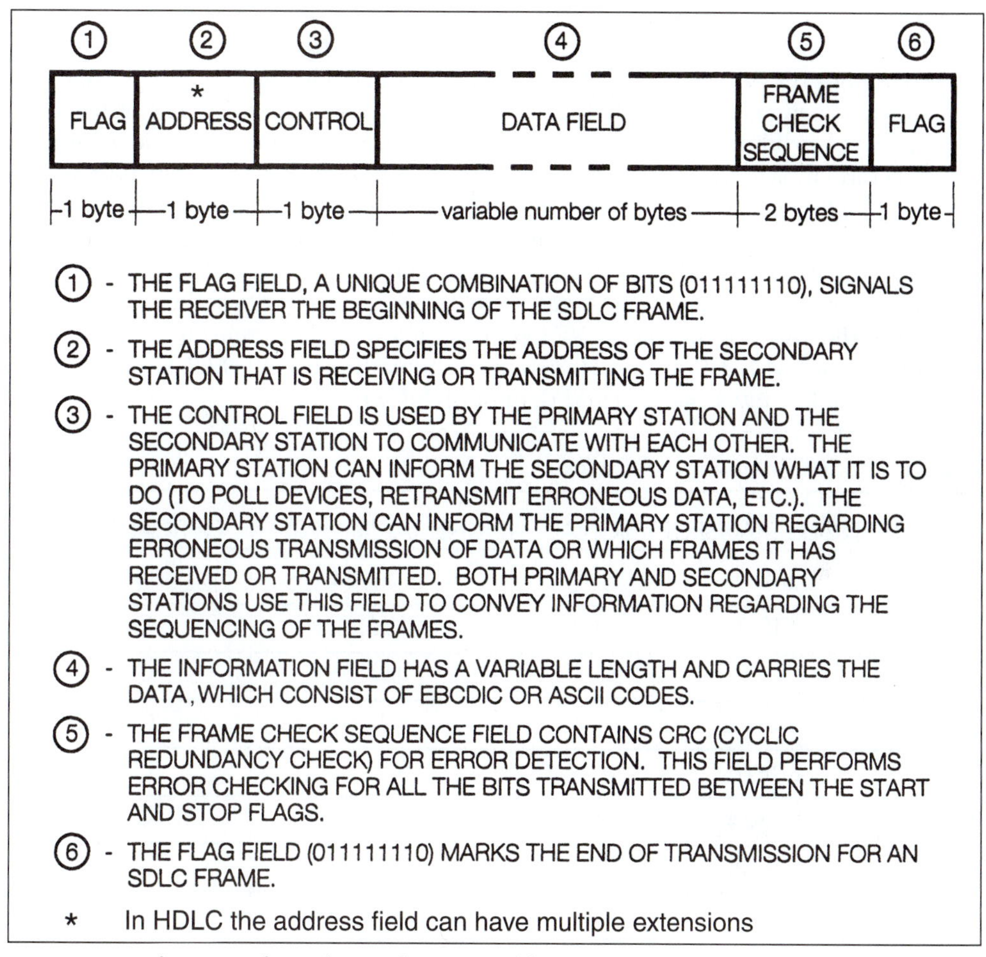

FIGURE 19-15(b) SDLC frame format (bit--oriented frame)

synchronous data link control (SDLC) A data link protocol standard developed by IBM as a part of systems network architecture (SNA). It is a bit-oriented, full duplex, synchronous protocol in which information is sent by frames. Each frame consists of multiple fields, each representing a specific function. Figure 19-15(b) illustrates the SDLC frame format. *See also* **synchronous communication** *and* **high-level data link control**.

Synchronous Optical Network (SONET) A broadband transmission technology that uses optical fiber to offer data transmission rates from 51.84 Mbps (OC-1) to 2.488 Gbps (OC-48). SONET is an American National Standard for a high-capacity optical telecommunications network. It is a synchronous digital transport system aimed at providing a simple, economical, and flexible telecommunications infrastructure. In SONET, the lowest level signal is called synchronous transport signal level 1 (ITS-1), which has a signal rate of 51.84 Mbps. The optical equivalent of ITS-1 is the optical carrier level 1 signal (OC-1). The higher level signals obtained by multiplexing lower level signals are represented by STN-N and OC-N, where *n* is an integer. The line rate of the higher level OC-N signal is *n* times the ITS-1, the line rate at the lowest level. The SONET standard allows only certain values of *n*. The values of *n* are: 1, 3, 9, 12, 18, 24, 36, and 48. Table 19-2 lists the SONET signal hierarchy and line rates. *See also* **add/drop multiplexer (ADM)**.

TABLE 19-2 SONET hierarchy and line rates

SYNCHRONOUS TRANSPORT SIGNAL	LINE RATE (MBPS)	OPTICAL CARRIER
STS-1	51.84	OC-1
STS-3	155.52	OC-3
STS-9	466.56	OC-9
STS-12	622.08	OC-12
STS-18	933.12	OC-18
STS-24	1244.16	OC-24
STS-36	1866.24	OC-36
STS-48	2488.32	OC-48
STS-96	4976	OC-96
STS-192	9953	OC-192

systems network architecture (SNA) A network architecture that serves as a blueprint for the hardware and software of a network consisting of IBM equipment. It is similar to the OSI reference model. *See also* Module 35.

MODULE

20 Definitions T

Tbs or Tb/s Terabits per second. *See* Module 31 for **Tera**.

T–carrier A time division, multiplexed, digital carrier that provides a bit rate of 1.544 Mbps and above. It is a pulse code modulation (PCM) system used by common carriers in providing long-distance telephone service. It can be used to transmit both voice and data. Figure 20-1 illustrates the North American digital multiplexing hierarchy (T-carrier system). *See also* **DS** *and* **E-Line**.

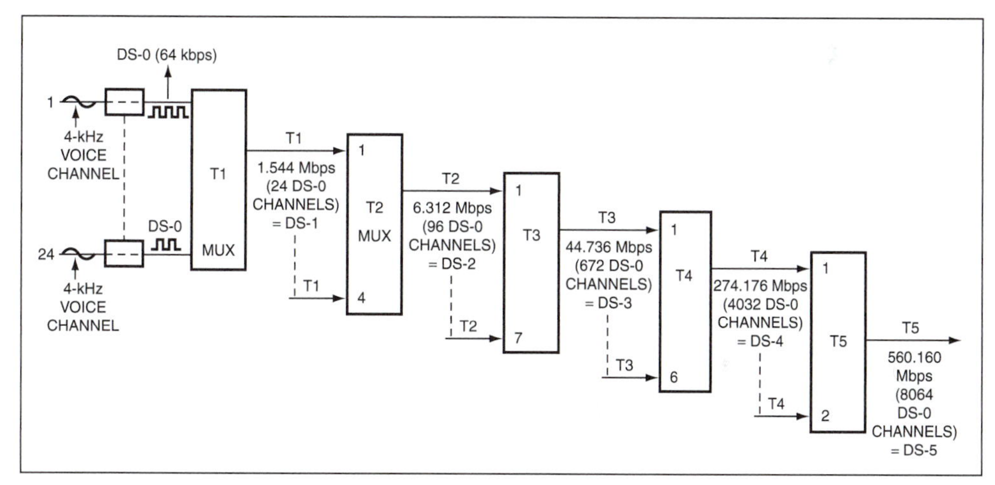

FIGURE 20-1 T–carrier system (North American digital multiplexing hierarchy)

T,R

1. Tip, ring
2. Transmit, receive
3. Transmitter, receiver

T1 A digital carrier facility used to transmit a DS1-formatted signal at 1.544 Mbps. The European standard is 2.048 Mbps. *See also* **E-Line** and Figure 20-1.

T1C, T1D A digital carrier facility used to transmit a DS1C-formatted digital signal at 3.152 Mbps.

T2 A digital carrier facility used to transmit a DS2-formatted digital signal at 6.312 Mbps.

T3 A digital carrier facility used to transmit a DS3-formatted digital signal at 44.736 Mbps.

T4 A digital carrier facility used to transmit a DS4-formatted digital signal at 274.176 Mbps.

tandem The connection of devices, circuits, or networks in series; i.e., the output of one is connected to the input of another.

tandem data circuit A data channel that passes through more than two DCE devices connected in series.

tandem office A telephone exchange used to switch traffic between central offices when direct interoffice trunks are not available. Also called *tandem exchange.*

tandem switching system A switched network system in which information can be transmitted through two or more switches.

tariff The rates, regulations, and rules that govern the provision of communication services.

TCAM *See* **telecommunications access method.**

TDM *See* **time-division multiplexing.**

TDMA *See* **time-division multiple access.**

tee coupler In fiber optics communications, a passive coupler that connects three ports. *See also* **tee network.**

tee network In fiber optics communications, a network that interconnects many terminals with the help of tee couplers (Figure 20-2).

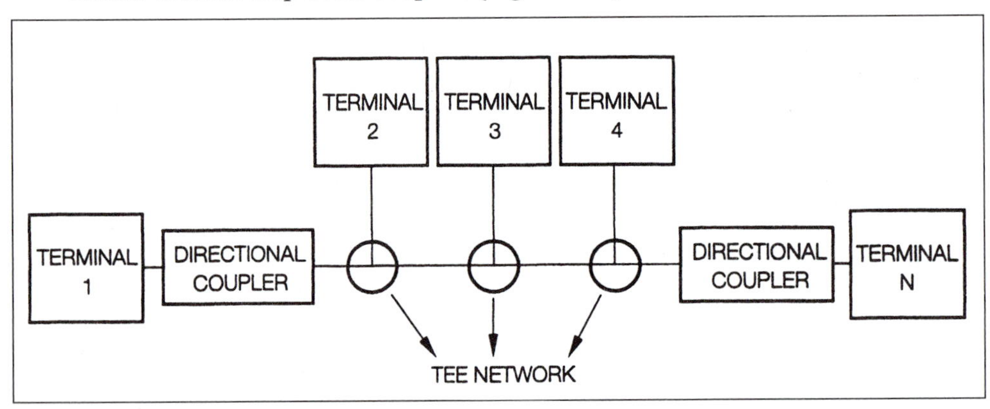

FIGURE 20-2 Tee network

TELCO An abbreviation for *tel*ephone *co*mpany.

telecommunications The process that allows the transmission of information (audio, video, data) in the form of electromagnetic signals from a transmitter to one or more receivers.

telecommunications access method (TCAM) An IBM software subsystem that supplements the conventional operating system to create data-communication-applications program and message control.

teleconference A conference in which participants from remote locations communicate to each other with the help of telecommunication lines.

telegraphy An aging branch of telecommunications that allows the transmission of information in the form of International Morse code signals over wire or radio. *See also* **Morse code** *and* Module 40.

telematics The technological services that have evolved due to the merger of telecommunications with computers and television.

Telnet A terminal emulator program that allows a user to log on to a remote computer.

telepak A discontinued service of common carriers. It used to provide leasing of wideband transmission channels between two points.

telephone exchange A switching center at which subscriber lines are terminated. Also called *central office.*

telephony Generic term describing voice communications.

teleprocessing Computer-based data processing that uses a transmission link to receive information from or transmit information to remote terminals.

Teletype

1. A trademark of the former Teletype Corporation.
2. A term for teleprinter equipment used in the earlier version of asynchronous communications.

teletypewriter (TTY) A generic term for a teleprinter that consists of a keyboard transmitter and a printing receiver.

teletypewriter exchange service (TWX) A public switched network service that interconnects teletypewriters. It is available in most countries around the world.

television The branch of telecommunications that deals with the transmission and reception of visual images in the form of electromagnetic signals. *See also* Module 41 for **International television standards**.

television receive only (TVRO) An antenna that receives a television signal but cannot transmit it.

telex A worldwide dial-up telegraph service that enables subscribers to communicate with each other with teletypewriter equipment on the public telegraph network.

terminal

1. A point in a telecommunications network at which data can enter or leave a telecommunications network.
2. A device capable of transmitting and receiving data over a communications network.

terminal emulation Imitation of a specific terminal by a microcomputer through software in order to communicate with other terminals or computers in a communications network.

Terminal Equipment Type 1 (TE1) In integrated services digital network (ISDN) technology, an ISDN-compatible terminal device such as digital telephone, data terminal, etc. *See also* Module 34.

Terminal Equipment Type 2 (TE2) In integrated services digital network (ISDN), a non-ISDN–compatible terminal device such as analog telephone, EIA/TIA-RS-232 interface, etc. *See also* Module 34.

text The information portion of a transmitted message. Generally surrounded by control characters, e.g., header, error check, end of text, etc.

thermal noise *See* **noise** and Table 14-1.

thin-film multiplexer A device consisting of thin films of optical materials that use electro-optic, electroacoustic, or magneto-optic effects to multiplex two or more optical signals.

thin-film optical modulator A device fabricated by depositing onto a semiconductor or a glass substrate thin films of optical materials that are capable of modulating an optical signal by electro-optic, electroacoustic, or magneto-optic effects.

thin-film optical switch An optical switching device that consists of thin layers of optical materials that use electro-optic, electroacoustic, or magneto-optic effects to perform switching functions similar to those of semiconductor logic gates.

thin-film optical waveguide An optical waveguide that consists of thin films of materials having different refractive indices and is used for the propagation of single-mode optical signals.

throughput The volume of information processed or communicated during a specific time period.

tie line (TL) A leased or private communication channel provided by the common carrier to connect two points without using a switching network.

time-division multiple access (TDMA) 1. A technique used in satellite communications that allows multiple earth stations to have sequential access to the total available transponder power and bandwidth. *See also* **Frequency-division multiple access (FDMA)** *and* Module 47 for **wireless technologies and standards**. 2. A time-division multiplexing technique used in cellular communications that allows sharing of one frequency (30 kHz) channel between several users by dividing it into three distinct time slots, thereby allowing each subscriber to get the channel for one-third of the total time. In TDMA, voice is digitized and compressed before it is transmitted, increasing the spectrum efficiency to three times that of AMPS. *See also* Module 47 for **wireless technologies and standards**.

time-division multiplexing (TDM) A multiplexing technique in which the bandwidth of the transmission path is divided into a number of time slots, each representing a separate channel (Figure 20-3). *See also* **frequency-division multiplexing (FDM), multiplexing,** *and* Module 47.

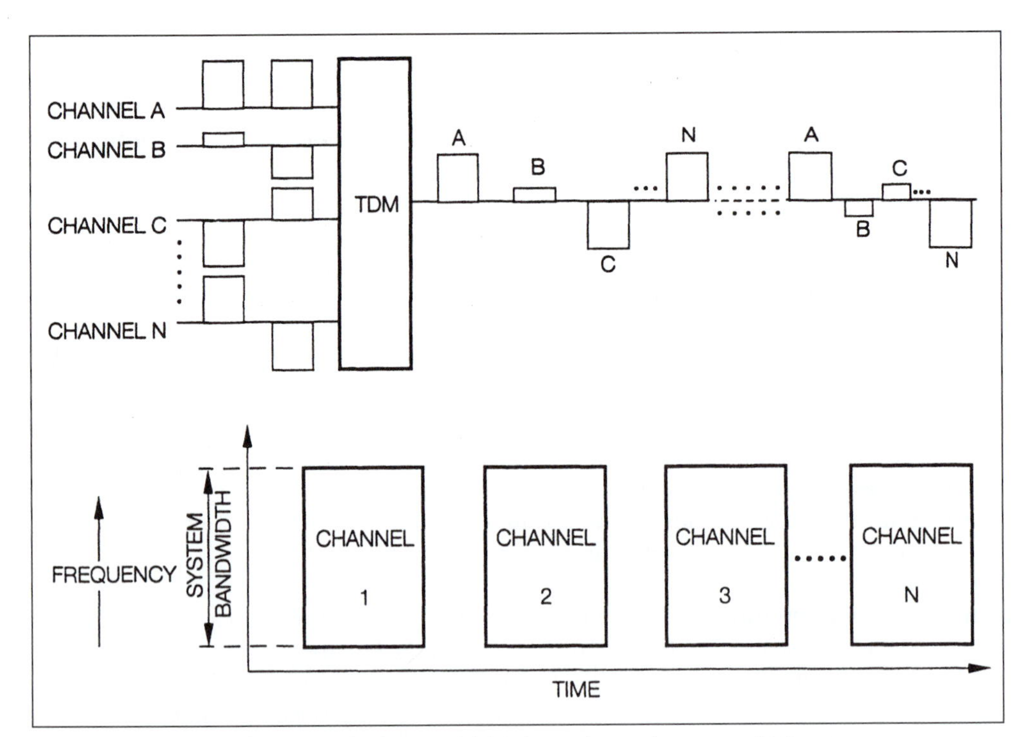

FIGURE 20-3 Time-division multiplexing: (a) N-channel time-division multiplexer; (b) frequency-time relationship for TDM

timeout A predefined time period of waiting before a specific function is initiated by a system.

time-sharing A method of operation in which a data processing facility is shared by multiple users at the same time. Each user gets service in a sequential manner but high-speed processing makes it appear that all tasks are processed simultaneously.

time-sharing option (TSO) An IBM programming package that runs under the operating system and provides a processing environment that is shared by many users at the same time.

tip The connecting part at the end of a telephone plug or the top spring of a jack. Any conductor connects to these contacts. The other contact of the plug is the ring. The terms tip and ring are used in a telephone network to identify the talk pair of a subscriber, the central office line, or the central office trunk. The use of the terms tip and ring goes back to the earliest days of telephony when all calls were handled manually by plugging cords into the jacks of a switchboard. Figure 20-4 illustrates tip and ring.

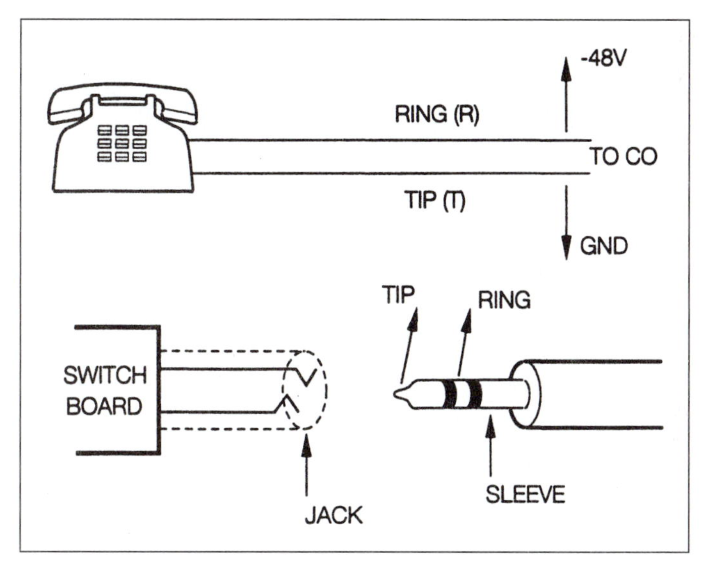

FIGURE 20-4 Tip and ring

token bus A local-area-access method in which all stations connected to a bus listen for a broadcast token. Any station wishing to gain access to the network must receive the token first and then transmit data. The next station to receive the token is not necessarily the next physical station on the bus. Bus access is controlled by the software. *See also* Module 48.

token passing A local-area-network-access method in which a token—a special bit pattern—is circulated among all network stations. The token gives a station the right to transmit over the network. It is generally used in ring topology.

token ring A local-area-network topology and access method in which a supervisory frame or token is sequentially passed from station to station. Any station wishing to gain access to the network must receive the token first and then transmit data. In a token ring, the next logical station to receive the token is also the next physical station. Contrast to *token bus*. *See also* **topology** *and* Module 48.

toll center A class 4 switching center where time- and distance-based information is collected. Usually there is one toll center per metropolitan area.

top-level domain (TLD) *See* **domain name**.

topology The physical and logical arrangement in which the nodes of a network are interconnected. Table 20-1 lists and compares the common topologies used in networks. *See also* Module 48 for **computer networks**.

TABLE 20-1 Topology comparisons

TOPOLOGY	ADVANTAGES	DISADVANTAGES
STAR	1. LOW FUTURE INCREMENTAL COST 2. FAILURE IN ONE STATION DOES NOT AFFECT REST OF NETWORK 3. REQUEST FOR SERVICE MAY BE BLOCKED AT THE SWITCH IN A PBX IN HEAVY TRAFFIC CONDITIONS 4. GOOD AVAILABILITY OF NETWORK MONITORING AND CONTROL SOFTWARE 5. CAPACITY OF CENTRAL NODE DETERMINES THROUGHPUT	1. HIGH COST INITIALLY 2. FAILURE OF THE CENTRAL NODE CAUSES THE WHOLE NETWORK TO FAIL. POOR RELIABILITY 3. LIMITATIONS IN DISTANCE BETWEEN THE CENTRAL NODE AND THE USER STATION 4. TWISTED-PAIR WIRE IS VULNERABLE TO ERRORS, LOW ERROR RATES POSSIBLE WITH COAXIAL AND FIBER CABLE
RING	1. NODES ARE EASY TO CONSTRUCT AND MAINTAIN 2. LOWER COST PER STATION THAN OTHER TOPOLOGIES 3. OVERALL GOOD RELIABILITY	1. WAITING TIME DEPENDS ON NUMBER OF NODES IN NETWORK 2. IF ONE STATION FAILS THE WHOLE NETWORK FAILS UNLESS BYPASS CIRCUITRY OR REDUNDANCY IS BUILT 3. IF THE NETWORK FAILS, RECOVERY IS DIFFICULT 4. ADDITION OF EACH STATION DIRECTLY AFFECTS PERFORMANCE. THROUGHPUT DECREASES DUE TO ADDITIONAL NODES 5. LIMITATIONS IN DISTANCE BETWEEN EACH STATION AND OVERALL
BUS	1. FAILURE OF ONE STATION DOES NOT AFFECT THE REST OF THE NETWORK, BREAK-IN CABLE MAY AFFECT ALL OR PART OF NETWORK 2. STATIONS MAY BE ADDED WITHOUT RECONFIGURING THE NETWORK 3. LOWER COST PER STATION THAN STAR TOPOLOGY AND HIGHER THAN RING TOPOLOGY 4. OVERALL GOOD RELIABILITY	1. LIMITATIONS IN DISTANCE BETWEEN STATIONS AND OVERALL. 2. HIGH ERROR RATE WITH TWISTED PAIR, LOW ERROR RATES POSSIBLE WITH COAXIAL AND FIBER CABLE 3. ADDITION OF EACH STATION DIRECTLY AFFECTS PERFORMANCE. THROUGHPUT DECREASES DUE TO ADDITIONAL NODES

total internal reflection The reflection of a light ray when it strikes an interface at an angle (with respect to the normal) greater than the critical angle (Figure 20-5).

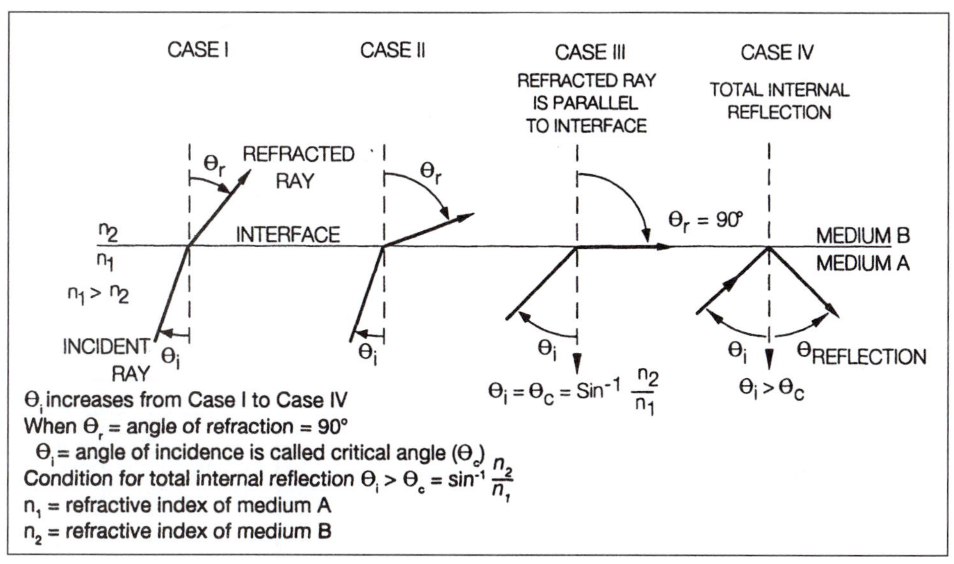

FIGURE 20-5 Total internal reflection

Touch-Tone Registered trademark of AT&T for push-button dialing. *See also* **dual tone multifrequency (DTMF)**.

traffic

1. The flow of information through a communication system.
2. Quantitative monitoring of total messages and their length, usually expressed in hundred call seconds (CCS).

traffic flow The measure of the traffic density, expressed in erlangs. *See also* Module 33 for **Traffic** formulas.

transceiver A device or a system that can receive as well as transmit information.

transistor-transistor logic (TTL) One of the IC (integrated circuit) digital logic families, used in the design of digital logic circuits. It uses combinations of bipolar transistors to form logical gates, which are the basic building blocks of digital circuitry. Figure 20-6 shows a TTL logic gate.

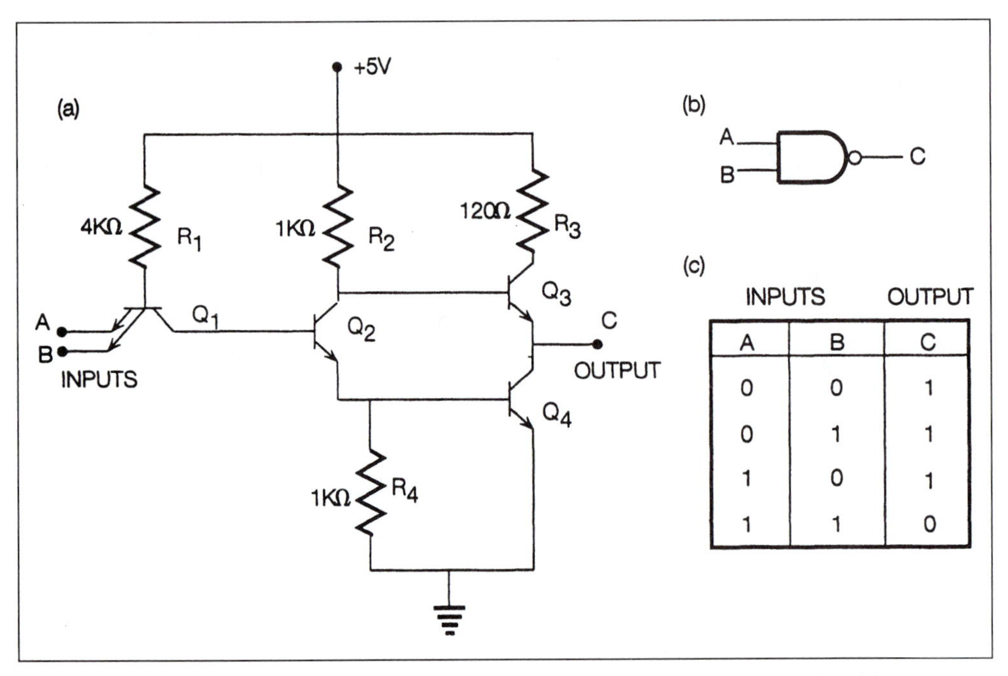

A	B	C
0	0	1
0	1	1
1	0	1
1	1	0

FIGURE 20-6 Transistor-transistor logic: (a) TTL NAND gate; (b) symbol; (c) truth table

transmission Sending information in the form of electromagnetic signals over transmission lines or waveguides (metallic or dielectric) or through air/space.

Transmission Control Protocol (TCP) A connection-oriented transport layer protocol that provides transport mechanisms for many Internet services, such as FTP, SMTP, and HTTP. *See* Figure 20-7.

TCP/IP	OSI MODEL	HARDWARE DEVICE IMPLEMENTATION
4. APPLICATION AND PROCESS (TELNET, FTP, SMTP)	7. APPLICATION	GATEWAY
	6. PRESENTATION	GATEWAY
	5. SESSION	GATEWAY
3. TRANSPORT (TCP, DNS, UDP)	4. TRANSPORT	GATEWAY
2. INTERNET (IP, DHCP, ICMP)	3. NETWORK	ROUTER
1. NETWORK INTERFACE: DATA LINK (ETHERNET, TOKEN RING, FDDI, ATM)	2. DATA LINK	BRIDGE/SWITCH
1. NETWORK INTERFACE: PHYSICAL (COPPER WIRE/OPTICAL FIBER)	1. PHYSICAL	REPEATER

IP = INTERNET PROTOCOL
DHCP = DYNAMIC HOST CONFIGURATION PROTOCOL
ICMP = INTERNET CONTROL MESSAGE PROTOCOL
UDP = USE DATAGRAM PROTOCOL
TCP = TRANSMISSION CONTROL PROTOCOL
SMTP = SIMPLE MAIL TRANSPORT PROTOCOL
FDDI = FIBER DATA DISTRIBUTION INTERFACE
ATM = ASYNCHRONDUS TRANSFER MODE

FIGURE 20-7 Comparison of TCP/IP suite with OSI model

transmission level point (TLP) The power of a transmission measured at any point in a transmission system. Usually expressed in db or dbm.

transmission line A material structure consisting of one or more conductors used for propagating electromagnetic signals from one point to another. *See also* **coaxial cable** and **two-wire line**.

transmission loss The loss of power that occurs during the propagation of an electromagnetic signal between two points in a medium. Usually expressed in decibels (db).

transmittance The ratio of the radiation transmitted through an object to the incident radiation.

Transmitter (XMTR) A device or equipment that prepares an information signal to propagate as an electromagnetic signal through a transmission medium. It is one of three major blocks of a typical communication system (transmitter, transmission medium or channel, and receiver). *See also* **channel**.

transparent A device, system, or function that appears not to exist but actually does. *See also* **real time** *and* **virtual**.

transponder A communication satellite's receiver-transmitter unit that receives an up-link signal, filters and amplifies it, and retransmits it at a different frequency as a down-link signal. Communication satellites contain several transponders to increase the number of transmission channels. Figure 20-8 illustrates the block diagram of a transponder. *See also* Module 46.

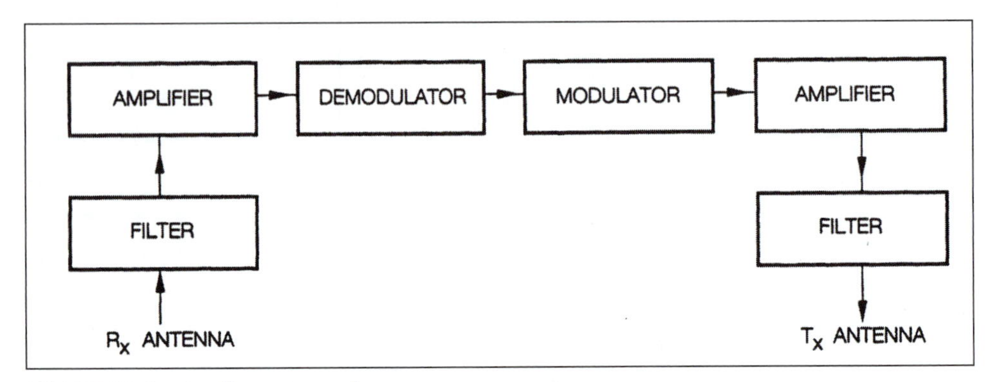

FIGURE 20-8 Satellite transponder

transport layer The OSI model's fourth software layer, responsible for the transmission of a message from one network user to another. This layer provides the appropriate service to the session layer. (*See* Module 35.)

tree A type of network topology in which the communication medium branches at some points along its length to connect stations. It resembles a tree.

tribit A group of three bits. It can represent eight distinct values, which are 000, 001, 010, 011, 100, 101, 110, and 111.

trunk A circuit that connects two switching centers in a telephone network.

trunk exchange An exchange that interconnects trunk lines.

TSO *See* **time-sharing option**.

TTL *See* **transistor-transistor logic**.

TTY *See* **teletypewriter**.

turnaround time In data communications, the time required to reverse the direction of transmission from transmit to receive or vice versa. It affects the throughput, especially when the propagation delay of the channel is lengthy.

TVRO *See* **television receive only**.

twisted pair A transmission line that consists of two insulated wires twisted together. Twisting the wire together reduces external interference. Contrast to *two-wire line or circuit.*

two-wire line A circuit consisting of two single wires in pair for half-duplex transmission. Also called *two-wire channel.*

TWX *See* **teletypewriter exchange service**.

MODULE

21 Definitions U

UART *U*niversal *a*synchronous *r*eceiver *t*ransmitter. A device that converts parallel data from a system into serial data for asynchronous transmission and converts it back to parallel at the receiver. Figure 21-1 illustrates the block diagram of a UART.

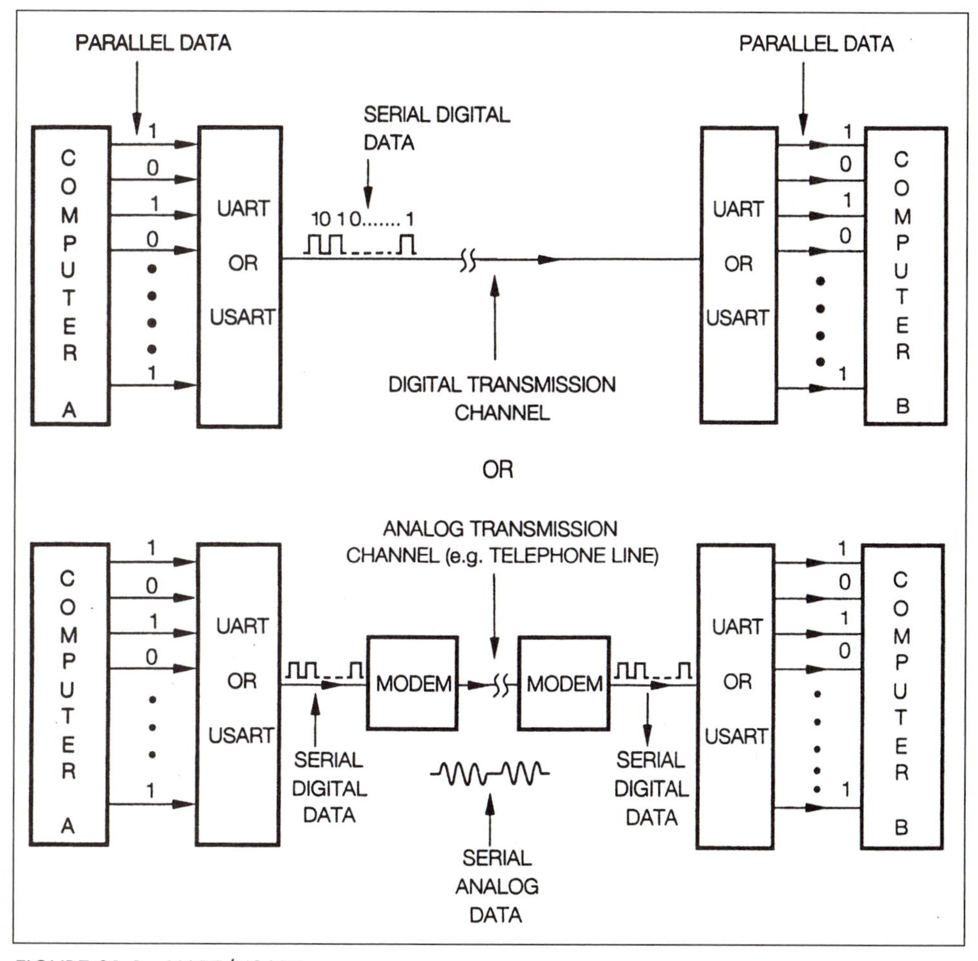

FIGURE 21-1 UART/USART

UG Underground.

UHF *See* **ultra high frequency**.

ultra high frequency The band of electromagnetic signal frequencies ranging from 300 MHz to 3 GHz.

ultrasonic frequency A sonic frequency lying above audio frequencies, ranging usually above 20 kHz. *See also* **audio frequencies**.

ultraviolet (UV) The band of electromagnetic frequencies ranging from 0.4 to 0.003 micron.

U-matic A type of video cassette recorder that uses 0.75-inch-wide recording tape.

unattended operation Transmission and/or reception that is designed to operate automatically and does not require a human operator.

unbalanced-to-ground A two-wire transmission line in which the impedance-to-ground on one conductor is measurably different from that on the other. *See also* **balanced line**.

unbundled Software, services, and training sold separately from the hardware by the manufacturer.

undistorted wave A periodic wave in which both the attenuation and the speed of propagation are the same for all sinusoidal components.

unicode A 16-bit code set developed as universal set by Unicode consortium (http://www.unicode.org) to represent characters of any language used in the world. *See also* Module 40 *for Data Codes*.

uniform resource locator (URL) Address of a website. *See also* **domain name**.

universal data link control (UDLC) A bit-oriented communication protocol developed by Sperry.

UNIX A portable operating system designed by AT&T.

upload The process of transmitting data from a user's computer to another computer connected to the network. In contrast to *download* process, in which data is transferred from a computer connected to the network to a user's computer.

up link That part of a satellite link in which signals are transmitted from an earth station to the satellite. *See also* **down link**.

USART *U*niversal *s*ynchronous/*a*synchronous *r*eceiver *t*ransmitter. A device that takes parallel data from a system and converts the data into serial data for synchronous/asynchronous transmission. At the receiver, it converts the data back to parallel form. *See also* **UART** *and* **USRT**.

USASCII *See* **American Standard Code for Information Interchange (ASCII)**.

USB *U*niversal *S*erial *B*us. It is an external bus standard that supports high-speed data transfer (12 Mbps) between a personal computer and peripheral devices such as digital cameras, printers, modems, mice, CD-ROM drives, digital scanners, etc. Most operating systems support the USB. See Figure 16-4.

User Datagram Protocol (UDP) A fast, connectionless transport-layer protocol that is faster than TCP but has less reliability. *See also* **Transmission Control Protocol (TCP)**.

user friendly Computer software or hardware that is easy to use.

USITA *U*nited *S*tates *I*ndependent *T*elephone *A*ssociation.

USRT *U*niversal *s*ynchronous *r*eceiver *t*ransmitter. A device that converts parallel data from a system into serial form for synchronous transmission and at the receiver converts it back to parallel form. *See also* **UART** *and* **USART**.

MODULE

22 Definitions V

V A set of ITU-TS (CCITT) standards that deal with data communication using the telephone system. Also called *V. Series Recommendations*. *See* Module 38.

value-added carrier A carrier that orders service from a common carrier and adds special features (e.g., electronic mail, protocol conversion, error detection and correction, temporary data storage, etc.) before resale to customers.

value-added network (VAN) A type of public network that provides computer-oriented special services (e.g., computerized switching, protocol conversion, store and forward message switching, electronic mail, etc.) in addition to basic voice and data transmission service.

VDSL Very high bit rate digital subscriber line. *See* Module 36.

vertical redundancy check (VRC) An error-checking technique in which a parity bit is added to each character. *See also* **parity check**.

very high frequency (VHF) That part of the electromagnet spectrum with frequencies in the range of 30 to 300 MHz.

very-large-scale integration (VLSI) An IC (integrated circuit) fabrication technique that allows thousands of logic gates on a single chip. Table 22-1 compares various types of IC fabrication technologies.

TABLE 22-1 Comparison of IC fabrication technologies

TECHNOLOGY TYPE	NUMBER OF LOGIC GATES/CHIP
SSI—Small-scale Integration	≃10
MSI—Medium-scale Integration	10–100
LSI—Large-scale Integration	0.1 k–9.99 k
VLSI—Very-large-scale Integration	10 k–99 k
ULSI—Ultra-large-scale Integration	100 k+

very low frequency (VLF) That part of the electromagnetic spectrum with frequencies in the range of 3 to 30 kHz.

very-small-aperature terminal (VSAT) A satellite earth station with a small antenna, used in a point-to-multiple-point communication network.

vestigial sideband The remaining part of a sideband that has been nearly suppressed with a gradual cutoff in the vicinity of the carrier frequency. *See also* **modulation**.

vestigial sideband transmission A technique of signal transmission in which one complete sideband and the corresponding vestigial sideband is transmitted.

VHS One of the two original formats for recording video information on VCR cassettes. The other is beta format.

video The conversion and/or display on a CRT of electronic signals that can be seen when converted.

video conferencing The blending of software, hardware, and telecommunication technologies to allow users to hold a virtual meeting on computers.

video signal In a television system the signal that conveys the whole of the intelligence present in an image. Figure 22-1 shows a composite video signal.

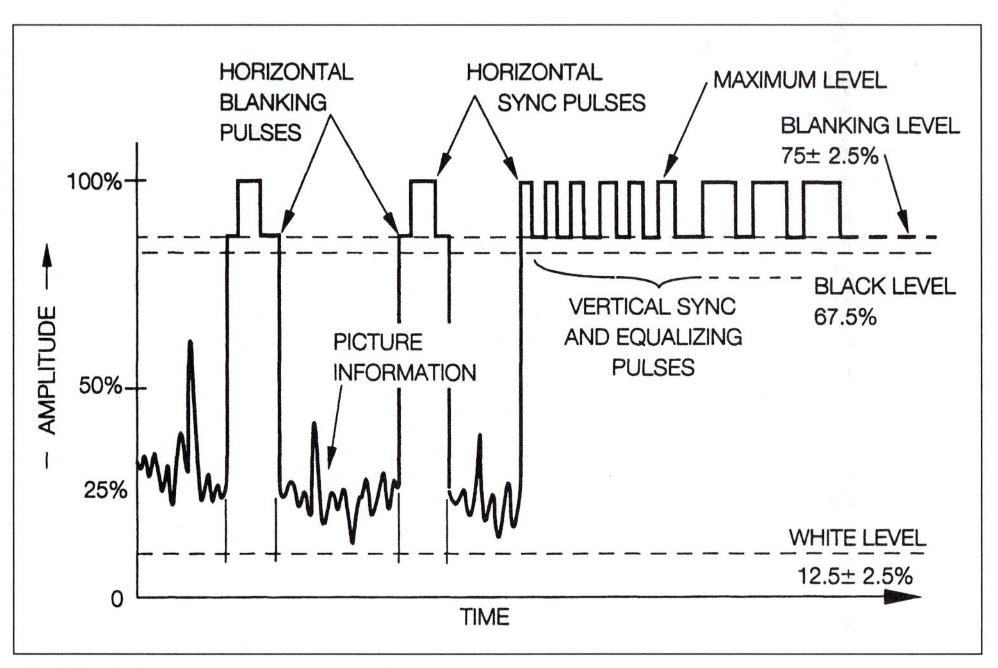

FIGURE 22-1 Composite video signals

videotex An interactive data communication service in which an unsophisticated user interacts with a remote data base over the public switched network.

virtual A device, system, or function that appears to exist but actually does not.

virtual channel identifier (VCI) *See* **asynchronous transfer mode (ATM)**.

virtual local area network (VLAN) A group of network devices that forms a single broadcast domain. VLANs logically group users together, regardless of their physical connection with respect to the network, with the objective of keeping traffic within the broadcast domain.

virtual path identifier (VPI) *See* **asynchronous transfer mode (ATM)**.

virtual private network (VPN) A secure private network that uses public transmission networks, such as the Internet, for communication.

virtual reality The use of software and hardware to create an artificial environment that mimics reality; has current and potential applications in the fields of medicine, education, research, etc.

virtual telecommunications access method (VTAM) An IBM mainframe communications software that provides resource sharing for the efficient use of a network.

virus A program or a code that loads itself on a computer and corrupts data. *See also* **worm**.

visible light That part of the electromagnetic spectrum that can be perceived by the human eye. It covers frequencies in the range of 0.7 to 0.4 micron.

VT100 A standard for dumb terminals.

voice digitization The process in which voice is converted from analog to digital form for storage or transmission.

voice frequency (VF) Any frequency within the audio frequency range that is used to transmit speech of commercial quality. Typically, it represents a range from 300 to 3400 Hz.

voice grade A telecommunication device, channel, or system capable of voice transmission (0–4 kHz).

voice-grade channel A telecommunications channel with a bandwidth of 4 kHz, primarily used for voice transmission but also suitable for the transmission of data or facsimile.

voice-over-IP The use of Internet Protocol (IP) to transmit digitized voice signals over long distances, over the Internet (Figure 22-2).

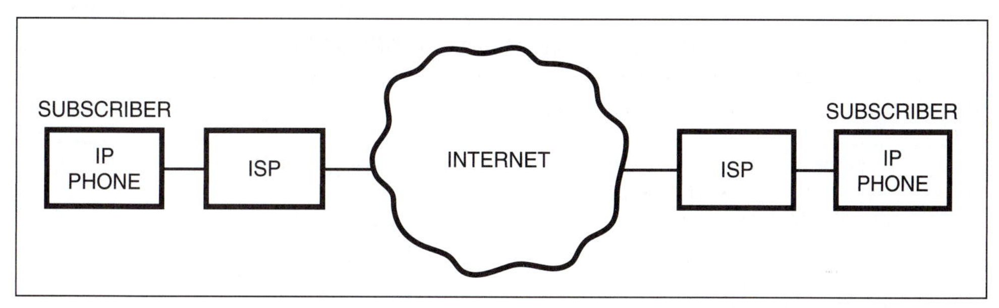

FIGURE 22-2 Voice-over-IP (Internet Telephony)

voice recognition A technique in which a human voice is converted into digital commands to activate a device or achieve a task.

voice store-and-forward system (VSFS) A computer-controlled system that allows voice messages to be created, edited, stored, forwarded, and transmitted to appropriate destinations with the help of a 12-button dial pad.

volatile storage A type of device that loses its contents when applied power is removed.

volt The unit of electromotive force and potential difference. It is that electromotive force or potential difference when applied across a conductor with a resistance of one ohm that produces a current of one ampere.

voltage A difference of electrical potential expressed in volts.

volume The specific amount of storage contained on storage media such as hard disk drives, CD-ROMs, or DVDs.

MODULE

23 Definitions W

WAN *See* **wide area network** *and* Module 48 *for computer networks.*

WARC *W*orld *A*dministrative *R*adio *C*onference. An ITU conference in which decisions are made regarding the allocation of international frequencies and geosynchronous satellite orbit locations.

WATS *See* **wide area telephone service**.

WATTIC *W*orld *A*dministrative *T*elegraph and *T*elephone *C*onference. An ITU conference in which provisions regarding the interconnecting and interworking of various international networks are discussed.

wave A vibrational disturbance in a medium or in quanta of energy, propagated through space, in which displacement may be a function of time or space or both.

waveform The shape of a wave represented graphically in amplitude and time.

waveguide Generally, a system of material boundaries capable of guiding electromagnetic signals. Specifically, a transmission line consisting of a hollow rectangular, elliptical, or circular metallic conductor within which electromagnetic waves may be propagated (Figure 23-1).

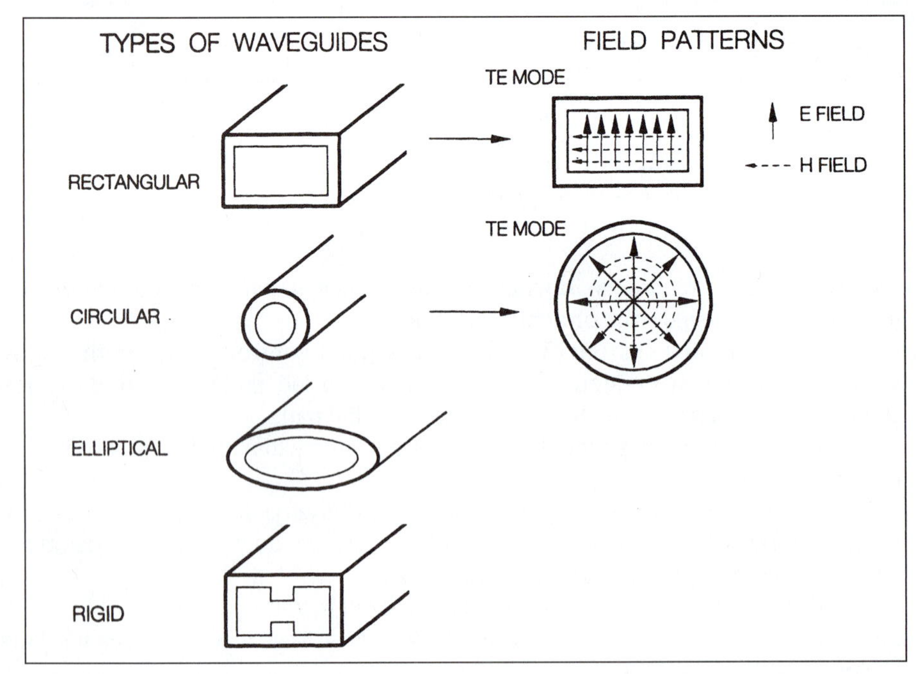

FIGURE 23-1 Waveguides

waveguide dispersion *See* **dispersion** *and* Module 27.

wavelength Of a periodic wave, the distance from any point on one wave to the corresponding point on the next wave, such as from crest to crest. It is also the distance traveled by the wave in one cycle (Figure 23-2). Mathematically, wavelength is given by:

$$\lambda = \frac{c}{f}$$

or

$$\lambda = \frac{v}{f}$$

Where: λ = wavelength of signal in meters
f = frequency of signal in Hz
v = velocity of wave in m/s (for air/free space $v = c = 3 \times 10^8$ m/s)

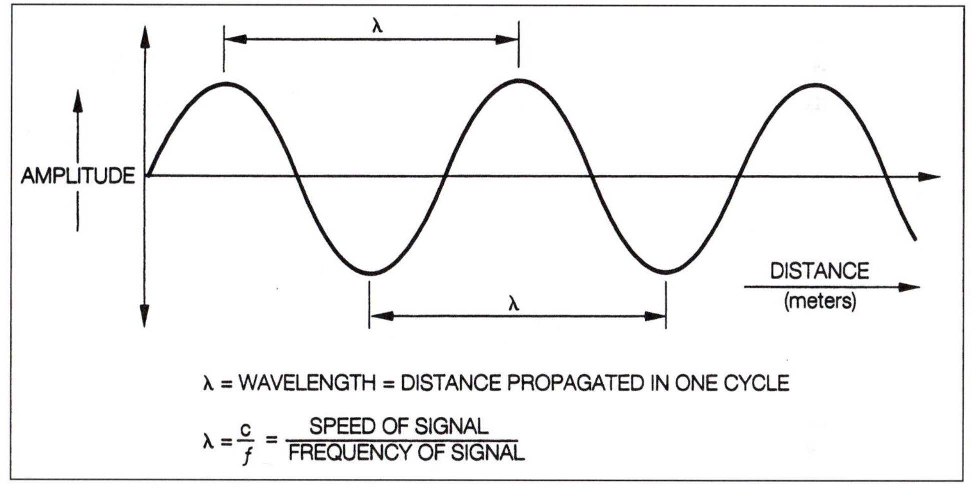

FIGURE 23-2 Wavelength

wavelength-division multiplexing (WDM) A type of multiplexing used in fiber optics communications to transmit two or more channels over a single fiber. Each channel is assigned a separate wavelength. It is similar to the frequency-division multiplexing (FDM) technique used in radio communications. Figure 23-3 illustrates the half-duplex and full-duplex WDM. *See also* Modules 27 *and* 28.

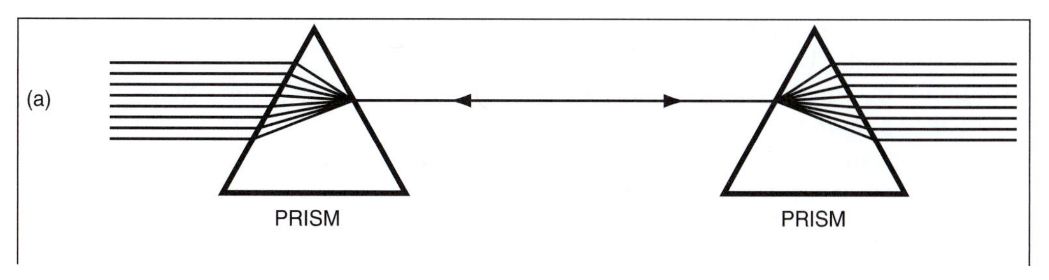

FIGURE 23-3 (a) Wavelength-division multiplexing; (b) half-duplex WDM; (c) full-duplex WDM

(continues)

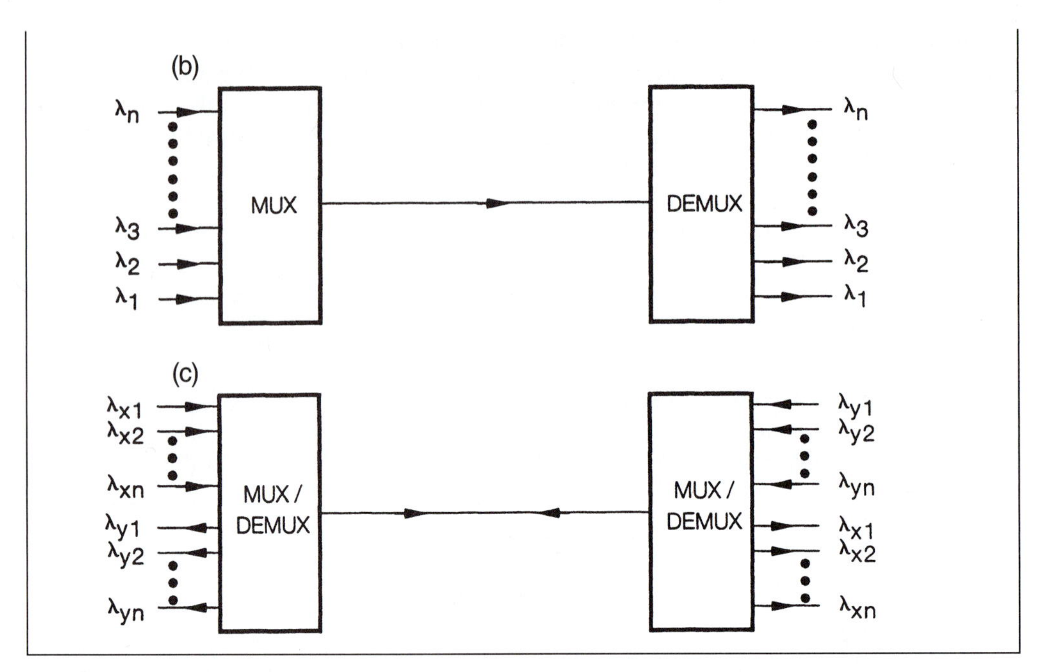

FIGURE 23-3 (continued)

web browser *See* **browser**.

Westar The name of a communication satellite owned by Western Union.

white noise Noise having a wide-frequency spectrum (resembling white light in this respect). It is one of the sources for the degradation of telecommunication signals. *See also* **noise**.

who A utility that lists all the users logged in to a network.

wide area network (WAN) A network that interconnects local area networks (LANs) or metropolitan area networks (MANs) that may be located at diverse geographic locations within a county or around the world. *See also* **local area network (LAN), topology,** *and* Module 48.

wide area telephone service (WATS) A service provided by a telephone company that allows customers to make (OUTWATS) or receive (INWATS) long-distance voice or data calls on a flat-rate or a measured-bulk-rate basis. INWATS or 800 service allows calls to be made to a central location from anywhere, with no cost to the calling party. OUTWATS allows calls to be placed from a central location. Generally, cost is based on hourly usage per WATS circuit and on distance based on zones from which or to which calls are placed. The United States has been divided into five zones.

wideband A signal whose bandwidth is broader than the voice-grade channel. Also called *broadband.*

wideband channel A telecommunication channel with a bandwidth greater than the voice-grade channel. Also called *broadband channel.*

Wi-Fi *Wi*reless *Fi*delity. It refers to IEEE 802.11 network. *See also* Module 47 *and* Module 48.

wireless Refers to radio communications, i.e., communication without wires.

wireless LAN (WLAN) A network configuration that uses infrared (IR) or radio frequency (RF) signals for interconnecting computers located in a building. *See also* Module 47 *and* Module 48.

word A group of characters processed by a computer as an entity.

word length The number of bits or bytes in a word, determined by optical or convenient size for processing, storage, or transmission. In computers, word lengths are determined by the size of registers.

workstation A term for personal computer, terminals, and other personal productivity devices used to perform a job.

worm A code that self-replicates and invades computers connected to a network. Compared to a virus it requires less human intervention for propagation.

world number plan *See* Module 42.

write To store information in a memory device or medium.

WWW World Wide Web. *See* **Internet**.

MODULE 24 Definitions X

X A set of ITU-TS (CCITT) standards that deal with data transmission over public networks. Also called *X. Series Recommendations. See also* Module 38.

X client A graphical program that runs on a host computer.

X rays Electromagnetic radiation of wavelengths less than 300 angstroms. Can penetrate solid objects.

X server The screen where a graphical program is remotely displayed.

xbar Crossbar.

XC Cross connect.

XD Ex-directory refers to a subscriber number that is not listed in the printed telephone directory. Also known as *unlisted number.*

XENIX Microsoft's trade name for a 16-bit microcomputer operating system, derived from UNIX.

xfr Transfer.

XMIT Transmit.

XMODEM A data link protocol used in asynchronous transmission for transferring files from one computer to another. It uses 128-byte data block and cyclic redundancy code (CRC) or checksum error-checking techniques. Figure 24-1 illustrates an XMODEM data block.

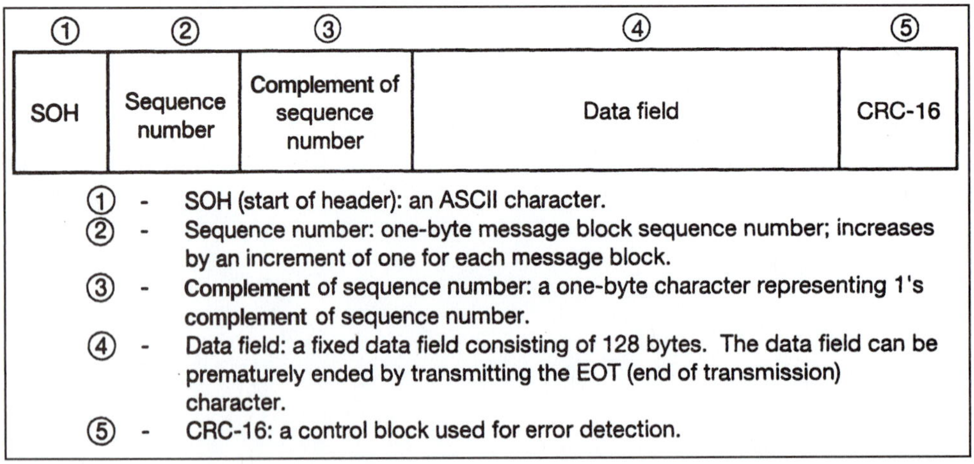

①	②	③	④	⑤
SOH	Sequence number	Complement of sequence number	Data field	CRC-16

① - SOH (start of header): an ASCII character.
② - Sequence number: one-byte message block sequence number; increases by an increment of one for each message block.
③ - Complement of sequence number: a one-byte character representing 1's complement of sequence number.
④ - Data field: a fixed data field consisting of 128 bytes. The data field can be prematurely ended by transmitting the EOT (end of transmission) character.
⑤ - CRC-16: a control block used for error detection.

FIGURE 24-1 XMODEM data block format

XON/OFF Transmitter ON/transmitter OFF. A method of controlling communications between a terminal and a host, with the help of status signals. The XOFF signal is activated when the terminal does not want the host to transmit more data. XON signal is activated when the terminal wants the host to transmit more data.

MODULE 25 Definitions Y

Y signal In a composite video signal, the part that conveys the brightness information about the original scene that was imaged.

yagi antenna A type of antenna that is highly directional, and is used for television signal reception. Figure 25-1 illustrates a yagi antenna. *See also* **antenna**.

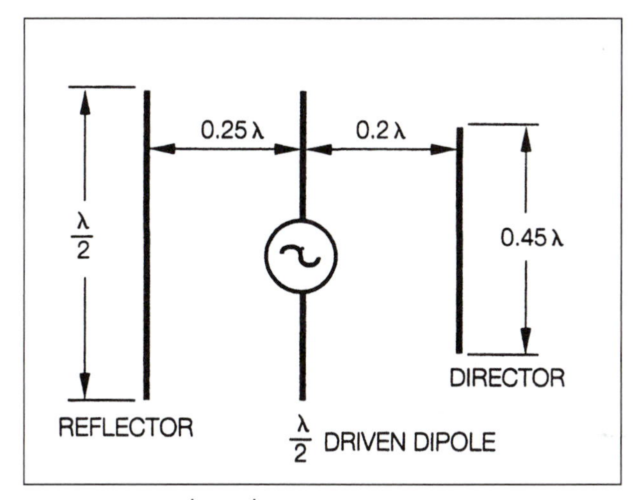

FIGURE 25-1 Three-element yagi antenna

yoke An assembly of one or more coils that carry current to produce a magnetic field for the deflection of an electron beam in a device. The most common application is around the neck of a CRT, as shown in Figure 25-2.

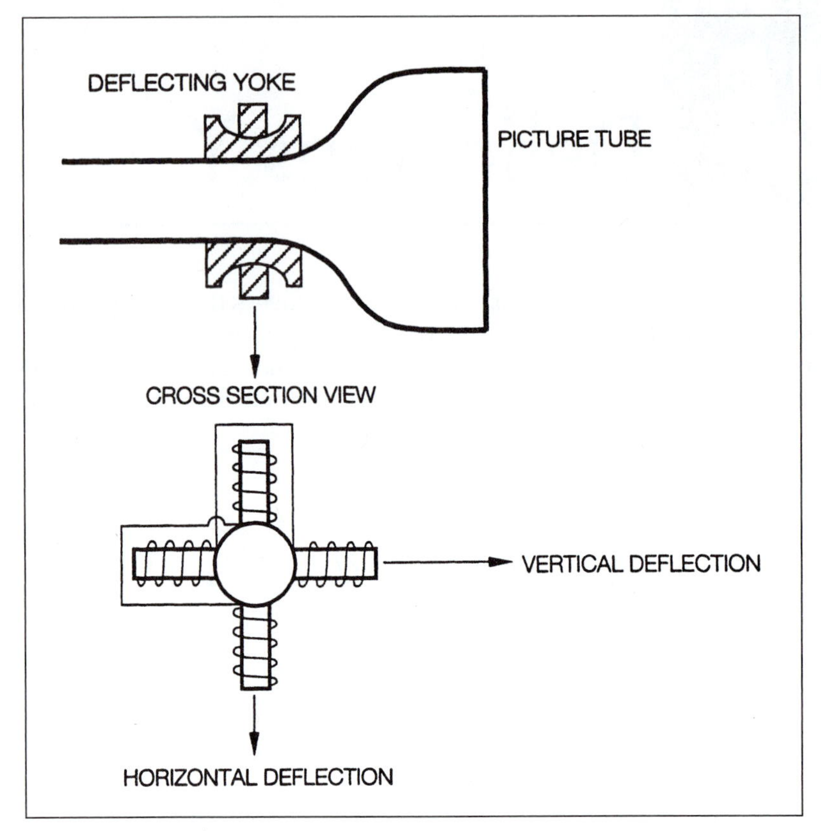

FIGURE 25-2 Yoke

MODULE

26 Definitions Z

Z modem A file transfer protocol that uses a 32-bit CRC for error checking.

zero bit insertion Also called zero bit stuffing, a technique used in bit-oriented protocols (e.g., in HDLC), which allows the transmission of bit patterns similar to opening and ending flags (01111110) in the information field. A "1" bit is inserted or stuffed by the transmitter after any succession of five consecutive 1 bits in the information field. At the receiver, this extra bit is removed to restore data to its original form. *See also* **High-level data link (HDLC) protocol**.

zero code suppression The insertion of a "1" (one) bit to eliminate the transmission of eight or more consecutive "0" (zero) bits. Used in digital telephony, T-1, and related facilities that require a certain number of "1" bits to keep individual channels of a multiplexer active.

zero transmission level point (0TLP) In a telephone network, a point used as a reference to measure signal and noise levels at other points in the system relative to it. In North America, 0TLP is located at the output of the toll exchange. For measuring signal levels with respect to 0TLP, the unit of dbm0 is used.

zone transfer The process of exchanging information between a master DNS server and a slave DNS server.

MODULE

27 Fiber Optics Communications

■ 27-1 INTRODUCTION TO FIBER OPTICS COMMUNICATIONS

The recent exponential growth in fiber optics (lightwave) technology has opened a new avenue for integrating voice, data, and video transmission. After two decades, the performance of lightwave transmission systems is still advancing rapidly. Fiber optics, which until 1970 was primarily a technology confined to research laboratories, has emerged as the medium of choice for short- and long-haul terrestrial and oceanic broadband links.

Many innovations in various fields of science have contributed to the development of modern fiber optics communications systems. Table 27-1 presents the developmental history of fiber optics communications.

TABLE 27-1 Fiber optics communications—development history

1790	–Optical telegraph system built by Claude Chappe
1870	–Demonstration by John Tyndall before the British Royal Society that light can be guided in a stream of water
1880	–Photophone developed by Alexander Graham Bell
1934	–Patent for optical telephone system received by Norman R. French
1950s	–Advances in semiconductor technology –Development of LASER
1960s	–Lightwave system feasibility study –Experimental lightwave systems –Development of optical fiber with loss in range 500–1000 db/km
1970s	–Development of LEDs (light-emitting diodes) –Advances in development of low-loss optical fiber. In 1970, fiber with attenuation values of less than 20 db/km manufactured by Corning Glass Works; improved to 4 db/km in 1972. –Interoffice trunk systems
1980s	–Development of optical fiber with loss < 1 db/km –Advances in development of laser diodes and LEDs –Fiber optics LANs and MANs –Completion of first fiber optics transoceanic link (TAT-8)

(continues)

TABLE 27-1 (continued)

1990s	–Long haul fiber optics transoceanic networks –Synchronous optical network (SONET) –Fiber Distributed Data Interface (FDDI) –Erbium doped fiber amplifier (EDFA) –Broadband integrated services digital network (B–ISDN) –Synchronous optical network (SONET) –Fiber to the curb (FTTC) –Fiber to the home (FTTH) –Fiber CATV
2000+	–Lightwave Internet backbone networks –Optical switches –Optical computing –Global oceanic links –Solitons –WDM, DWDM, & UDWDM

A typical fiber optics communications system consists of three basic components:

1. optical transmitter
2. optical fiber
3. optical receiver

Figures 27-1 a and b illustrate the block diagram and components of a typical system. The transmission of information over a distance using optical fiber usually requires several steps. First, the information is converted into an electrical signal, if it is not yet in that form. Second, the electrical information signal is changed into an optical signal with the help of an optical source. Third, the optical signal is transmitted through the optical fiber. Fourth, the optical signal is detected and converted into an electrical signal with the help of an optical detector. Finally, signal processing is done.

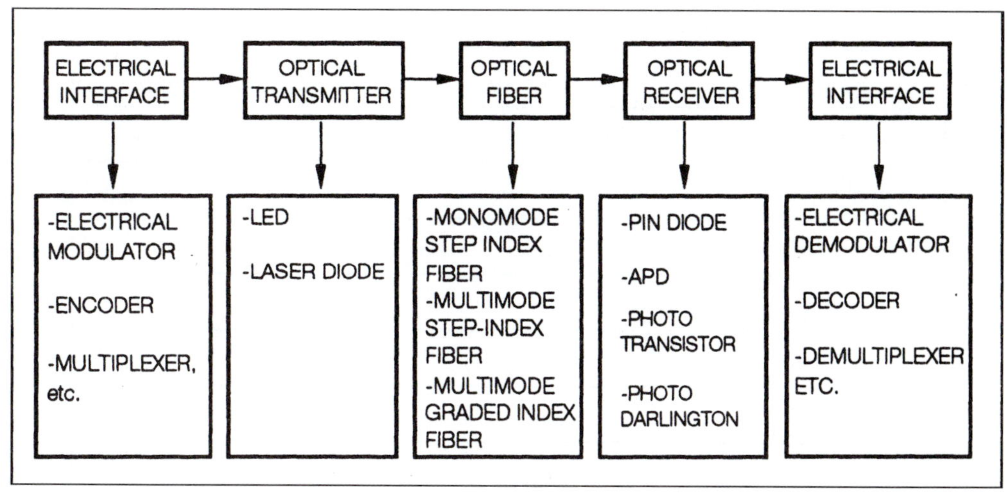

FIGURE 27-1(a) Fiber optics communication system

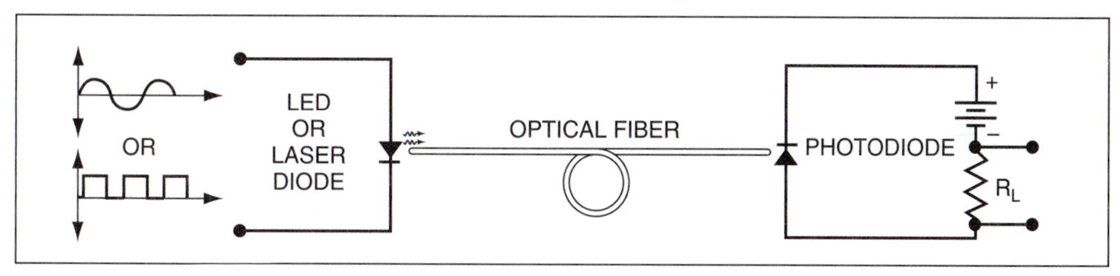

FIGURE 27-1(b) Components of a fiber optics system

Fiber optics communication offers numerous advantages over conventional copper cable and microwave communications techniques. Table 27-2 compares characteristics of optical fiber with copper wires and Table 27-3 lists the advantages of fiber optics. The major advantage of lightwave communications is its potentially enormous bandwidth. For example, for an optical signal with a wavelength of 1 micrometer, theoretically a bandwidth of 300 THz is possible, using which 75 billion voice channels (BW = 4kHz) can be transmitted over an optical fiber with multiplexing techniques. But at present, the maximum data transmission rate for lightwave systems is under 10 Gbps because the present state of technology does not permit the fabrication of optical sources, optical fibers, and optical detectors that can operate in THz range.

TABLE 27-2 Comparison of optical fiber with copper wires

CHARACTERISTICS	TWISTED PAIR	COAXIAL	FIBER
Data rate (Mbps/km)	16	500	1000+
Signal security	No	No	Yes
Static problems	Yes	Yes	No
Grounding problems	Yes	Yes	No
Bit error rate (BER)	10^{-6}	10^{-6}	10^{-9}
Size	Large	Large	Small
Signal radiation	Yes	Yes	No

TABLE 27-3 Advantages of fiber optics communication

LARGE BANDWIDTH
Theoretically, a bandwidth on the order of 100 THz is possible. However, the present state of technology permits data rates in the range of 10–100 Gbps.

SMALL SIZE AND LIGHT WEIGHT
Optical fibers have a very small diameter (on the order of tenths of a micrometer) compared with bulky copper cables.

DIELECTRIC CONSTRUCTION
No ground loops required. There is no induction of electromagnetic fields.

EMI and RFI IMMUNITY
There is no generation and conduction of electrical or electromagnetic noise or interference.

LOW TRANSMISSION LOSS
With low-loss fibers, long-haul systems can be designed with a minimum number of repeaters.

SIGNAL SECURITY
Optical fibers do not radiate any energy. Therefore it is very difficult to make unauthorized taps in a noninvasive manner. This feature makes the fiber a highly secure medium.

HIGH RELIABILITY & DURABILITY
Optical fibers are durable and reliable. They have no inherent degradation mechanisms and no susceptibility to corrosion or oxidation. They have a high temperature tolerance and can be used in an explosive or nuclear environment.

■ 27-2 BASIC FIBER CONSTRUCTION

Figure 27-2 shows the typical construction of an optical fiber. The optical fiber can be made of silica and plastic. An optical fiber is composed of two layers: core and cladding. The core is the central portion of fiber through which light propagates from one end to another. It has a higher refractive index than the cladding which surrounds it. The difference in the refractive indices of core and cladding defines an optical interface, which keeps the signal within the core (i.e., which light propagates through the core due to total internal reflection of light).

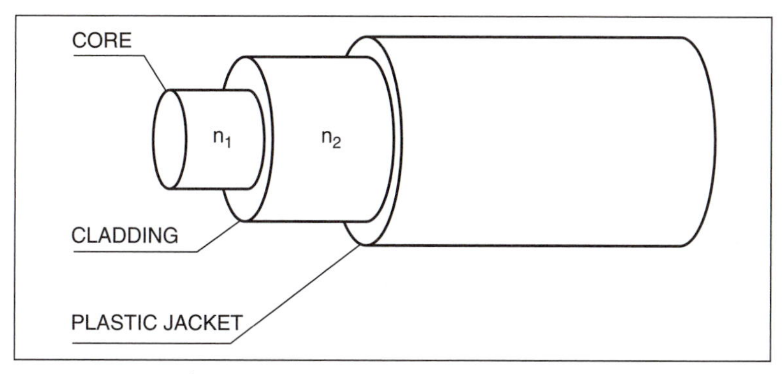

FIGURE 27-2 Fiber construction

■ 27-3 TYPES OF OPTICAL FIBERS

Optical fibers are classified into three types:

1. Mono or single-mode step-index fiber
2. Multimode step-index fiber
3. Multimode graded-index fiber

A mono or single-mode step-index fiber supports only one mode. It has a very narrow core, which limits the wavelength of light that can pass through it. It has very low signal attenuation and dispersion, making it suitable for long-haul links that require a high bandwidth. A multimode step-index fiber has a larger core and is capable of transmitting many modes. It is generally used for low-bandwidth, low-cost, short-haul applications. The multimode graded-index fiber also has a larger core and can transmit a signal by many modes. It is suitable for high bandwidth and medium-haul applications. Table 27-4 presents a summary of characteristics and applications for different types of optical fibers.

TABLE 27-4 Characteristics, types, and applications of optical fibers

TYPES OF OPTICAL FIBER										
TYPE OF FIBER	FIBER DIAGRAM / TYPES OF RAY PATHS	REFRACTIVE INDEX PROFILE n = REFRACTIVE INDEX r = DISTANCE	INPUT PULSE/ OUTPUT PULSE	TYPE OF SOURCE REQUIRED	≈ NUMERICAL APERTURE (NA)	≈ CORE DIAMETER (µm)	≈ ATTENUATION (db/km)	≈ DISPERSION (ns-km)	NUMBER OF NODES (N) OR V-number	APPLICATIONS
MONOMODE STEP-INDEX FIBER	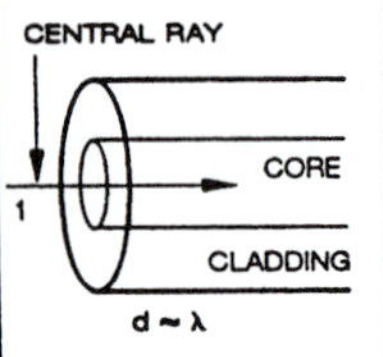 CENTRAL RAY, 1, CORE, CLADDING, $d \sim \lambda$	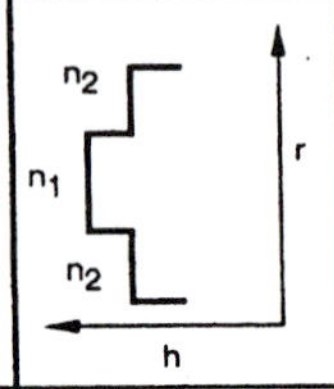 n_2, n_1, n_2, r, h	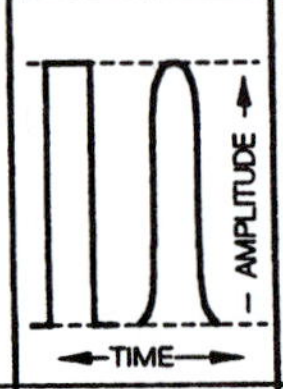 AMPLITUDE, TIME	LASER DIODE (LD)	0.1	5–10	0.01–0.2	0.005–0.5	$V = \frac{2\pi a}{\lambda_0} NA$ < 2.405	LONG-HAUL LINKS, eg., TAT-8, TAT-9, SONET
MULTIMODE STEP-INDEX FIBER	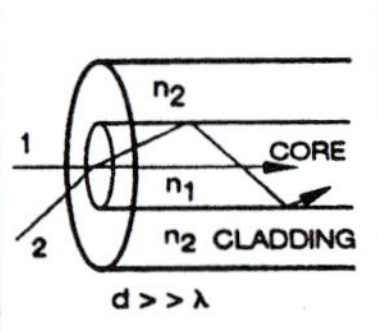 1, 2, n_2, n_1, CORE, n_2 CLADDING, $d >> \lambda$	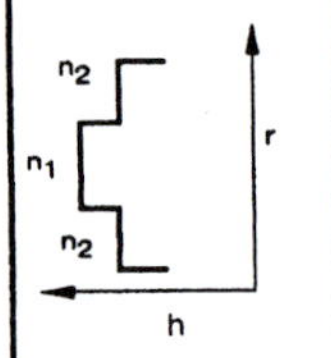 n_2, n_1, n_2, r, h	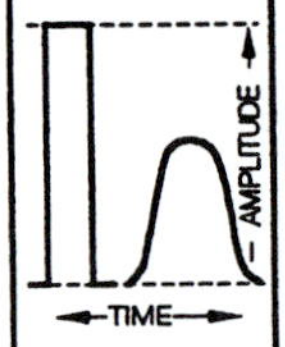 AMPLITUDE, TIME	LASER DIODE (LD) OR LED	0.24	50–200	5	15	$V = \frac{2\pi a}{\lambda_0} NA$	SHORT-HAUL DATA LINKS
MULTIMODE GRADED INDEX FIBER	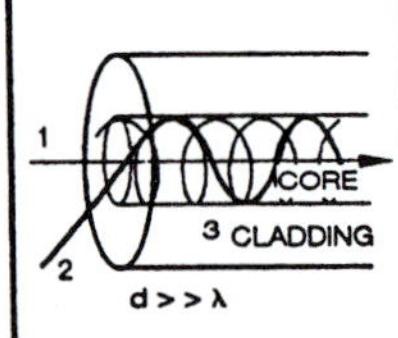 1, 2, 3, CORE, CLADDING, $d >> \lambda$	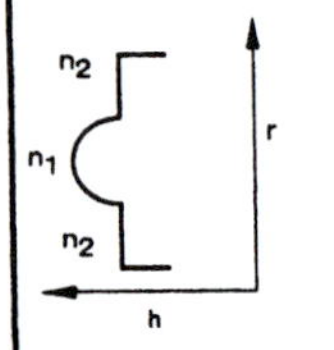 n_2, n_1, n_2, r, h	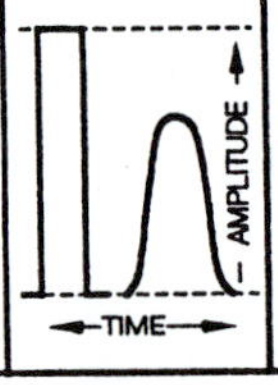 AMPLITUDE, TIME	LASER DIODE (LD) OR LED	0.2	50–100	5	0.5–1	$N = \left(\frac{2\pi a n_1}{\lambda_0}\right)^2 \Delta \left(\frac{\alpha}{\alpha + 2}\right)$	DATA LINKS, TELEPHONE TRUNKS

1 = CENTRAL RAY
2 = MERIDIONAL RAY
3 = SKEW RAY

n_1 = CORE REFRACTIVE INDEX
n_2 = CLADDING REFRACTIVE INDEX

a = CORE RADIUS, λ_0 = WAVELENGTH
$NA = (N_1^2 - N_2^2)^{1/2}$ Δ = DEVIATION INDEX
α = POWER LAW COEFFICIENT

NUMBER OF MODES ARE PROPORTIONAL TO V-number PARAMETER

■ 27-4 PRINCIPLES OF LIGHT TRANSMISSION, ATTENUATION, AND DISPERSION MECHANISMS IN OPTICAL FIBER

Table 27-5 lists the principles of optical signal transmission through a fiber. As the optical signal propagates through the optical fiber it undergoes attenuation (a decrease in signal strength due to various types of losses) and dispersion (pulse spreading) as illustrated in Figure 27-4. These two transmission characteristics dictate the design of a fiber optic link. Attenuation controls the maximum length of the link or maximum inter-repeater distance, and the maximum data transmission rate or bandwidth is determined by dispersion. Attenuation in optical fiber is wavelength dependent. Figure 27-5 illustrates the relationship between attenuation and wavelength for a silica fiber. The three low-loss windows are used in fiber optics links. The first window (800–900 nm) is used for shorter links. The second window (1200–1300 nm) is used for longer links. The third window (1500–1600 nm) offers the lowest attenuation and is used for the longest links. Systems operating at wavelengths near 1550 nm can transmit typically 40 percent farther

TABLE 27-5 Principles of light transmission in fiber

Conditions for the total internal reflection of a light signal in an optical fiber:

1. The refractive index of the core must be greater than the refractive index of the cladding.
2. The angle of incidence at the core-cladding interface must be greater than the critical angle.

In order to propagate through the fiber by total internal reflection, the optical signal must be launched within the acceptance cone. An optical signal launched outside the acceptance cone will enter the cladding as a leaky mode. Figure 27-3 illustrates the acceptance cone. The maximum acceptance angle (also called the half-cone angle) is given by:

$$\Theta_0 = \sin^{-1}\sqrt{\frac{n_1^2 - n_2^2}{n_0}}$$

$$\text{if } n_0 = 1 \text{ then } \Theta_0 = \sin^{-1} \text{NA}$$

Where: n_0 = refractive index of the launching medium
n_1 = refractive index of the fiber core
n_2 = refractive index of the fiber cladding
$(n_1^2 - n_2^2)^{1/2}$ = numerical aperture (NA)

The numerical aperture of a fiber is a figure of merit that indicates the angular range of incident light that can be guided by a fiber. It dictates the number of modes propagating within the fiber, which has consequent effects on both fiber attenuation and fiber dispersion.

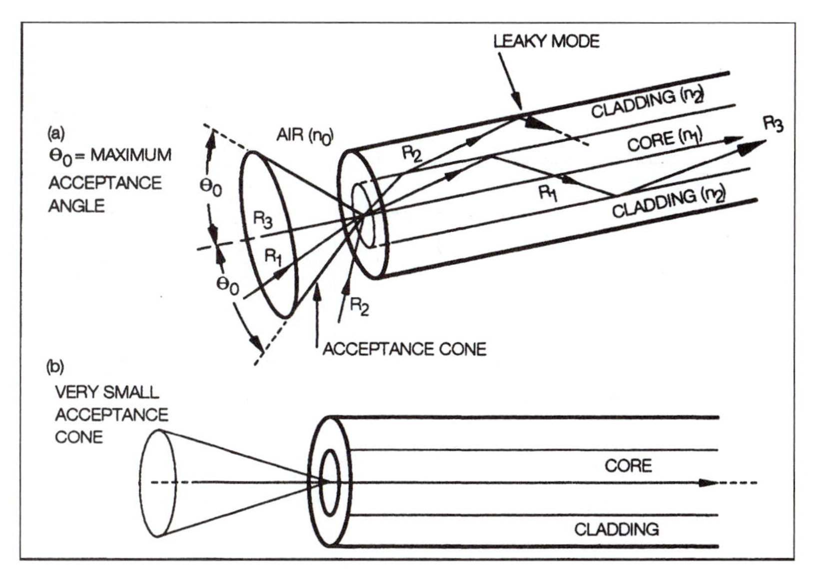

FIGURE 27-3 (a) Multimode step-index fiber; (b) monomode step-index fiber

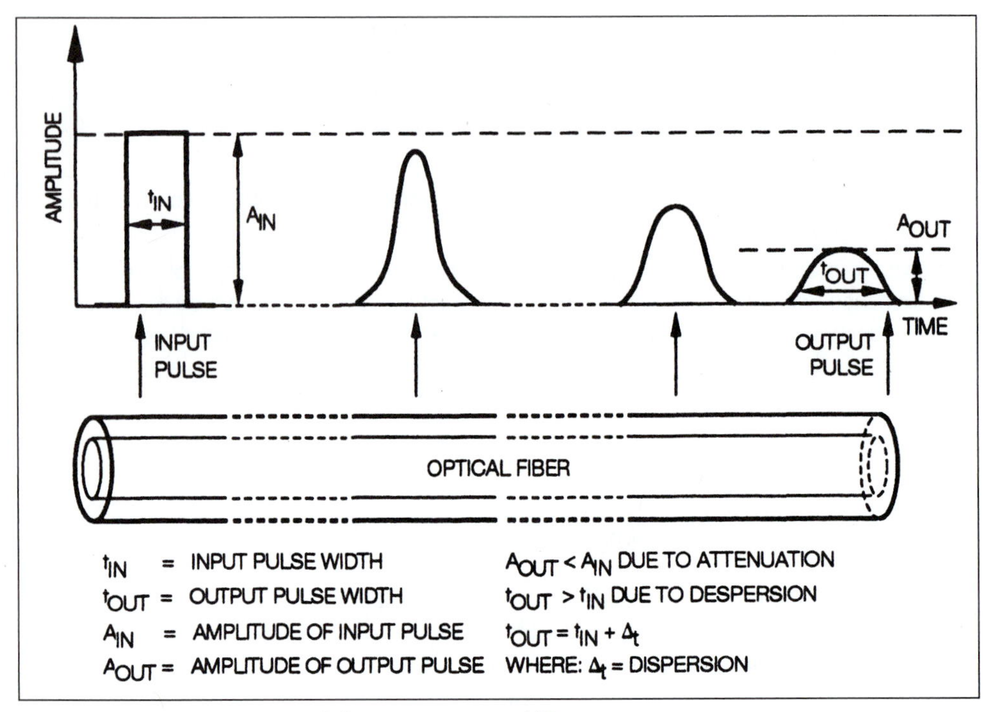

FIGURE 27-4 Attenuation and dispersion in optical fiber

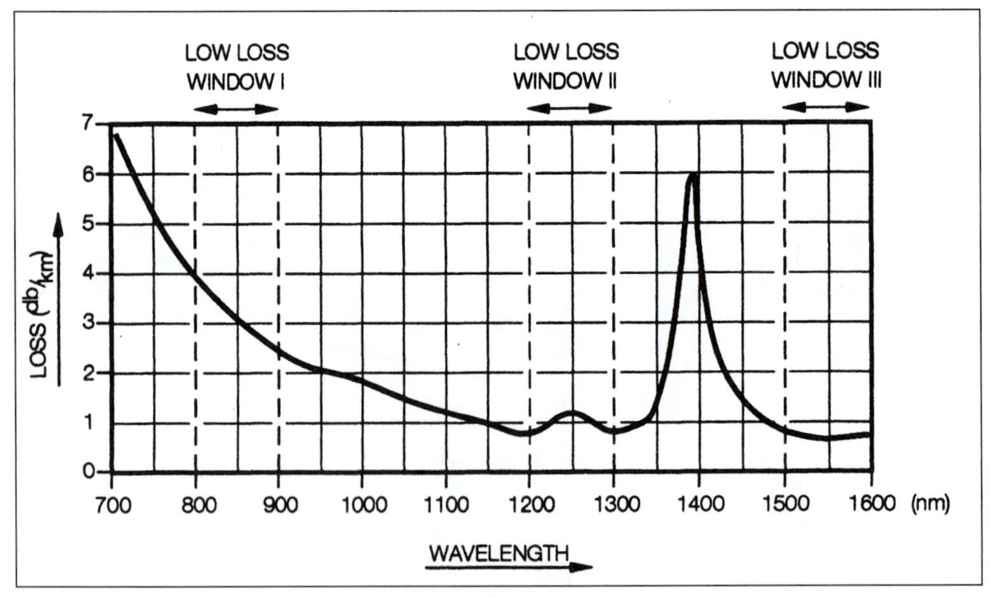

FIGURE 27-5 Attenuation versus wavelength relationship for silica fiber

than at the 1310 window, provided the signal is not too degraded by dispersion. Typical attenuation for silica fiber is less than 1 db/km for 1330 nm and 1500 nm wavelength windows, and for plastic (acetylic-core) fiber typical loss is 65 db/km at 570 nm and 160 db/km at 650 nm. Due to high attenuation, plastic fibers can be used only for short distance links typically less than 100 m long. Table 27-6 lists attenuation mechanisms in optical fiber.

TABLE 27-6 Attenuation mechanisms in optical fiber

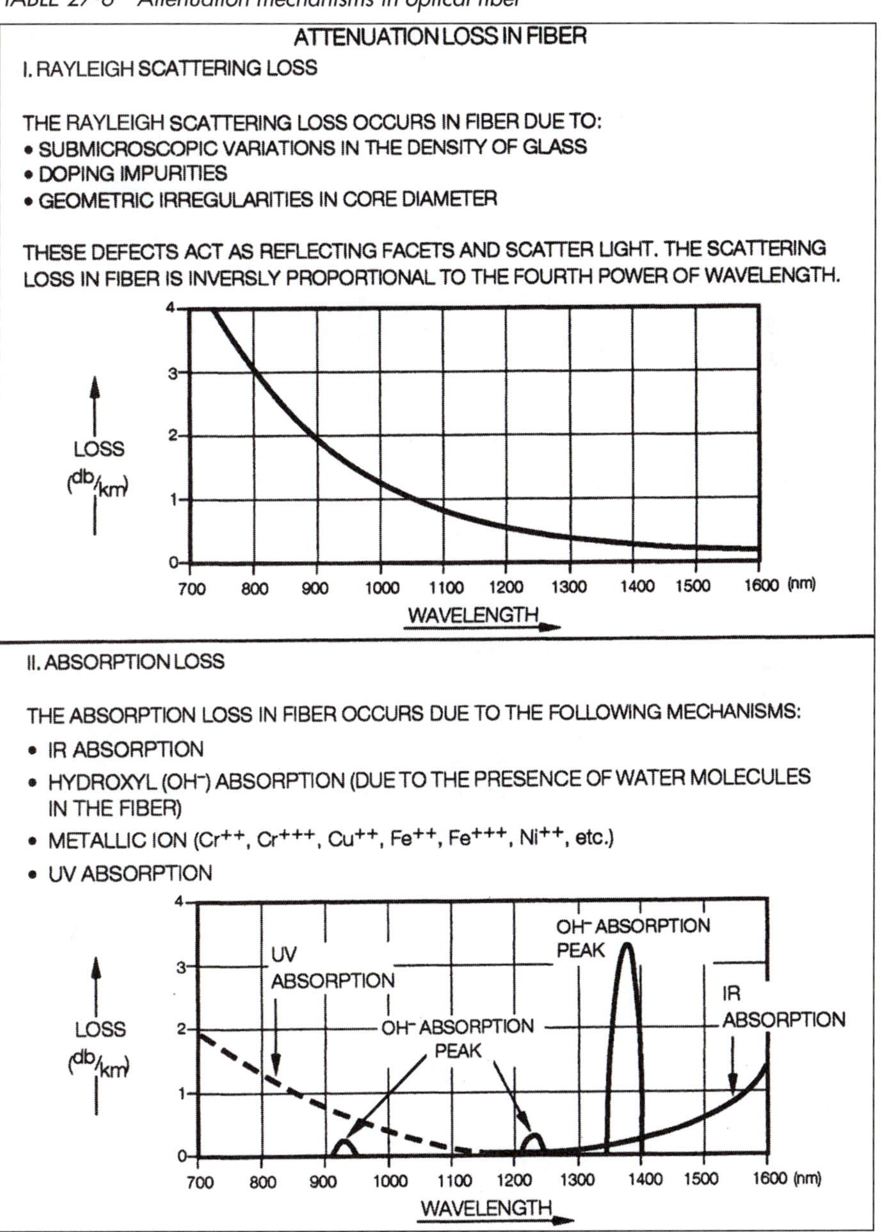

ATTENUATION LOSS IN FIBER

I. RAYLEIGH SCATTERING LOSS

THE RAYLEIGH SCATTERING LOSS OCCURS IN FIBER DUE TO:
- SUBMICROSCOPIC VARIATIONS IN THE DENSITY OF GLASS
- DOPING IMPURITIES
- GEOMETRIC IRREGULARITIES IN CORE DIAMETER

THESE DEFECTS ACT AS REFLECTING FACETS AND SCATTER LIGHT. THE SCATTERING LOSS IN FIBER IS INVERSLY PROPORTIONAL TO THE FOURTH POWER OF WAVELENGTH.

II. ABSORPTION LOSS

THE ABSORPTION LOSS IN FIBER OCCURS DUE TO THE FOLLOWING MECHANISMS:
- IR ABSORPTION
- HYDROXYL (OH^-) ABSORPTION (DUE TO THE PRESENCE OF WATER MOLECULES IN THE FIBER)
- METALLIC ION (Cr^{++}, Cr^{+++}, Cu^{++}, Fe^{++}, Fe^{+++}, Ni^{++}, etc.)
- UV ABSORPTION

(continues)

TABLE 27-6 (continued)

III. LEAKY-MODE LOSS

THE LEAKY-MODE LOSS OCCURS IN A FIBER CLOSE TO THE TRANSMITTER WHEN LIGHT ENTERS THE FIBER OUTSIDE THE ACCEPTANCE CONE. IT GETS REFRACTED INTO THE CLADDING AS A LEAKY MODE. THIS LEAKY MODE MAY TRAVEL WITHIN THE CLADDING AND ATTENUATE OR MAY RE-ENTER THE CORE AT SOME POINT ALONG THE LENGTH OF THE FIBER AND INTERFERE WITH OTHER MODES INSIDE THE FIBER AND CAUSE ATTENUATION.

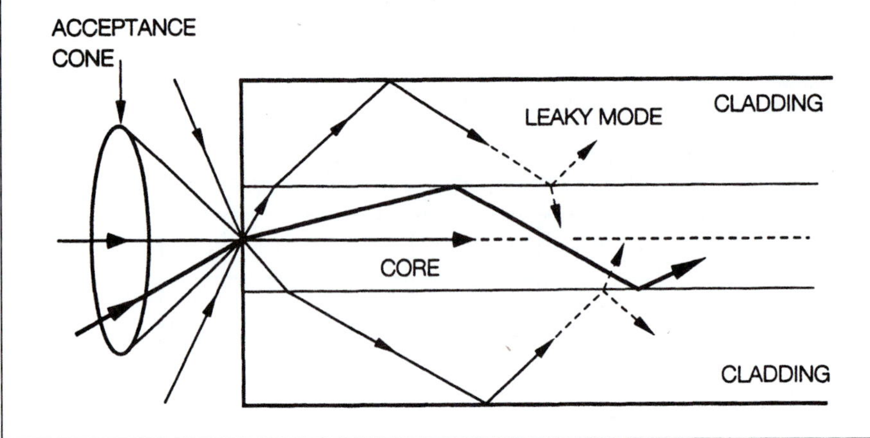

IV. MODE-COUPLING LOSS

THE MODE-COUPLING LOSS IN FIBER OCCURS WHEN POWER LAUNCHED INTO A PROPAGATION MODE WITHIN THE CORE TURNS INTO A LEAKY OR RADIATING MODE DUE TO:

1. (A) VARIATIONS IN GLASS CORE DIAMETER
   (B) VARIATIONS IN CORE CLADDING INTERFACE
   (C) PRESENCE OF BUBBLES OR IMPURITY CLUSTERS IN THE CORE

2. (A) IMPERFECT SPLICE
   (B) MISALIGNED CONNECTORS

TABLE 27-6 (continued)

V. BENDING LOSS

THERE ARE TWO TYPES OF BENDING LOSSES: MICROBENDING AND CONSTANT RADIUS BENDING.

(1) MICROBENDING LOSS: MICROBENDS ARE FORMED IN THE CORE AS A RESULT OF SLIGHTLY DIFFERENT THERMAL CONTRACTION RATES BETWEEN THE CORE MATERIAL AND THE CLADDING MATERIAL DURING THE MANUFACTURING PROCESS. THESE MICROBENDS SCATTER LIGHT, CAUSING ATTENUATION.

(2) CONSTANT RADIUS BENDING LOSS: IT OCCURS WHEN A FIBER IS BENT SO THAT THE ANGLE OF INCIDENCE AT THE CORE CLADDING INTERFACE BECOMES SMALLER THAN THE CRITICAL ANGLE AND SOME MODES ARE REFRACTED INTO THE CLADDING, CAUSING ATTENUATION.

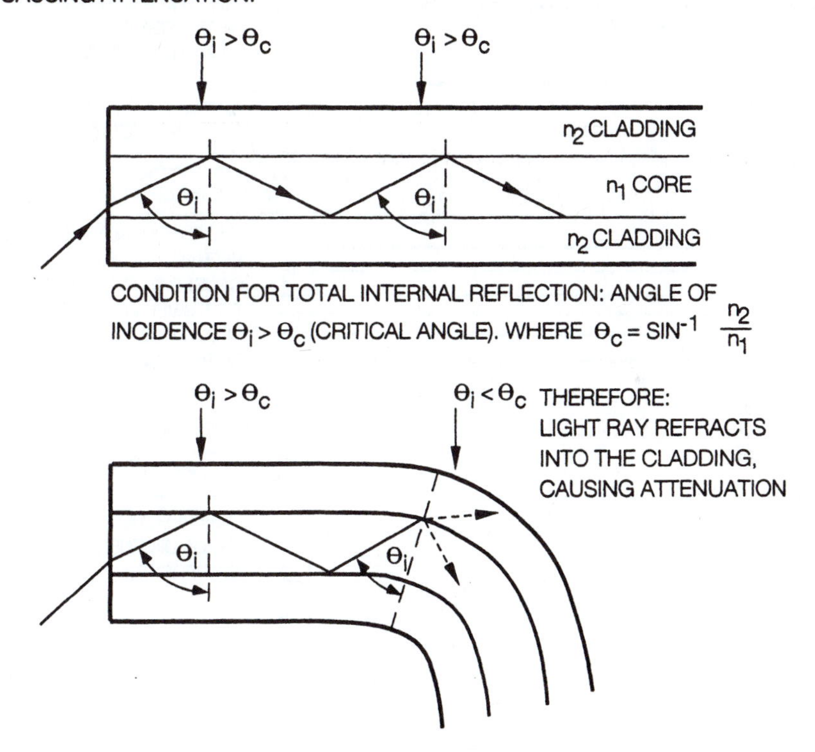

As mentioned earlier, dispersion dictates the bit rate or the bandwidth of the fiber. The higher the dispersion, the lower the potential bit rate or bandwidth. Table 27-7 presents a summary of various mechanisms that contribute to overall dispersion in optical fiber. The chromatic dispersion is the major type of dispersion in single-mode fibers. In order to minimize dispersion the optical source should have narrow spectral width. The wavelength at which dispersion theoretically becomes zero and bandwidth is maximized is called zero-wavelength. Conventional single-mode fibers have zero-dispersion wave-

TABLE 27-7 Dispersion mechanisms in optical fiber

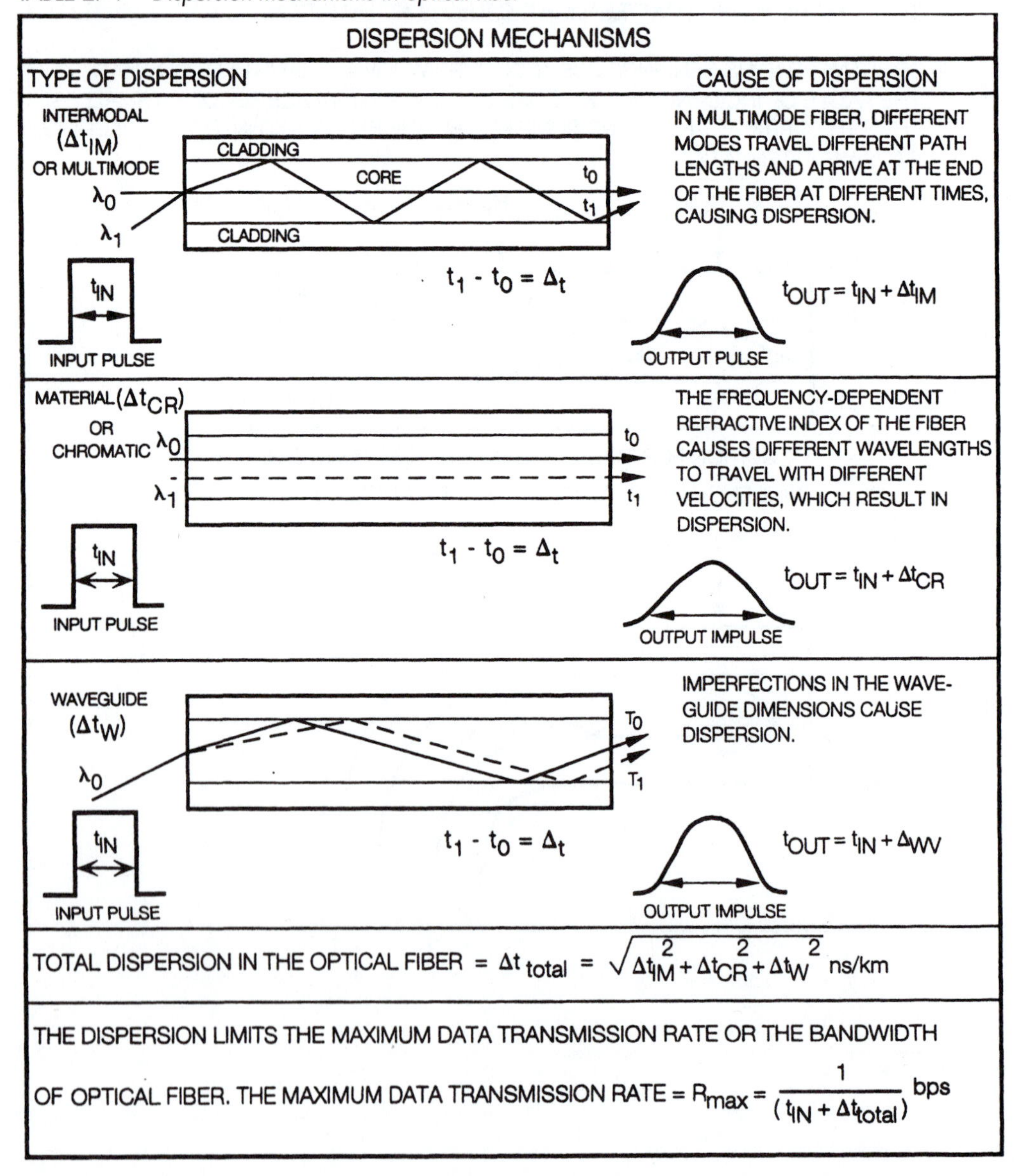

lengths near 1310 nm and relatively high dispersion in the low-loss 1550 nm window. In the mid-1980s, fiber designers developed methods for shifting the zero-dispersion wavelength from the 1310 nm to the 1550 nm window by balancing the two major components of dispersion: material and waveguide dispersion. Material dispersion is an intrinsic property of the type of glass and cannot be changed without radically altering the type of glass. Waveguide dispersion depends on fiber design and can be tailored by changing the fiber's refractive index profile. This change is made during manufacture by injecting appropriate dopants into glass in varying concentrations radially from the center of the fiber. This balancing of dispersion results in a fiber called dispersion-shifted fiber, which has zero-dispersion wavelength at 1550 nm and high negative dispersion at 1310 nm.

■ 27-5 OPTICAL SOURCES

The major requirements for an optical transmitter are as follows:

1. Small emission area (compatible with the core of an optical fiber).
2. Narrow spectral bandwidth (should emit narrow range of wavelengths).
3. Emission wavelengths must be compatible with the detector and fiber.
4. Large modulation bandwidth.
5. Highly directional output.
6. Linear transfer characteristics.
7. High reliability.
8. Long life.
9. Low cost.
10. High tolerance to change in ambient conditions.

The light-emitting diodes (LEDs) and laser diodes satisfy these requirements and therefore are widely used in fiber optics communication systems. Table 27-8 compares the characteristics of LEDs and laser diodes.

TABLE 27-8 Optical sources: LEDs versus laser diodes

	LEDs vs	LASER DIODES
Spectral width	Large 30–100 nm	Narrow 1–5 nm
Modulation bandwidth	1 Gbps	6–10 Gbps
Insertion loss	15–20 db	3 db
Output power	1–5 mW	5–30 mW
Life expectancy	100 million hours	1 million hours
Temperature sensitivity	Tolerant	Sensitive
Beam divergence	Large	Narrow
Cost	Low	High

■ 27-6 OPTICAL DETECTORS

The major requirements for an optical detector are as follows:

1. Small detection area (compatible with the core of an optical fiber).
2. Large signal-to-noise ratio (SNR).
3. High efficiency.
4. High sensitivity at optical-fiber and optical-transmitter-compatible wavelengths.
5. Short response time large detection bandwidth.
6. High reliability.
7. Low cost.

Table 27-9 compares the various optical detectors used in fiber optics communication systems.

The most widely used optical detectors are the PIN [P-type, I (intrinsic), N-type] diode and the avalanche photo diode (APD). Table 27-9 compares the characteristics of the PIN diode with the APD.

TABLE 27-9 Optical detectors: PIN diode versus avalanche photodiode

	PIN (P-TYPE, INTRINSIC, N-TYPE)	APD
Sensitivity	Low	High
Cost	Low	High
Temperature sensitivity	Tolerant	Sensitive
Bias voltage	Low 10–50 V	High 100–300 V

■ 27-7 FIBER OPTICS SYSTEM DESIGN AND CONDITIONS FOR SYSTEM VIABILITY

In order to design a fiber optics system, the following factors should be considered:

1. Distance (short haul or long haul)
2. Cost
3. Environment (land-based or oceanic)
4. Flexibility
5. Reliability (MTBF, MTTR)
6. Bandwidth requirements
7. Type of signal (analog or digital)

Figure 27-6 illustrates the flowchart for the fiber optic system design process. Table 27-10 presents the condition for the viability of a fiber optics system.

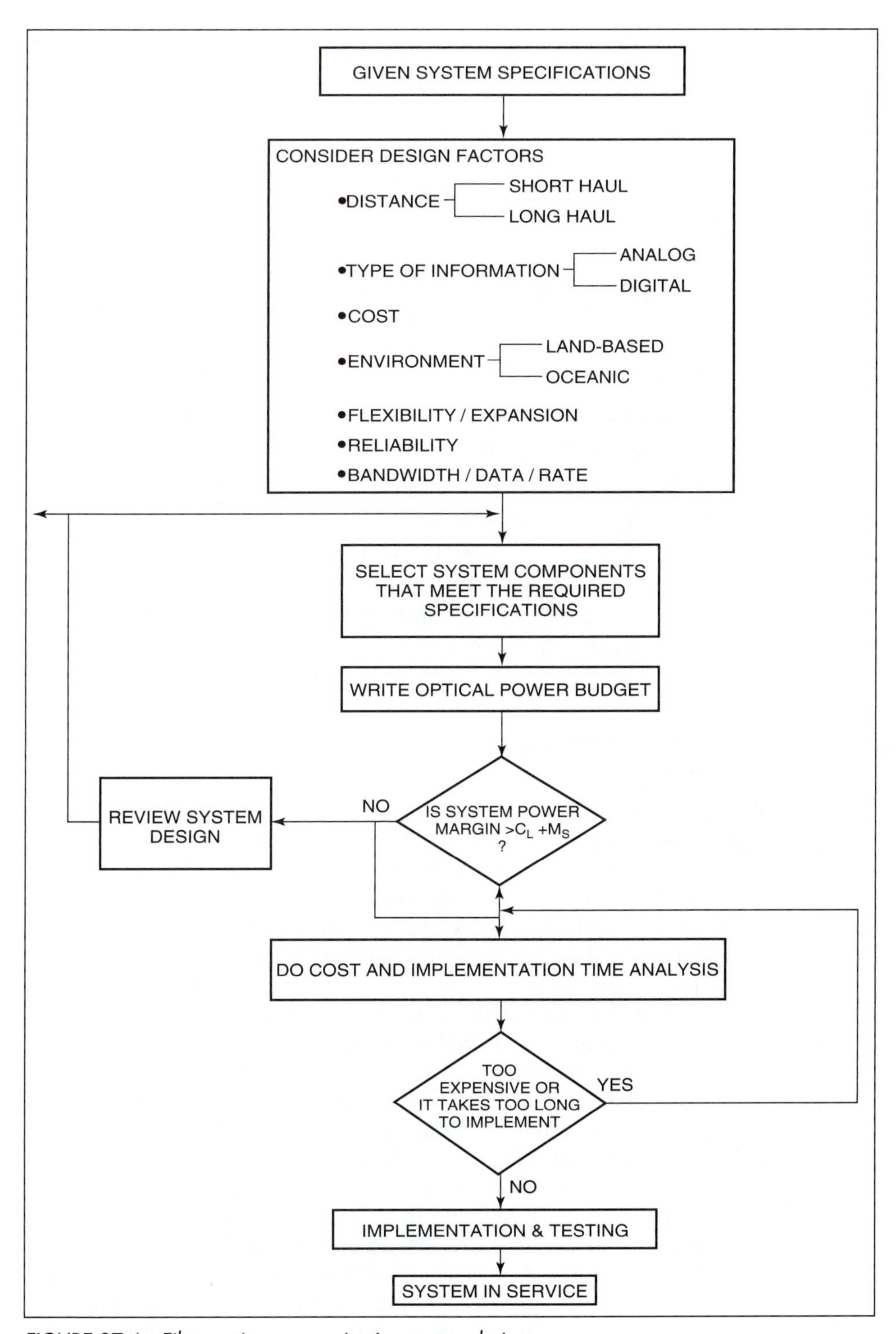

FIGURE 27-6 Fiber optics communication system design

TABLE 27-10 Condition for the viability of a fiber optics system

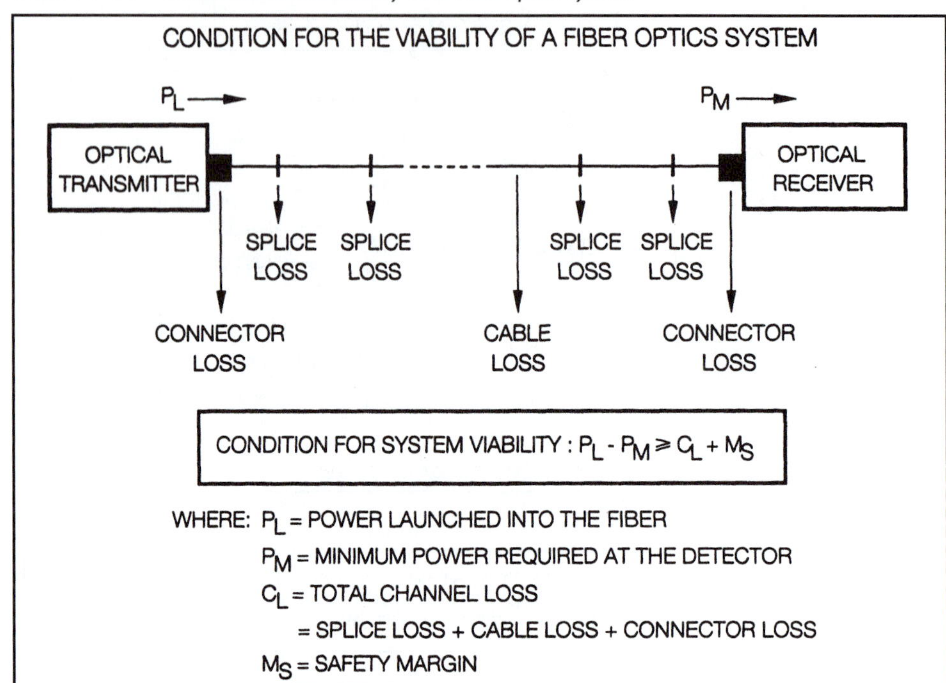

■ 27-8 OPTICAL AMPLIFIERS/REPEATERS

Although optical fibers offer low transmission loss, regeneration is still necessary to compensate for transmission loss over long links. In a conventional system this requires the following: photon-to-electron conversion, electrical amplification, re-timing and pulse-shaping, and finally electron-to-photon conversion. This conversion of an information signal from the electrical domain to the optical domain and vice versa often creates a bottleneck and may restrict both the operating bandwidth and the quality of the signal. In many applications it is advantageous to use direct optical amplification. Optical amplifiers operate in the optical domain with no interconversion of photons to electrons. Recent advances in the field of optical amplifiers have been so rapid that new components are expected to replace optoelectronics repeaters in many existing applications. These include (a) power amplifiers, which boost transmitter power and increase span length; (b) preamplifers, which enhance receiver sensitivity; and (c) repeaters, which boost the signal periodically in the long distance links.

The major types of optical amplifiers are semiconductor laser amplifiers (SC), erbium doped fiber amplifiers (EDFAs), and fiber Raman amplifiers (FRAs). Of these three, erbium doped fiber amplifiers are the most promising. EDFAs operate at a low-loss 1550 nm wavelength window, and offer high gain (> 40 db), high output power (>100 mW), immunity to cross talk among wavelength-multiplexed channels, and low insertion loss. EDFAs coupled with dispersion-shifted fibers will make it possible to transmit a 1550 nm bit stream thousands of kilometers without regeneration.

■ 27-9 APPLICATIONS

The first major application of fiber optics technology was in telephone systems. In the 1970s, multimode graded-index fiber links operating at 45 Mbps using the first low-loss window of fiber (800–900 nm) connected the central exchanges and the toll exchanges. In the 1980s, single-mode fiber links were used for telephone links. These links operated at 405 Mbps and used the second low-loss window of fiber (1300–1400 nm). Presently, systems are available consisting of single-mode dispersion-shifted fiber and erbium doped fiber amplifiers, which can transmit multi-Gbps bit streams at the third low-loss window (1500–1600 nm) over thousands of kilometers without any regeneration.

Recent advances in the application of fiber include fiber to the curb (FTTC), fiber to the home (FTTH), fiber CATV, fiber in local area networks (LANs), and synchronous optical network (SONET). FTTC and FTTH are the network architectures that will bring fiber to the subscriber's home, offering a large bandwidth for a number of interactive services. In local area networks, the fiber distributed data interface provides a data rate of 100 Mbps.

SONET (synchronous optical network) is an American National Standard for a high-capacity optical telecommunications network. It is a synchronous digital transport system aimed at providing a simple, economical, and flexible telecommunications infrastructure. In SONET, the lowest level signal is called synchronous transport signal level 1 (ITS-1), and it has a signal rate of 51.84 Mbps. The optical equivalent of ITS-1 is the optical carrier level 1 signal (OC-1). The higher level signals obtained by multiplexing lower level signals are represented by STN-N and OC-N, where n is an integer. The line rate of the higher level OC-N signal is n times the ITS-l, the line rate at the lowest level. The SONET standard allows only certain values of n. The values of n are: 1, 3, 9, 12, 18, 24, 36, 48, 96, and 192. Table 27-11 lists the SONET signal hierarchy and line rates.

TABLE 27-11 SONET hierarchy and line rates

SYNCHRONOUS TRANSPORT SIGNAL	LINE RATE (Mbps)	OPTICAL CARRIER
STS-1	51.84	OC-1
STS-3	155.52	OC-3
STS-9	466.56	OC-9
STS-12	622.08	OC-12
STS-18	933.12	OC-18
STS-24	1244.16	OC-24
STS-36	1866.24	OC-36
STS-48	2488.32	OC-48
STS-96	4976	OC-96
STS-192	9953	OC-192

Oceanic fiber optics systems have revolutionized global telecommunications. Oceanic lightwave systems have been in service since 1988 across the Atlantic Ocean, and since 1989 across the Pacific Ocean. These high-capacity digital communications systems have brought about a revolution in available system capacity and service quality compared to prior analog coaxial systems. On the drawing board are systems with a capacity of 100+ Gbps. Oceanic fiber optics systems are transforming the world into a global village.

i. Transatlantic Telephone Cable-8 (TAT-8)

The first transatlantic telephone cable, TAT-8, was installed in 1988. Connecting the U.S. to the U.K. and France, it has a capacity of 40,000 simultaneous voice communications circuits. Since 1988 more than 300 undersea fiber optics projects have been planned and presently are in various stages of development. In 1994, a record high 46 new fiber optics undersea links were placed into service in international and domestic telephone networks worldwide. Table 27-12 lists the characteristics of TAT-8, and Table 27-13 lists major oceanic links.

TABLE 27-12 Transatlantic telephone cable #8 (TAT-8)

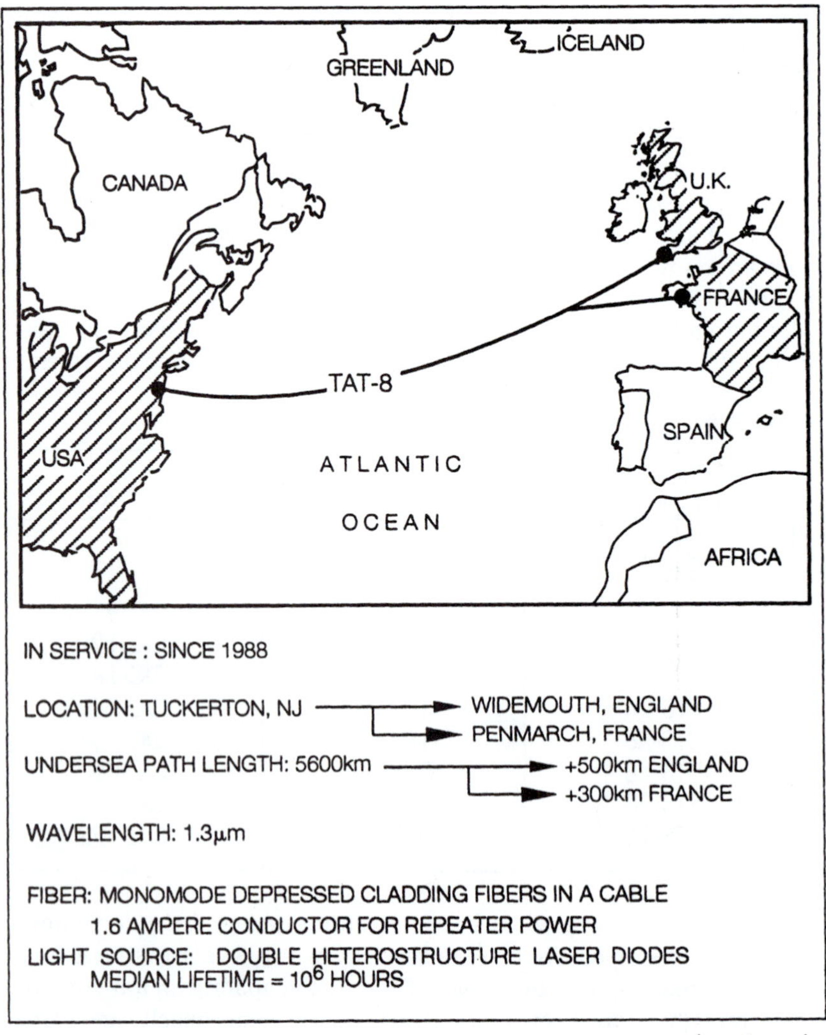

(continues)

TABLE 27-12 (continued)

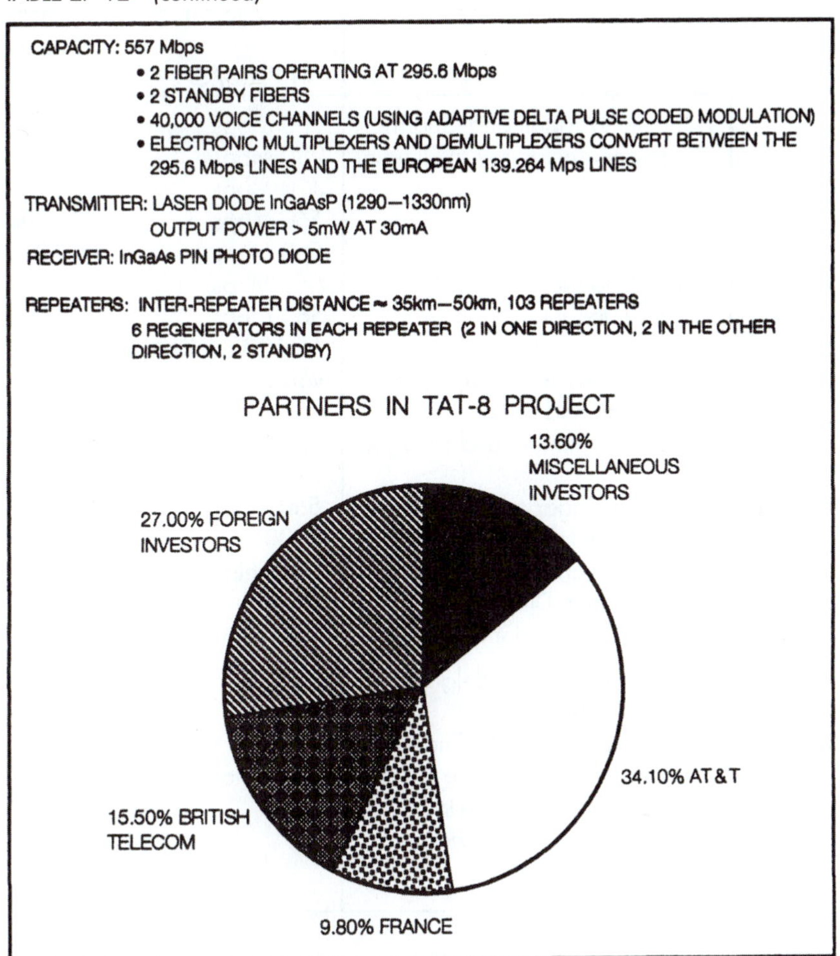

CAPACITY: 557 Mbps

- 2 FIBER PAIRS OPERATING AT 295.6 Mbps
- 2 STANDBY FIBERS
- 40,000 VOICE CHANNELS (USING ADAPTIVE DELTA PULSE CODED MODULATION)
- ELECTRONIC MULTIPLEXERS AND DEMULTIPLEXERS CONVERT BETWEEN THE 295.6 Mbps LINES AND THE EUROPEAN 139.264 Mps LINES

TRANSMITTER: LASER DIODE InGaAsP (1290–1330nm)
OUTPUT POWER > 5mW AT 30mA

RECEIVER: InGaAs PIN PHOTO DIODE

REPEATERS: INTER-REPEATER DISTANCE ~ 35km–50km, 103 REPEATERS
6 REGENERATORS IN EACH REPEATER (2 IN ONE DIRECTION, 2 IN THE OTHER DIRECTION, 2 STANDBY)

TABLE 27-13 Major fiber oceanic systems

CABLE	ROUTE	BIT RATE/CAPACITY	IN-SERVICE DATE
TAT-8	U.S. to U.K. and France	280 Mbps	1988
TAT-9	U.S. to U.K., Spain and France	560 Mbps	1991
PTAT-l	U.S. and Bermuda to U.K. and Ireland	420 Mbps	1989
TAT-12	U.S. to U.K. and France	5 Gbps (60,000 voice circuits)	1995
TAT-13	U.S. to France	5 Gbps (60,000 voice circuits)	1996
PTAT-2	U.S. to U.K.	420 Mbps	1992

(continues)

TABLE 27-13 (continued)

CABLE	ROUTE	BIT RATE/CAPACITY	IN-SERVICE DATE
TPC-3	Honolulu to Japan and Guam	280 Mbps	1988
G-P-T	Guam and Philippines to Taiwan	280 Mbps	1989
NPC	Hong Kong to Philippines	420 Mbps	1990
TASMAN-2	New Zealand to Australia	560 Mbps	1991
TPC-4	U.S. and Canada to Japan	560 Mbps	1992
PAC RIM EAST	Honolulu to New Zealand	560 Mbps	1993
PAC RIM WEST	Australia to Guam	560 Mbps	1996
Americas-I	United States to the U.S. Virgin Islands, Trinidad, Venezuela, and Brazil	15,100 equivalent voice circuits	1994
Unisur	Brazil to Argentina and Paraguay	15,000 equivalent voice circuits	1994
Sea-Me-We-2	France to Singapore, with landing points in Italy, Algeria, Tunisia, Egypt, Cyprus, Turkey, Saudi Arabia, Djibouti, India, Sri Lanka, and Indonesia	560 Mbps 8000 equivalent voice channels (digital)	1994
RIOJA	Spain with England, Belgium, and the Netherlands	2.5 Gbps	1994
FLAG	U.K. to Japan with landing points in Spain, Italy, Egypt, the United Arab Emirates (UAE), India, Thailand, Malaysia, Hong Kong, and Korea	120,000 equivalent voice channels	1996

ii. Americas-1

Americas-1 links the United States, the U.S. Virgin Islands, Brazil, Trinidad, and Venezuela. Joining two continents, the Americas-1 submarine system complements Latin America's existing and planned land and undersea fiber optic cable system to provide regional and international networking.

The system supplies varied transmission rates to accommodate each connected country's anticipated communications needs. The optically amplified route between Florida and St. Thomas contains two fiber pairs operating at 2.5 Gbps. These fiber pairs are capable of handling 320,000 simultaneous telephone calls.

The cable segment into Chaguaramas, Trinidad, operates at 560 Mbps over three fiber pairs to transmit approximately 120,000 simultaneous telephone calls. The routes into Brazil and Venezuela are connected by deep-water branching units. These units operate at 560 Mbps over two fiber pairs and can handle 80,000 simultaneous voice channels or equivalent voice and data information.

In Brazil, Americas-1 connects Unisur, a 1,715 km system that links Florianopolis, Brazil; Los Toninas, Argentina; and Maldonado, Uruguay. In St. Thomas, Americas-1 connects to Taino-Carib, a 187 km undersea cable that provides intraregional communications service to Tortola and Puerto Rico. In St. Thomas, Americas-1 also connects to Columbus-II, a 12,000 km international system with landing points in Portugal, Spain, Italy, Mexico, St. Thomas, and the United States.

iii. TAT-12/13

The TAT-12/13 oceanic system is a fiber optic ring network that links the United States, England, and France. The ring network contains two transatlantic submarine fiber-cable pairs operating at 5 Gbps and uses optically amplified repeaters. The TAT-12 system connects the United States with England and France. Approximately 4,000 km in length, TAT-12 uses 85 undersea repeaters. The TAT-13 system also connects the United States to France.

iv. Fiber Link Around the Globe (FLAG)

Fiber link Around the Globe (FLAG) is an 18,190-km international link between the United Kingdom and Japan. The landing points along the route include Spain, Italy, Egypt, the United Arab Emirates (U.A.E.), India, Thailand, Malaysia, Hong Kong, and Korea.

The FLAG cable contains two fiber pairs each operating at 5 Gbps. The use of synchronous digital hierarchy technology allows transmission of 120,000 digital channels, each operating at 64 kbps.

v. Southeast Asia-Middle East-Western Europe-2 (SEA-ME-WE-2)

Sprawled over about 18,000 km via 137 submerged repeaters and six branching units, the SEA-ME-WE-2 oceanic network consists of seven segments that connect three continents and 13 countries. The network connects France to Singapore with landing points in Italy, Algeria, Tunisia, Egypt, Cyprus, Turkey, Saudi Arabia, Djibouti, India, Sri Lanka, and Indonesia. The system has a capacity of 560 megabits per second over two fiber pairs. One fiber pair is dedicated for traffic between Europe and Asia and the other for regional traffic. The system capacity can be increased four to five times using voice-compression technology.

The exponential growth in fiber optics technology during the past two decades has significantly changed global telecommunications. Today fiber is the medium of choice for short- and long-haul broadband applications. Numerous new applications of fiber are proving to be the impetus for technological developments in the field. With rapid advances in optoelectronic devices, low-loss zero-dispersion fibers, optical amplifiers—coupled with innovations in optical computing and switching—fiber optics is transforming the 20th century electronic era of telecommunications into an optical era of the 21st century.

Useful Web Resources for Fiber Optics Communications

RESOURCE/PUBLICATION	PUBLISHER	URL
• *Applied Optics* • *Journal of Optical Society of America*	Optical Society of America	http://www.osa.org
• *Fiber Optics and Integrated Optics*	Taylor and Francis	http://www.taylorandfrancis.com
• *Fiber Optics and Communication*	Information Gatekeeper	http://www.igigroup.com
• *Fiber Optic Product News* • *Fiber Optic Market Place*	Cahners	http://www.fpnmag.com http://www.fiberoptic.com
• *IEEE Communication Magazine*	Institute of Electrical and Electronics Engineers (IEEE)	http://www.comsoc.org
• *IEEE Lasers and Electro-Optics Society (LEOS) Newsletter*	Institute of Electrical and Electronics Engineers (IEEE)	http://www.ieee.org/leos
• *IEEE/OSA Journal of Lightwave Technology*	Institute of Electrical and Electronics Engineers (IEEE)	http://www.opera.ieee.org
• *IEEE Spectrum*	Institute of Electrical and Electronics Engineers (IEEE)	http://www.spectrum.ieee.org
• *IEEE Transactions on Communications*	Institute of Electrical and Electronics Engineers (IEEE)	http://www.opera.ieee.org
Web Tutorials on key communication technologies	International Engineering Consortium	http://www.iec.org
Lasers and Optronics	Cahners	http://www.lasersoptrmag.com
Laser Focus World	PennWell Publishing Company	http://www.Optoelectronics-world.com
Lightwave	PennWell Publishing Company	http://www.light-wave.com
Optical Fiber Technology	Academic Press	http://www.academicpress.com
Optics and Photonics News	Optical Society of America	http://www.osa.org
Optics Communications	Elsevier Science	http://www.elsevier.com
Photonics Spectra	Laurin Publishing Company	http://www.photonics.com

MODULE 28 Optical Amplifiers*

■ 28-1 OPTICAL AMPLIFIERS

The exponential growth in lightwave technology during the past two decades has significantly changed global telecommunications. Today, fiber is the medium of choice for short- and long- haul applications [1]. Approximately 60+ million kilometers of fiber have been installed worldwide. The double-digit growth rate continues unabated, as nations strengthen and extend their infrastructure, and as societies enter the Information Age [2]. The two developments that have revolutionized lightwave communications are advances in rare-earth optical amplifiers, and the use of solitons for data transmission. Optical amplifiers have matured into one of the most significant advances in fiber optic technology since fiber itself. With the advent of erbium doped fiber amplifiers (EDFA), transmission capacities have increased from hundreds to tens of thousands of gigabit-kilometer per second [3]. Optical solitons that exploit both the nonlinearity and dispersion of the fiber medium to maintain a constant pulse shape offer a potential for high-capacity, ultra-long-distance systems. Researchers at AT&T have reported achieving error-free soliton data transmission at 10 Gbits/sec over a distance of 1,000,000+ km using a recirculating EDFA loop [4].

Although optical fibers offer low transmission loss, regeneration is still necessary to compensate for transmission loss for long links. In a conventional system, regeneration requires photon-to-electron conversion, electrical amplification, re-timing and pulse-shaping, and finally electron-to-photon conversion. This conversion of an information signal from the electrical domain to the optical domain and vice versa often creates a bottleneck and may restrict both the operating bandwidth and the quality of the signal. Optical amplifiers operate in the optical domain with no inter-conversion of photons to electrons. Recent advances in the field of optical amplifiers have been very rapid, and these amplifiers are replacing optoelectronic repeaters in many existing applications [5]. Table 28-1 compares amplifiers with regenerators. The use of optical amplifiers in long-haul systems can be divided into three different applications: (a) A power or booster amplifier that boosts the transmitter power and increases the span length. Architecturally, this amplifier is located immediately after the transmitter. A receiver pre-amplifier boosts the signal level within the sensitivity range of the source; (b) an in-line amplifier or repeater that compensates for attenuation; (c) a distributed amplifier provides a continuous pump that is absorbed down the entire fiber span [6]. Figure 28-1 illustrates the application of optical amplifiers.

TABLE 28-1 Amplifier–regenerator comparison

AMPLIFIER	REGENERATOR
Dispersion Accumulated	Dispersion Reset
Noise Accumulated	Bit Error Rate (BER) Accumulated
Multichannel	Single Channel
Modulation/Bit-rate Transparent	Modulation/Bit-rate Specific

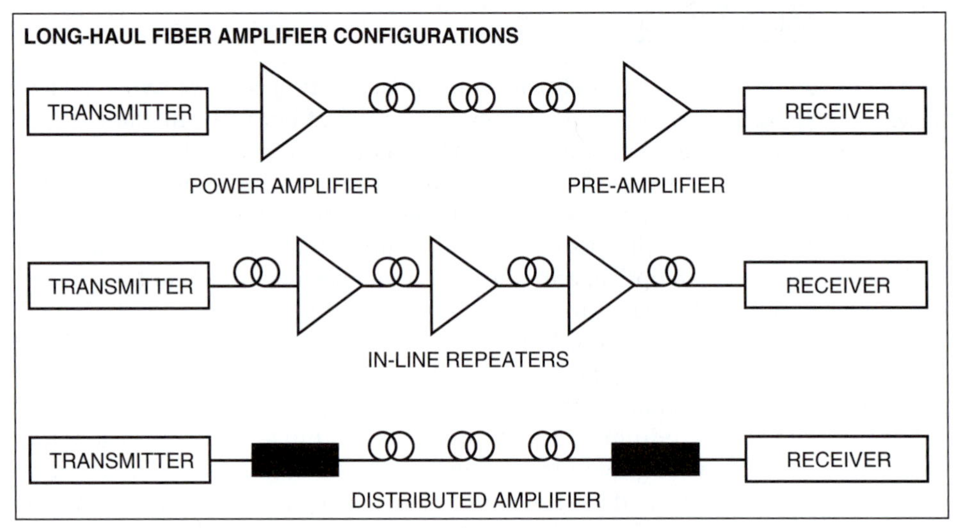

FIGURE 28-1 Optical amplifier

The major type of optical amplifiers are semiconductor laser amplifiers (SCA), rare-earth doped fiber amplifiers, and Raman and Brillouin amplifiers. Table 28-2 presents the development history of optical amplifiers.

TABLE 28-2 Optical amplifiers—development history

1950s	– Development of LASER. – Advances in semiconductor technology.
1960s	– Enhancements in detector's sensitivity with gas lasers as optical amplifier. – 1963: C. Koester and E. Snitzer investigated the idea of using rare-earth ions such as neodymium for optical amplification. – 1969: G. C. Holst and E. Snitzer demonstrated the improvement in detector sensitivity through the use of optical amplifier.
1970s	– Advances in the development of low-loss fiber. In 1970 Corning Glass manufactured fiber with attenuation values of less than 20 db/km. This was improved to 4 db/km in 1972.

(continues)

TABLE 28-2 Optical amplifiers—development history (continued)

	– Development of LEDs. – Discovery of scattered Raman effect in fiber. – Rapid progress in semiconductor lasers.
1980s	– Investigations in Raman amplifiers were hindered due to lack of appropriate pump sources. – Investigations in semiconductor diode amplification led to more research problems than solutions. Many experts dismissed optical amplifiers in favor of electronic regeneration or more sensitive generation techniques.
1985–1986	– David Payne and co-workers at University of Southampton, England, showed that ions of the rare-earth element erbium could exhibit optical gain in fiber at the second low attenuation window of 1500 nm.
1987	– Emmanuel Desurvire and co-workers reported a 22-db gain for EDFA using 100 mW of green light.
1989	– Researchers at NTT, Japan, used newly developed InGaAsP laser diode to pump EDFA with a wavelength of 1.48 microns.
1990–1991	– researches at Bell laboratories reported data rates of 5 Gbits/sec and 2.4 Gbits/sec for distances of 14,000 km and 21,000 km respectively.
1992–1993	– AT&T and NTT reported a joint test bed experiment in which 10 Gbits/sec signals were transmitted over a distance of 9000 km using 274 optical fiber amplifiers.
1993–1994	– Demonstration of 10 Gbits/sec error-free soliton data transmission using EDFA loop over a distance of 1,000,000+ km.
1994–1995	– EDFAs are deployed in lightwave systems.
2000+	– Depoyment of EDFAs in DWDM, UDWDM, SONET/SDH, LANs, CATV, and fiber sensor networks.

■ 28-2 SEMICONDUCTOR LASER AMPLIFIER

A laser diode amplifier can be described as a "poor laser," one that uses low injection current or low facet reflectivity or both. Laser diode amplifiers are classified into two types: traveling-wave amplifier (Figure 28-2) and Fabry-Perot amplifier (Figure 28-3). In a traveling-wave amplifier the reflectivity of facets is made as small as possible. If the gain per unit length is high enough, the device amplifies an incident signal sufficiently in a single pass through the active region. There is no, or very little, resonance effect visible in the output spectrum. The traveling-wave type of laser diode amplifier is used in wavelength-division multiplexing (WDM). For network applications, traveling-wave amplifiers are much more interesting because of their flatter overall gain spectrum and greater ease of controlling their operating points. In a Fabry-Perot amplifier, the reflectivity of facets is kept high enough for light to undergo multiple reflections within the cavity. Typical values of power reflectivity are up to 30 percent [7].

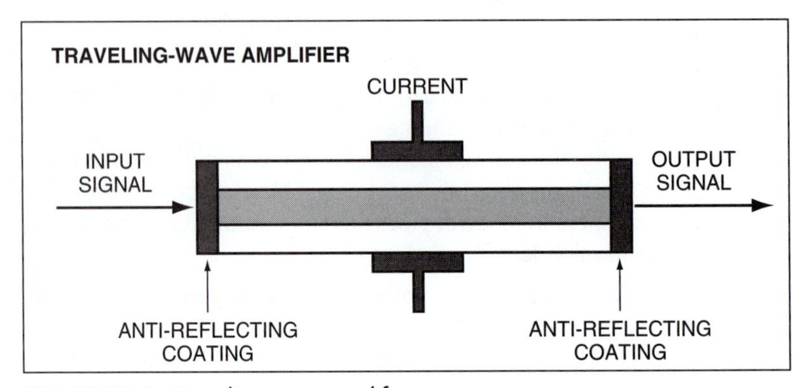

FIGURE 28-2 Traveling-wave amplifier

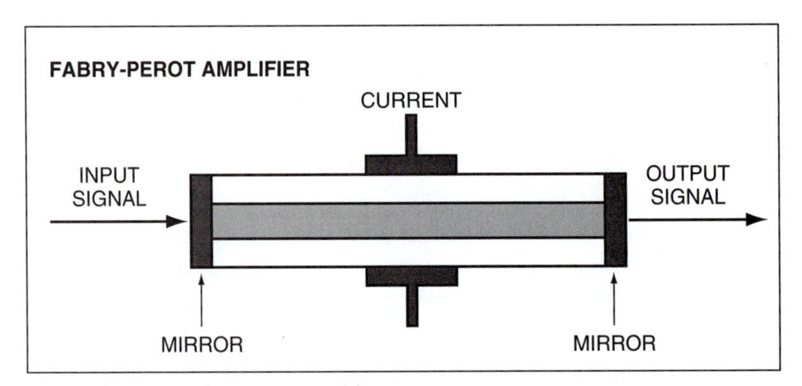

FIGURE 28-3 Fabry-Perot amplifier

In principle these devices work, but in practice limitations such as low gain, polarization dependent gain, high crosstalk, and inter-modulation distortion have kept laser diode amplifiers from becoming widely accepted [8].

■ 28-3 DOPED-FIBER AMPLIFIERS (DFAs)

Optical amplifiers, in the form of optical fibers doped with rare-earth ions, have been recognized for many years as interesting devices both for signal amplification and as sources of radiation [9,10]. The rare-earth doped fibers function similarly to a laser cavity. That is, rare-earth ions in the silica core are excited to a higher energy state with an optical pump. The ions produce stimulated emission when they drop from a metastable state to their stable state. Laser power levels of more than 500 mW have been reported.

The primary wavelengths of interest in rare-earth doped fibers are 1060, 1320, and 1550 nm. The principal dopants used in these fibers include neodymium, erbium, and ytterbium. The composition and mix of the dopants define the lasing wavelengths, and can be tailored to optimize performance. Typical doping levels range from tens to thousands of parts per million in the silica core. The precise concentration of doping levels also depends on the intended length of the final doped fiber. A "lumped" amplifier will have higher doping concentration than a "distributed" amplifier.

The dopant praseodymium can also be used for 1060 nm lasing fibers, and can be doped in fluoride glass materials for 1300 nm lasing. However, these fluoride glasses are not readily compatible with silica fibers and their conventional manufacturing systems. Rare-earth doped fibers have achieved significant success during the past few years, mostly in fiber amplifiers. The fiber amplifier provides a cost-effective solution to conventional repeaters; it produces higher power that travels longer distances at lower costs [11].

The rare-earth doped fiber amplifiers offer the following advantages:

- High gain
- High saturation output power
- Insensitive to polarization
- Low splice losses and reflections
- High speeds (100 GHz), bit-rate transparent
- Low inter-modulation distortion, crosstalk
- Operation in saturation above 10 kHz

28-4 ERBIUM DOPED FIBER AMPLIFIER (EDFA)

Erbium doped fiber amplifiers (EDFAs) are at the forefront of doped amplifier technology, with hundreds of systems operating or planned. Erbium amplifiers accept an incoming optical signal at the 1550 nm wavelength and increase its optical power level by up to 40 db. Amplification is accomplished by combining the incoming signal, along with a "pump" signal, into a length of fiber that has its core doped with trace amounts of the rare-earth element erbium. When two signals are combined in the core of doped fiber, the high-powered laser pump signal excites the erbium atoms and they emit light at the 1550 nm incoming signal wavelength. The amplification or power gain that results from 15 to 40 db can add an additional 160 kilometers to inter-repeater distance. Figure 28-4 shows the block diagram of an erbium doped fiber amplifier, and Figure 28-5 illustrates a typical EDFA long-haul transmission system.

Figure 28-6 illustrates the energy level diagram for trivalent erbium. The energy level diagram reveals the three most important aspects that make the trivalent erbium a perfect candidate for a 1500 nm amplifier. First, the transition from the meta-stable state to ground state is compatible with the third low attenuation window (1500 nm) of optical fiber. Second, the lifetime to decay to the ground is remarkably long (10 ms) so that enormous amounts of energy remain stored in the inverted erbium ion at that level. Third, some of the upper levels correspond to wavelengths where laser diodes exist to do the pumping. Without pump sources as practical as semiconductor laser diodes, EDFA could never have been considered for deployment in an actual system. In 1989, researchers in Japan were the first to use a newly developed InGaAsP laser diode to pump the EDFA near a wavelength of 1.48 microns [12,13]. The EDFAs offer the following advantages:

- Low pump power: only a few milliwatts of pump power is required to generate a gain of a factor of thousands.
- Gain is intrinsically insensitive to light polarization.
- Temperature is stable.
- Gain is immune to interference or "crosstalk."
- Can operate in a regime of minimum spontaneous emission noise
- Can be spliced to fiber without coupling loss and end reflections

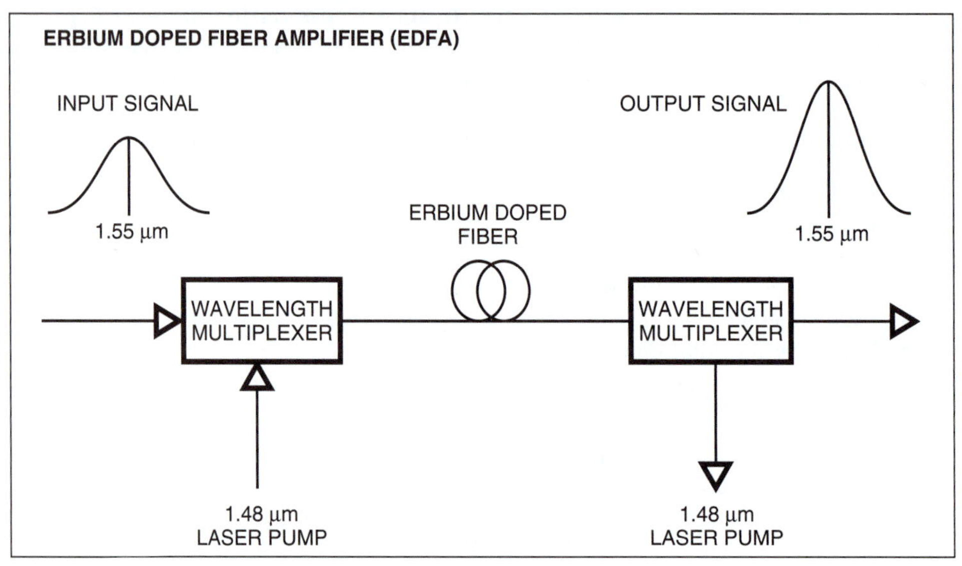

FIGURE 28-4 Block diagram of erbium doped amplifier

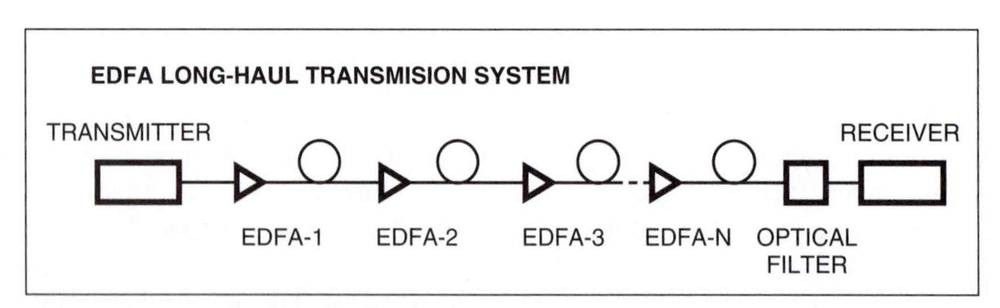

FIGURE 28-5 EDFA long-haul transmission system

The EDFAs have the following limitations:

- Gain non-uniformity; i.e., gain spectrum is not uniform over 30 nm usable wavelength
- The amplification wavelength is fixed due to the nature of erbium-ion energy levels. EDFA works only for signals around 1.5 microns.

The limitation of gain non-uniformity can be corrected by a variety of equalization techniques. But the EDFA cannot be tuned to wavelengths other than 1.5 microns. To amplify signals at 1.3 microns, recent work has focused on praseodymium doped fiber amplifiers [14].

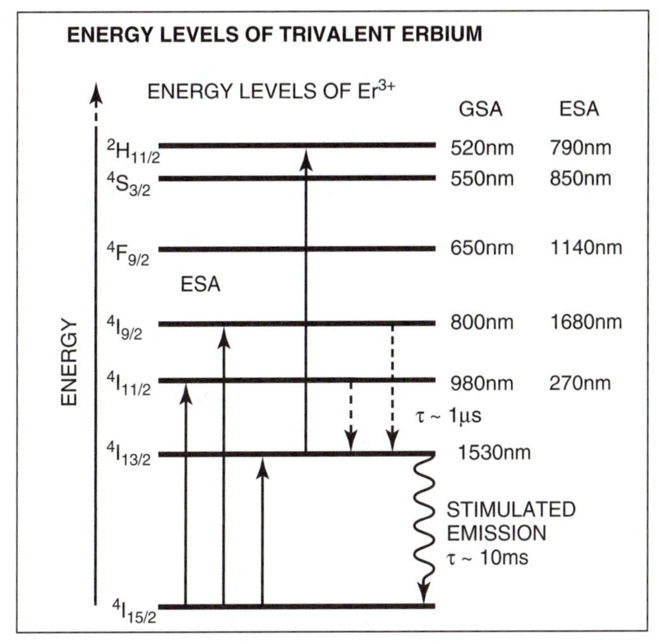

FIGURE 28-6 Energy levels of trivalent erbium

The EDFAs have been recognized as the "ideal" for optical signal amplification in the 1.5 micron band. The EDFA characteristics are determined by many factors, which include (a) material parameters [glass composition, co-dopant elements, concentration profile, etc.], (b) waveguide parameters [V-value, numerical aperture, cutoff wavelength, spot size, etc.], and (c) working parameters [signal/pump wavelength and power, pump scheme, active fiber length, etc.] [15].

■ 28-5 PRASEODYMIUM DOPED FLUORIDE-FIBER AMPLIFIER (PDFFAs)

In this 1300 nm amplifier, trivalent praseodymium is used to dope fluoride glass fiber. The PDFFA operating near 1.3 microns is important because the majority of installed lightwave communication systems operate in this wavelength region. A fluoride glass, rather than erbium doped silica fiber, is required because the radiative electron transition that produces the 1.3 micron gain is totally quenched in silica. The lower energy phonon vibrations of fluoride glasses have a longer meta-stable lifetime that allows relatively efficient amplification. While trivalent praseodymium in ZBLAN (zinc barium lanthanum aluminum sodium) fiber is presently the best fiber amplifier in the 1300 nm region, the intrinsic performance of PDFFAs is significantly poorer than that of EDFAs. Table 28-3 compares the performance of EDFA with PDFFA [16].

TABLE 28-3 EDFA–PDFFA comparison

PUMP WAVELENGTH (nm)	SATURATED POWER (dbm)	MINIMUM NOISE FIGURE (db)	PUMP-POWER REQUIREMENTS FOR 30 DB GAIN (mW)	3-DB OPTICAL BANDWIDTH IN HIGH SATURATION (nm)
EDFA				
980	> +18	3	<10	> 40
1480	> +21	5	10+	> 40
1047	> +24	3	10+	> 40
PDFFA				
1017	> +18	5	100	> 25
1047	> +24	5	100+	> 25

■ 28-6 RAMAN AND BRILLOUIN AMPLIFIERS

Long before erbium doped fibers became practical as the basis for lightwave amplifiers, other forms of fiber amplifiers were being considered and implemented that did not involve optical pumping at ionic level, but instead involved processes at the molecular level. These amplifiers used one of two such physical phenomenon, either stimulated Raman scattering (SRS) or stimulated Brillouin scattering (SBS) [17,18]. The SRS and SBS effects are much weaker than the gain that stimulated emission provides in a doped amplifier. But, the cost and reliability aspects of doped fiber amplifiers, and the availability of appropriate pump sources, have opened a window of opportunity for scientists to take another look at the Raman amplifier for 1310 nm and 1550 nm operation.

Unlike erbium or praseodymium types, the Raman type is a distributed amplifier. The communication fiber itself is the amplification cavity, i.e., the farther the signal travels, the more amplification the signal gets [19].

Figure 28-7 illustrates a Raman amplifier scheme proposed by W. Hicks. The Raman multi-oscillator starts with a neodymium doped fiber pumped by a multi-stripe laser diode at around 0.8 micron. The laser-diode pump beam travels in a large pump core surrounding the laser core, coupling in substantial laser-diode power that is absorbed by the neodymium doped core. Several generations of Raman-Strokes lines are stimulated, the last of which is used for pumping the communications signals. This last line is removed by a fine filter that also serves to time the various stroke lines. The system incorporates means to control and stabilize Brillouin offspring; this action prevents undue broadening of the Raman pump and stabilizes the power level [20].

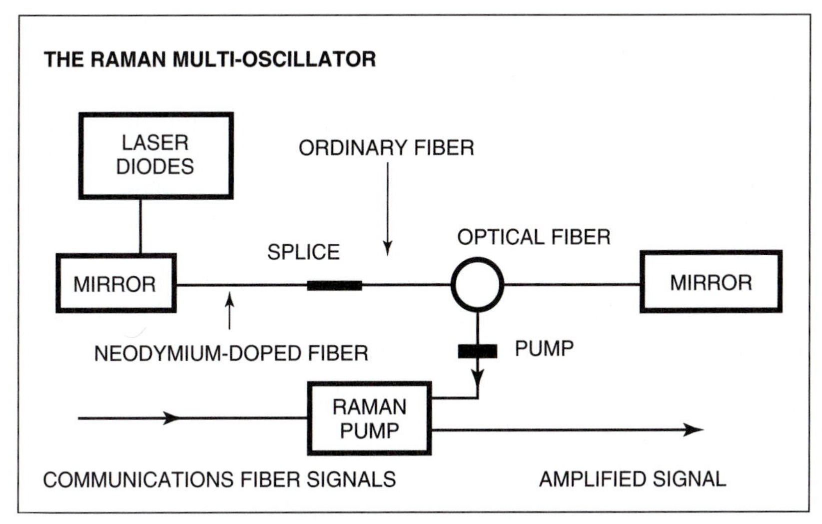

FIGURE 28-7 Raman amplifier

With wavelength-division multiplexing (WDM), Raman optical amplifier networks could possibly carry 100–500 separate channels on the same fiber. With the use of separate optical channels, a Raman network could feasibly run a terabit of data, spread out on many channels [21].

■ 28-7 SOLITONS

A soliton is a pulse with a certain ratio of intensity and width, whose shape is maintained by the balance of negative group dispersion and the self-phase modulation arising from the Kerr nonlinearity. To a communications engineer, solitons are a dream come true: pulses that propagate entirely without distortion. They were observed in water as long ago as 1838, and described by a mathematical theory in 1895. But not until 1973 was it predicted that optical solitons could propagate in fiber. In 1973, Akira Hasegawa of Bell Telephone Laboratories realized that optical fibers exhibit both the optical Kerr effect (an intensity-dependent refractive index) and at wavelengths longer than 1.3 microns have negative group velocity dispersion. Thus he predicted soliton transmission in optical fibers [22].

Optical solitons were first observed in fiber by Linn Mollenauer and co-workers at Bell Laboratories in 1980. In 1985, Mollenauer and colleagues first investigated soliton regeneration with a Raman fiber amplifier. In a Raman amplifier the gain is distributed over the length of fiber; this feature enables adiabatic regeneration of soliton pulses. Based on this principle, in 1988 soliton propagation over more than 4000 km was demonstrated using a recirculating loop. In 1990 and 1991, research groups at AT&T and NTT investigated the transmission of actual soliton data in EDFA systems. In 1992, AT&T researchers reported modulating two soliton channels at 10 Gbits/sec and transmitting over 11,000 km with low BER. In 1993, Nakazawa and colleagues at NTT laboratories achieved error-free soliton data transmission at 10 Gbits/sec over a distance of 1,000,000 km, using a recirculating EDFA loop with a 500 km circumference [23]. Similar results were reported in 1994 by French researchers. The workers at the Center National deludes des Telecommunications (CENT) research laboratories of France Telecom in Paris achieved 1,000,000 km of error-free, 10 Gbits/sec transmission using soliton technology and a 350-km recirculating EDFA loop [24].

Recent advances in optical amplifiers are having a profound impact on short- and long-haul lightwave systems. Erbium doped fiber amplifier (EDFA) are changing the designs of many networks, from transoceanic to "fiber-to-the-home (FTTH)." EDFAs allow increased design flexibility, tremendous cost savings, and greatly increased capacity and reliability.

The rapid advances in optical amplifiers and low-loss dispersion-shifted fibers coupled with multichannel WDM soliton transmission technology are transforming lightwave systems into broadband, integrated-service, end-to-end optical networks.

REFERENCES

1 Khan, Ahmed S, "Oceanic Optical Systems Transform World Into a Global Village," *Annual Review of Communications 1994–95*, vol. 48, pp. 649–651.

2 Li, Tingye, "Optical Amplifiers Transform Lightwave Communications," *Photonics Spectra*, January 1995, p. 115.

3 Desurvire, Emmanuel, "Lightwave Communications: The Fifth Generation," *Scientific American*, January 1992, pp. 114–121.

4 Desurvire, Emmanuel, "The Golden Age of Optical Fiber Amplifiers," *Physics Today*, January 1994, pp. 20–27.

5 Khan, Ahmed S., "Fiber Optics Communications: An Introduction to Technology and Applications," *Annual Review of Communications 1993–94*, vol. 47, pp. 503–508.

6 Warr, Michael, "Optical fiber amplifiers: Everybody wants one," *Telephony*, March 18, 1991, pp. 51–52.

7 Green, Paul E., "Fiber Optic Networks," Englewood Cliffs, N.J. Prentice-Hall, Inc., 1993, pp. 219–244.

8 Palais, Joseph, "Fiber Optic Communications," Englewood Cliffs, N.J. Prentice-Hall, Inc., 1992, pp. 161–163.

9 Koester, C.J., and Snitzer, E., "Amplification in a fibre laser," *Appl. Opt*, 1963, 3, pp. 1182–1186.

10 Snitzer, E., "Neodymium glass laser," *Proc. 3rd Int. Conf. on Solid State Lasers*, Paris, 1963, pp. 999–1019.

11 McCann, B.P., "Specialty optical fibers resolve challenging application problems," *Lightwave*, November 1994, pp. 48–52.

12 ibid. Green, Paul E., pp. 237–238.

13 ibid. Desurvire, Emmanuel, "The Golden Age of Optical Fiber Amplifiers," p. 23.

14 Haus, Hermann A., "Molding light into solitons," *IEEE Spectrum*, March 1993, pp. 48–53.

15 Chongcheng, Fan, et al., "Theoretical and Experimental Investigations on Erbium-doped Fiber Amplifiers," *Fiber and Integrated Optics*, volume 13, 1994, p. 247.

16 Poole, S., "Fiber amplifier use expands as functionality increases," *Laser Focus World*, October 1994, pp. 111–118.

17 ibid. Green, Paul E., pp. 243–244.

18 Lawton, George, "Major advances in optical amplifiers attract worldwide attention," *Lightwave*, October 1994, pp. 11–12.

19 Kotelly, George., "Raman unit boosts 1310 and 1500 nm," *Lightwave*, October 1994, p. 1.

20 ibid. Kotelly, George, p. 20.

21 ibid. Kotelly, George, p. 23.

22 ibid. Haus, Hermann A., pp. 48–49.

23 ibid. Desurvire Emmanuel, "The Golden Age of Optical Fiber Amplifiers," pp. 25–26.

24 Hars, Adele, "Soliton pulses travel one million kilometer error free," *Lightwave*, October 1994, pp. 8–10.

MODULE 29

Fundamentals of Electromagnetic Signals

ELECTROMAGNETIC WAVES

In telecommunication systems, information is transmitted from source (transmitter) to destination (receiver) in the form of electromagnetic waves that pass through a channel. That channel could be free space, air, transmission lines, or metallic or dielectric waveguides (optical fiber).

Electromagnetic waves are generated by accelerating charges. The radiated wave consists of oscillating electric and magnetic fields that are at right angles to each other; the direction of propagation is illustrated in Figure 29-1.

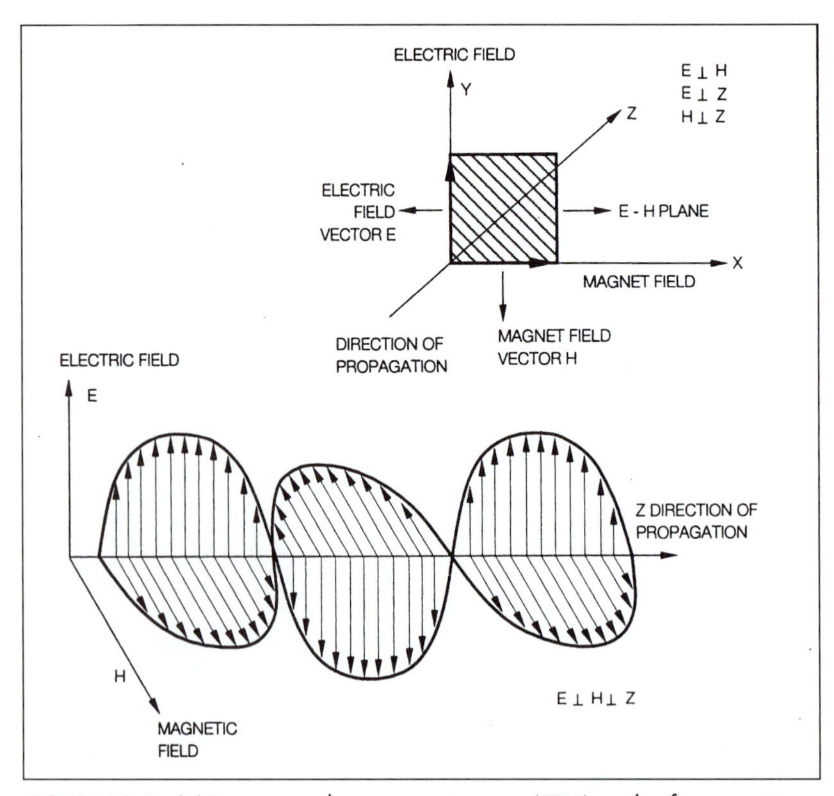

FIGURE 29-1 (a) Transverse electromagnetic wave (TEM) mode of propagation; (b) vectorial representation of TEM

The existence of electromagnetic waves was predicted in 1863 by Scottish physicist James Clerk Maxwell in the form of a theory. He described all electric and magnetic phenomena using four equations, which are presently known as Maxwell's equations. These equations predict the generation of electromagnetic waves by accelerating electric charges. The produced electromagnetic waves would propagate through space at the speed of light. Electromagnetic waves were experimentally generated and detected in 1887 by Heinrich Hertz. He produced electromagnetic waves with the help of oscillating circuits and detected them with circuits tuned to the same frequency. In the 1890s, Marconi demonstrated the use of electromagnetic waves for long-distance communications. He invented and developed the wireless telegraph, which initiated the radio revolution.

Velocity of Propagation: The velocity of propagation of electromagnetic waves in a medium is given by:

$$v = \frac{1}{\sqrt{\epsilon\mu}}$$

Where: ϵ = permittivity of the medium
μ = permeability of the medium

For air or free space, the velocity of propagation equals:

$$v = \frac{1}{\sqrt{\epsilon_o\mu_o}}$$

ϵ_o = permittivity of free space
$= 8.85 \times 10^{-12}\ C^2/N \cdot m^2$

μ_o = permeability of free space
$= 4\pi \times 10^{-7}\ T \cdot m/A$

$$v = \frac{1}{\sqrt{8.85 \times 10^{-12} \times 4\pi \times 10^{-7}}}$$

$v = 3 \times 10^8$ meters/second = c

c = velocity of propagation in air or free space.

The velocity of propagation can also be determined by:

$$v = \frac{c}{\sqrt{\epsilon_r}}$$

or

$$v = \frac{c}{n}$$

Where: ϵ_r = relative permittivity of the dielectric medium is also called the relative dielectric constant and

$n = \sqrt{\epsilon_r}$ = refractive index of the medium

For a transmission line, the velocity of the propagation of electromagnetic waves can be determined in terms of line inductance and capacitance:

$$v = \frac{1}{\sqrt{LC}}$$

Where: L = the inductance of a transmission line (H/unit length)
C = the capacitance of a transmission line (F/unit length)

The speed of an electromagnetic wave in a given medium is equal to the product of frequency and wavelength.

$$v = \lambda f$$

For air or free space:

$$C = \lambda f = 3 \times 10^8 \text{ m/s}$$

Energy in Electromagnetic Waves: Electromagnetic waves transport energy from one point to another point in space. The energy is associated with the magnitudes of the oscillating electric and magnetic fields. The energy density contained in an electric field is given by:

$$\mu_e = \frac{1}{2} \epsilon_o E^2$$

Where: E = the instantaneous value of the electric field
ϵ_o = permittivity of free space

The energy density contained in a magnetic field is given by:

$$\mu_b = \frac{1}{2} B^2/\mu_o$$

Where: B = the instantaneous value of the magnetic field
μ_o = permeability of free space

The total energy stored per unit volume is given by:

$$\mu_{total} = \mu_e + \mu_b$$

$$\mu_{total} = \frac{1}{2} \epsilon_0 E^2 + \frac{1}{2} B^2/\mu_0$$

$$\mu_{total} = \sqrt{\frac{e_0}{\mu_0}} EB$$

The total energy density in terms of the electric field only is given by:

$$\mu_{total} = \epsilon_0 E_2$$

The total energy density in terms of the magnetic field only is given by:

$$\mu_{total} = \frac{B^2}{\mu_0}$$

The instantaneous energy transported per unit time per unit area perpendicular to the direction of wave propagation is represented by a vector called the *Poynting vector* (S). The direction of the vector is the direction of wave propagation and the magnitude of the vector is given by:

$$S = \epsilon_0 c E^2 = c\frac{B^2}{\mu_0} = \frac{EB}{\mu_0} \text{ (J/s.m}^2\text{)}$$

The average value of the Poynting vector, i.e., the average energy transported per unit area per unit time, is given by:

$$\overline{S} = \frac{1}{2}\epsilon_0 c E_0^2 = \frac{1}{2}\frac{c}{\mu_0}B_0^2 = E_0B_0/2\mu_0$$

Where: B = maximum value of the magnetic field
E = maximum value of the electric field

The ratio of electric and magnetic field values is equal to the free space impedance:

$$\frac{E}{H} = \sqrt{\frac{\epsilon_0}{\mu_0}} = \eta = 377\ \Omega$$

For quantum waves (IR, visible, UV, X rays, etc.), the energy (E) of radiation is frequency dependent, and is given by:

$$E = hf = \frac{hc}{\lambda} \text{ joules (J)}$$

$$E = \frac{hf}{e} \text{ electron volts (eV)}$$

Where: h = Planck's constant = 6.625×10^{-34} J-s
f = frequency of radiation (Hz)
λ = wavelength of radiation (m)
e = charge of an electron = 1.6×10^{-19}C

Modes of Propagation: A mode of propagation refers to a specific relationship between the orientation of the electric field, the magnetic field, and the direction of the propagation vectors of the electromagnetic wave. In various media, the electromagnetic waves travel by different modes of propagation. The basic mode of propagation is called the TEM or transverse electromagnetic mode, in which the electric and the magnetic fields are perpendicular to each other and to the direction of propagation as illustrated in Figure 29-1. The other modes of propagation are transverse electric (TE), transverse magnetic (TM), and hybrid modes (HE and EH). Table 29-1 lists different modes of propagation that exist in various types of transmission media.

TABLE 29-1 Modes of electromagnetic wave propagation in transmission media

TRANSMISSION MEDIUM	MODE OF PROPAGATION
Air/free space	TEM
Transmission lines –Two Wire line – Coaxial Cable	TEM
Waveguides – Metallic – Dielectric (optical fibers)	TE, TM, and HYBRID (HE, EH)
MODE DEFINITION	
TEM: Transverse Electromagnetic ($E \perp H$, $E \perp z$, $H \perp z$)	E = electric field vector
TE: Transverse Electric ($E \perp H$, $E \perp z$, $H \not\perp z$)	H = magnetic field vector
TM: Transverse Magnetic ($E \perp H$, $H \perp z$, $E \not\perp z$)	z = direction of propagation vector
HE: Hybrid Mode ($E \perp H$, $E \not\perp z$, $H \not\perp z$ $H > E$)	$\perp$ = perpendicular to
EH: Hybrid Mode ($E \perp H$, $E \not\perp z$, $H \not\perp z$, $E > H$)	$\not\perp$ = not perpendicular to
	(*See also* **modes of propagation**)

Electromagnetic Spectrum: Electromagnetic waves can be generated and detected over a wide range of frequencies. Depending on frequency or wavelength, electromagnetic waves are categorized on the basis of wavelengths or frequencies in an array called the electromagnetic spectrum (as illustrated in Figure 5-3).

MODULE 30

Digital Logic: A Summary

Digital logic is the foundation of digital computers. In digital logic, only two states are possible (Table 30-1). Digital logic is based on the binary number system (Table 30-2) and Boolean algebra (Table 30-3). The circuits that perform Boolean logic are called logic gates. Logic gates are the building blocks of digital circuitry. A combination of various logic gates can implement the decision-making function. Complex decision-making functions can be implemented by combining logic gates with a memory unit. Table 30-4 lists various types of logic gates, their symbols, and their truth tables. (A truth table defines the relationship between the input(s) and outputs of a logic gate.)

TABLE 30-1 Logic states

Binary Levels				
Binary	0	low	0v	off
Levels	1	high	5v	on

TABLE 30-2 Binary number system

	BINARY NUMBERS			
HEXA-DECIMAL	COUNT	DECIMAL NUMBER	BINARY NUMBER	COMMENTS
			$2^n \ldots 2^3\ 2^2\ 2^1\ 2^0$	
0	zero	0	0 0 0 0	The binary number system has only two digits: 0 and 1. A digit is called a bit.
1	one	1	0 0 0 1	
2	two	2	0 0 1 0	
3	three	3	0 0 1 1	The binary numbers belong to the base 2 number system.
4	four	4	0 1 0 0	
5	five	5	0 1 0 1	
6	six	6	0 1 1 0	$2^n \ldots 2^3\ 2^2\ 2^1\ 2^0$
7	seven	7	0 1 1 1	

(continues)

TABLE 30-2 (continued)

	BINARY NUMBERS			
HEXA DECIMAL	COUNT	DECIMAL NUMBER	BINARY NUMBER	COMMENTS
8	eight	8	1 0 0 0	highest decimal number represented in binary
9	nine	9	1 0 0 1	$= 2^n - 1$
A	ten	10	1 0 1 0	
B	eleven	11	1 0 1 1	where n = number of bits
C	twelve	12	1 1 0 0	2^n = possible combinations with n-bits
D	thirteen	13	1 1 0 1	
E	fourteen	14	1 1 1 0	
F	fifteen	15	1 1 1 1	

TABLE 30-3 Basic laws of Boolean algebra

LOGICAL ADDITION	LOGICAL MULTIPLICATION	COMPLEMENT RULES
$0 + 0 = 0$	$0 \cdot 0 = 0$	$\overline{0} = 1$
$0 + 1 = 1$	$0 \cdot 1 = 0$	$\overline{1} = 0$
$1 + 0 = 1$	$1 \cdot 0 = 0$	
$1 + 1 = 1$	$1 \cdot 1 = 1$	

BOOLEAN ALGEBRA RULES

1. $0 + x = x$
2. $1 + x = 1$
3. $x + x = x$
4. $x + \overline{x} = 1$
5. $0 \cdot x = 0$
6. $1 \cdot x = x$
7. $x \cdot x = x$
8. $x \cdot \overline{x} = 0$
9. $\overline{\overline{x}} = x$
10. $x + y = y + x$
11. $x \cdot y = y \cdot x$
12. $x + (y + z) = (x + y) + z$
13. $x(yz) = (xy)z$
14. $x(y + z) = xy + xz$
15. $x + xz = x$
16. $x(x + y) = x$
17. $(x + y)(x + z) = x + yz$
18. $x + \overline{x}y = x + y$
19. $xy + yz + \overline{y}z = xy + z$

DE MORGAN THEOREMS

$$\overline{x + y} = \overline{x} \cdot \overline{y}$$
$$\overline{x \cdot y} = \overline{x} + \overline{y}$$

TABLE 30-4 Logic gates

LOGIC GATES

TYPE | SYMBOL | TRUTH TABLE

INVERTER

X	A
0	1
1	0

AND

A	B	X
0	0	0
0	1	0
1	0	0
1	1	1

OR

A	B	X
0	0	0
0	1	1
1	0	1
1	1	1

NAND

A	B	X
0	0	1
0	1	1
1	0	1
1	1	0

NOR

A	B	X
0	0	1
0	1	0
1	0	0
1	1	0

EXCLUSIVE-OR

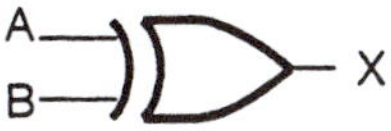

A	B	X
0	0	0
0	1	1
1	0	1
1	1	0

EXCLUSIVE-NOR

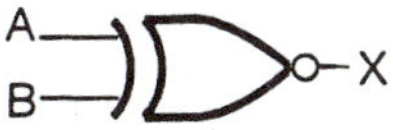

A	B	X
0	0	1
0	1	0
1	0	0
1	1	1

MODULE 31

Units and Physical Constants

TABLE 31-1 Prefixes for power of ten

PREFIX	SYMBOL	VALUE
yotta	Y	1,000,000,000,000,000,000,000,000 = 10^{24}
zetta	Z	1,000,000,000,000,000,000,000 = 10^{21}
exa	E	1,000,000,000,000,000,000 = 10^{18}
penta	P	1,000,000,000,000,000 = 10^{15}
tera	T	1,000,000,000,000 = 10^{12}
giga	G	1,000,000,000 = 10^{9}
mega	M	1,000,000 = 10^{6}
kilo	k	1,000 = 10^{3}
hecto	h	100 = 10^{2}
deca	da	10 = 10^{1}
deci	d	0.1 = 10^{-1}
centi	c	0.01 = 10^{-2}
milli	m	0.001 = 10^{-3}
micro	μ	0.000 001 = 10^{-6}
nano	n	0.000 000 001 = 10^{-9}
pico	p	0.000 000 000 001 = 10^{-12}
femto	f	0.000 000 000 000 001 = 10^{-15}
atto	a	0.000 000 000 000 000 001 = 10^{-18}
zepto	z	0.000 000 000 000 000 000 001 = 10^{-21}
yacto	y	0.000 000 000 000 000 000 000 001 = 10^{-24}

TABLE 31-2 Greek alphabet

	CAPITAL LETTERS	SMALL LETTERS
Alpha	Α	α
Beta	Β	β
Gamma	Γ	γ
Delta	Δ	δ
Epsilon	Ε	ε
Zeta	Ζ	ζ
Eta	Η	η
Theta	Θ	θ
Iota	Ι	ι
Kappa	Κ	κ
Lambda	Λ	λ
Mu	Μ	μ
Nu	Ν	ν
Xi	Ξ	ξ
Omicron	Ο	o
Pi	Π	π
Rho	Ρ	ρ
Sigma	Σ	σ
Tau	Τ	τ
Upsilon	Υ	υ
Phi	Φ	ϕ
Chi	Χ	χ
Psi	Ψ	ψ
Omega	Ω	ω

TABLE 31-3 Mathematical symbols

SYMBOL	MEANING
α	is proportional to
$=$	is equal to
$\approx$	is approximately equal to
$\neq$	is not equal to
$>$	is greater than
$>>$	is much greater than
$<$	is less than
$<<$	is much less than
$\leqslant$	is less than or equal to
$\geqslant$	is greater than or equal to
Σ	sum of
$\bar{x}$	average value of x
$\Delta \boldsymbol{x}$	change in x
$\perp$	perpendicular to

TABLE 31-4 Some SI (Systeme International d'Unites) Units

QUANTITY	UNIT	ABBREVIATION	VALUE
Capacitance	farad	F	$A^2 \cdot s^4/(kg \cdot m^2)$
Electric charge	coulomb	C	$A \cdot s$
Electric potential	volt	V	$kg \cdot m^2/(A \cdot s^3)$
Electrical resistance	ohm	Ω	$kg \cdot m^2/(A^2 \cdot s^3)$
Energy and work	joule	J	$kg \cdot m^2/s^2$
Force	newton	N	$kg \cdot /s^2$
Frequency	hertz	Hz	s^{-1}
Inductance	henry	H	$kg \cdot m^2/(s^2 \cdot A^2)$
Magnetic field strength	tesla	T	$kg/(A \cdot s^2)$
Magnetic flux	weber	Wb	$kg \cdot m^2/s^2$
Power	watt	W	$kg \cdot m^2/s^3$
Pressure	pascal	Pa	$kg/(m \cdot s^2)$

TABLE 31-5 Some physical constants

QUANTITY	SYMBOL	VALUE
Atomic mass unit (1 u)		1.6605×10^{-27} kg $= 931.51$ MeV/c^2
Avogadro's number	N_A	6.02×10^{23} mol^{-1}
Boltzmann's constant	k	1.38×10^{-23} J/K
Charge on electron	e	1.60×10^{-19} C
Electron rest mass	m_e	9.11×10^{-31} kg $= 0.000549$ u
Gas constant	R	8.315 J/mol • K = 1.99 cal/mol • K
Gravitational constant	G	6.67×10^{-11} N • m^2/kg^2
Neutron rest mass	m_n	1.6750×10^{-27} kg $= 1.008665$ u
Permeability of free space	μ_o	$4\pi \times 10^{-7}$ T • m/A
Permittivity of free space	$\epsilon_o = (1/c^2 \mu_o)$	8.85×10^{-12} C^2/N • m^2
Planck's constant	h	6.63×10^{-34} J • s
Proton rest mass	m_p	1.6726×10^{-27} kg $= 1.00728$ u
Speed of light in vacuum	c	3.00×10^8 m/s
Stefan-Boltzmann constant	σ	5.67×10^{-8} W/m^2 • K^4
Bohr radius	a	5.292×10^{-11} m
Deutron mass	m_d	3.344×10^{-27} kg
Rydberg constant	R	0.01097 nm^{-1}

TABLE 31-6 Unit conversion factors

Acceleration

1 m/s^2 = 3.28 ft/s^2 = 100 cm/s^2
1 ft/s^2 = 0.3048 m/s^2 = 30.48 cm/s^2

Angle

1 radian (rad) = 57.30° = 57°18′
1° = 0.01745 rad
1 rev/min (rpm) = 0.1047 rad/s

Area

1 m^2 = 10^4 cm^2 = 10.76 ft^2
1 ft^2 = 0.0929 m^2 = 144 in^2
1 in^2 = 6.452 cm^2

Energy

1 J = 0.738 ft • 1b=10^7 ergs
1 cal = 4.186 J
1 kcal = 4.18 × 10^3 J = 3.97 Btu
1 Btu = 252 cal = 1.054 × 10^3 J
1 eV = 1.6 × 10^{-19} J
931.5 MeV is equivalent to 1 u
1 kwh = 3.60 × 10^6 J = 860 kcal
1 ft • lb = 1.36 J = 1.29 × 10^{-3} Btu
= 3.24 × 10^{-4} kcal

Force

1 N =10^5 dyne = 0.2248 lb
1 lb = 4.448 N
1 dyne = 10^{-5} N = 2.248 × 10^{-6} lb

Length

1 in = 2.54 cm
1 m = 39.37 in = 3.281 ft
1 ft = 0.3048 m
12 in = 1 ft
3 ft = 1 yd
1 yd = 0.9144 m
1 km = 0.621 mi
1 mi = 1.609 km
1 mi = 5280 ft
1 Å = 10^{-10} m
1 μm = 1μ = 10^{-6} m = 10^4 Å
1 lightyear = 9.461 × 10^{15} m

Mass

1000 kg = 1 t (metric ton)
1 slug = 14.59 kg
1μ = 1.66 × 10^{-27} kg
(μ = atomic mass unit)

Power

1 hp 1 = 550 ft • lb/s = 0.746 kW
1 W = 1 J/s = 0.738 ft • lb/s
= 3.42 Btu/h
1 Btu/h = 0.293 W

Pressure

1 bar = 10^5 N/m^2 = 14.50 lb/in^2
1 atm = 760 mm Hg = 76.0 cm Hg
1 atm = 14.7 lb/in^2
= 1.013 × 10^5 N/m^2
1 Pa = 1 N/m^2 = 1.45 × 10^{-4} lb/in^2

Time

1 year = 365 days = 3.16 × 10^7 s
1 day = 24 h = 1.44 × 10^3 min
= 8.64 × 10^4 s

Velocity

1 mi/h = 1.47 ft/s = 0.447 m/s
= 1.61 km/h
1 m/s = 100 cm/s = 3.281 ft/s
1 mi/min = 60 mi/h = 88 ft/s

Volume

1 m^3 = 10^6 cm^3 = 6.102 × 10^4 in^3
1 ft^3 = 1728 in^3 = 2.83 × 10^{-2} m^3
1 liter = 1000 cm^3 = 1.0576 qt
= 0.0353 ft^3
1 ft^3 = 7.481 gal = 28.32 liters
= 2.832 × 10^{-2} m^3
1 gal = 3.786 liters = 231 in^3

MODULE

32 Symbols for Electronic Devices

NAME	TAG	SYMBOL
CROSSING OF PATHS (NOT CONNECTED)		
JUNCTION OF PATHS CONNECTED		
CIRCULAR WAVEGUIDE		
RECTANGULAR WAVEGUIDE		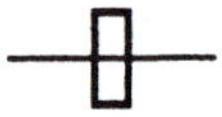
RESISTOR	R	
VARIABLE RESISTOR	R	
TAPPED RESISTOR	R	
THERMISTOR	T	
CAPACITOR	C	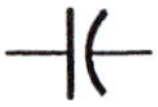
VARIABLE CAPACITOR	C	
INDUCTOR	L	
VARIABLE INDUCTOR	L	
MAGNETIC CORE INDUCTOR	L	

NAME	TAG	SYMBOL
MULTICELL BATTERY	DC V	
AC GENERATOR	DC V	
CURRENT SOURCE	I/S	
METER		
CRYSTAL	X	
NEG-POS-NEG (NPN)	Q	
POS-NEG-POS (PNP)	Q	
N-CHANNEL JFET	NJFET	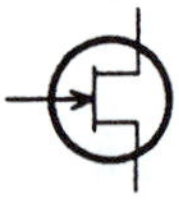
P-CHANNEL JFET	PJFET	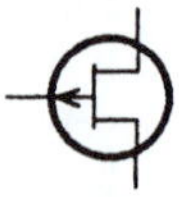
N-CHANNEL MOSFET	NMOSFET	
P-CHANNEL MOSFET	PMOSFET	

NAME	TAG	SYMBOL	NAME	TAG	SYMBOL
TRANSFORMER	T	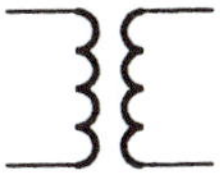	ENHANCEMENT MOSFET	EMOSFET	
DC BATTERY	DC V	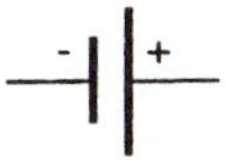	BASIC OPERATIONAL AMPLIFIER	AMP	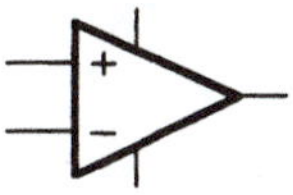
AMPLIFIER WITH ADJUSTABLE GAIN			AND GATE		
FUSE	F		OR GATE		
SWITCH	S		NAND GATE		
PUSH-BUTTON SWITCH (NORMALLY OPEN)	SW		NOR GATE		
PUSH-BUTTON SWITCH (NORMALLY CLOSED)			INVERTER		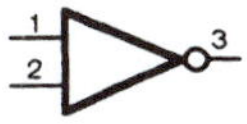
LAMP	LAMP		XOR		
SPEAKER	SPEAKER	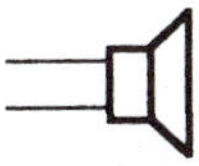	XNOR		
SUMMING POINT	S		SET/RESET FLIP-FLOP		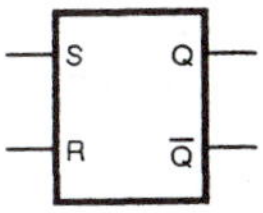
ANTENNA			J-K FLIP-FLOP		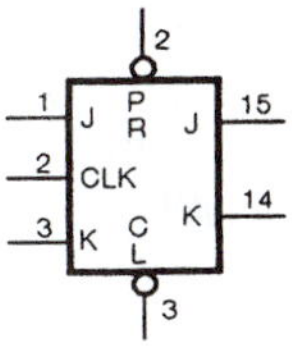
LOOP ANTENNA		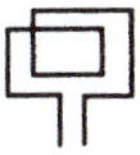	D FLIP-FLOP		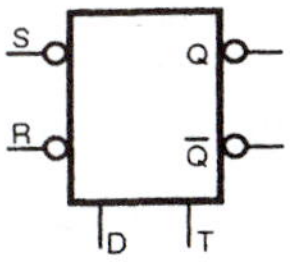
DIPOLE ANTENNA			MAGNETIC AMPLIFIER		
CIRCUIT BREAKER	CB		SILICON CONTROL RECTIFIER	SCR	
JACK (FEMALE)			TRIAC		
JACK (MALE)					
ZENNER DIODE	D				

NAME	TAG	SYMBOL	NAME	TAG	SYMBOL
TUNNEL DIODE	D		VACUUM TRIODE TUBE	V	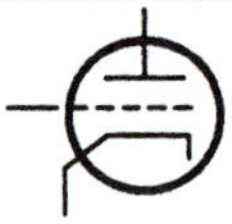
LIGHT-EMITTING DIODE	LED		VACUUM TETRODE TUBE	V	
PHOTODIODE		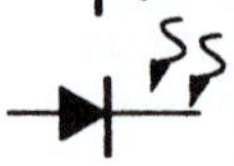	CATHODE RAY TUBE	CRT	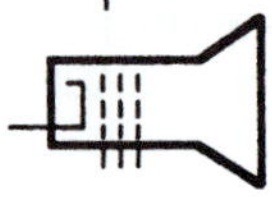
VACUUM PENTODE TUBE	V	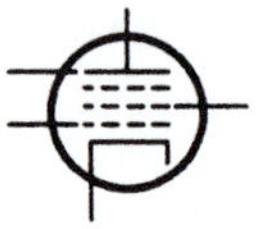	TWIN TRIODE TUBE	V	
THYRATON TUBE	V	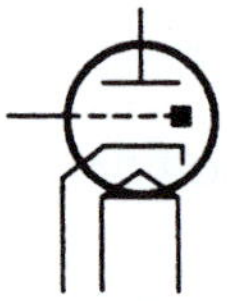	SYNCHRO		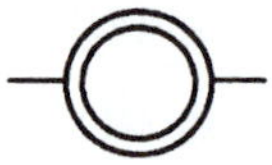
VACUUM DIODE	V	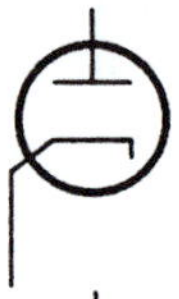	PICK-UP HEAD	PU	
DOUBLE-DIODE TUBE GAS-FILLED	V		NORMALLY OPEN RELAY	N.O.	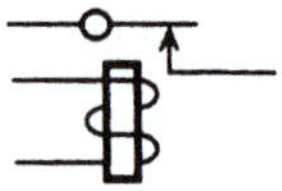
DIRECTLY HEATED CATHODE TUBE (DIODE)	V		NORMALLY CLOSED RELAY	N.C.	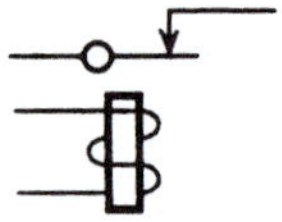
INDIRECTLY HEATED CATHODE TUBE (DIODE)	V	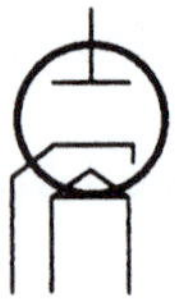	SINGLE POLE DOUBLE-THROW RELAY	SPDT	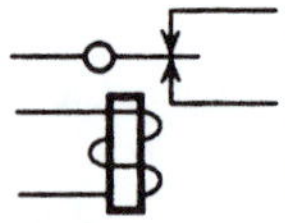
			SHIELDED CONDUCTOR WITH GROUND		
SINGLE POLE SINGLE-THROW SWITCH	SPST	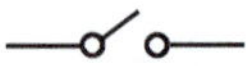	SINGLE POLE DOUBLE-THROW SWITCH	SPDT	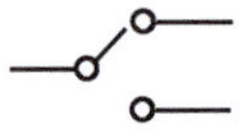

MODULE 33 Telecommunications Formulas

AMPLITUDE MODULATION

Instantaneous voltage for an AM signal

$$e_{mod} = E_c \sin \omega_c t - \frac{mE_c}{2}\cos(\omega_c + \omega_m)t + \frac{mE_c}{2}\cos(\omega_c - \omega_m)t$$

Where: E_m = information signal voltage
E_c = carrier signal voltage

$$m = \frac{E_m}{E_c}$$

Where: m is called modulation index.

$$\omega_c = 2\pi f_c$$

Where: f_c = carrier frequency

$$\omega_m = 2\pi f_m$$

Where: f_m = modulating signal frequency

Total voltage of an AM signal

$$E_T = \text{carrier voltage} + \text{USB voltage} + \text{LSB voltage}$$

$$E_{total} = E_c + \frac{mE_c}{2} + \frac{mE_c}{2}$$

For a complex AM signal

$$E_{total} = E_c\sqrt{1+\left(\frac{1}{2}\right)(m_1^2 + m_2^2 + \ldots + m_n^2)}$$

Where: $m_1^2 \ldots m_n^n$ = modulation indices of information frequencies

(continues)

Total power of an AM signal

Total power = carrier power + USB power + LSB power

$$P_T = P_c + P_{USB} + P_{LSB}$$

$$P_T = P_c + \frac{m^2P_c}{4} + \frac{m^2P_c}{4} = P_c\left(1 + \frac{m^2}{2}\right)$$

Where: P_c = carrier power = $\frac{E_c^2}{R}$

For a complex AM signal

$$P_T = P_c\left[1 = \left(\frac{1}{2}\right)(m_1^2 + m_2^2 + \ldots + m_n^2)\right]$$

Bandwidth for double sideband (DSB) AM channel

$$BW\,(DSB) = 2f_m$$

Where: f^m = highest information frequency

Bandwidth for a single sideband (SSB) AM channel

$$BW\,(SSB) = f_m$$

Instantaneous voltage for a frequency modulated signal

$$\begin{aligned} e_{FM} =\ & E_cJ_om_f\sin \omega_ct \\ & + E_cJ_1m_f[\sin(\omega_c + \omega_m)t - \sin(\omega_c - \omega_m)t] \\ & + E_cJ_2m_f[\sin(\omega_c + 2\omega_m)t - \sin(\omega_c - 2\omega_m)t] \\ & + E_cJ_3m_f[\sin(\omega_c + 3\omega_m)t - \sin(\omega_c - 3\omega_m)t] \\ & + E_cJ_4m_f[\sin(\omega_c + 4\omega_m)t - \sin(\omega_c - 4\omega_m)t] \\ & + E_cJ_n \ldots \end{aligned}$$

Where: m_f = modulation index = $\frac{\Delta f}{f_m}$

Δf = maximum frequency deviation

f_m = highest information signal

$$J_0J_1J_2J_3J_4, \ldots J_n$$

are the Bessels' functions that represent the amplitude of the carrier and relative amplitudes of the side frequencies. The FM spectrum contains an infinite number of side-frequency pairs whose amplitudes depend on the modulation index. In determining the bandwidth (according to Carson's rule) of the FM spectrum, only those side frequencies are considered that support 96 percent of the radiated power.

(continues)

Bandwidth of an FM channel

$$BW(FM) = 2(\Delta f + f_m)$$

This is called Carson's rule.

Total power of an FM signal

$$P_{total} = P_c[J_0^2 + (J_1^2 + J_2^2 + J_3^2 + \ldots + J_n^2)W$$

Where: P_c = carrier power = $\frac{E_c^2}{R}$

NOISE

Thermal noise average power

$$P_n = TB(\text{watts})$$

Where: k = Boltzmann's constant = 1.38×10^{-23} joules/kelvin
B = bandwidth (Hz)
T = temperature of conductor in kelvins

Thermal noise voltage (RMS)

$$V_n = \sqrt{4kTRB}$$

Where: R = resistance of conductor (ohms)

Thermal noise current (RMS)

$$I_n = \sqrt{4kTGB}$$

Where: G = conductance (siemens)

Shot noise current (RMS) for a vacuum diode

$$I_n = \sqrt{2I_{dc}eB}$$

Where: I_{dc} = dc current through diode
e = charge of electron (1.6×10^{-19}C)

For a semiconductor diode

$$I_n = \sqrt{2\,(I + I_0)\,eB}$$

Where: I = direct current across pn junction
I_0 = reverse saturation current

(continues)

Thermal noise for the equivalent noise resistance of an amplifier

$$V_{neq} = \sqrt{4\,(R_i + R_n)\,kTB}$$

Where: R_i = input resistance of the amplifier
R_n = noise resistance of the amplifier

Signal-to-noise ratio (SNR)

$$SNR = \frac{P_s}{P_n} = \frac{\text{signal power}}{\text{noise power}}$$

$$(SNR)_{db} = 10\log\left(\frac{P_s}{P_n}\right)$$

or

$$SNR = 10\log\left(\frac{V_s^2}{V_n^2}\right)$$

or

$$SNR = 20\log\left(\frac{V_s}{V_n}\right)$$

Where: V_s = signal voltage (volts)
V_n = noise voltage (volts)

Noise factor (F)

$$F = \frac{\text{signal-to-noise ratio at the input of circuit}}{\text{signal-to-noise ratio at the output of circuit}}$$

$$= \frac{P_{si}/P_{ni}}{P_{so}/P_{no}}$$

Where: P_{si} = signal power at input
P_{so} = signal power at output
P_{ni} = noise power at input
P_{no} = noise power at output

Noise factor for cascaded circuits

$$F_{cc} = F_1 + \frac{F_2 - 1}{G_1} + \frac{F_3 - 1}{G_1 G_2} + \frac{F_4 - 1}{G_1 G_2 G_3} + \ldots$$

Noise figure (NF)

$$NF = 10\log F$$

SPEED/VELOCITY, FREQUENCY, AND WAVELENGTH

$$V = \lambda f$$

$$\lambda = v/f \quad \text{or} \quad \lambda = c/f$$

$$f = v/\lambda \quad \text{or} \quad f = c/\lambda$$

For air or free space: $c = \lambda f = 3 \times 10^8$ m/s

Velocity of propagation: $v = \dfrac{c}{\sqrt{(\epsilon_r)}}$ or $v = \dfrac{c}{n}$

Where: ϵ_r = relative permittivity of the dielectric medium, also called the relative dielectric constant

$n = \sqrt{\epsilon_r}$ = refractive index of the medium

TRANSMISSION LINES

Phase velocity

$$V_p = \frac{c}{\sqrt{\epsilon_r}}$$

$$V_p = \frac{1}{\sqrt{LC}}$$

Where: $c = 3 \times 10^8$ m/s
ϵ_r = relative dielectric constant
L = inductance/unit length
C = capacitance/unit length

Characteristic impedance of coaxial cable

$$Z_0 = \frac{60}{\sqrt{\epsilon_r}} \ln\left(\frac{D}{d}\right) \text{ ohms}$$

Where: D = diameter of the outer conductor
d = diameter of the inner conductor
ϵ_r = relative dielectric constant

Characteristic impedance of two-wire open line

$$Z_0 = \frac{120}{\sqrt{\epsilon_r}} \ln\left(\frac{2D}{d}\right) \text{ ohms}$$

Where: D = distance between two conductors
d = diameter of the conductor
ϵ_r = relative dielectric constant

(continues)

Characteristic impedance of a transmission line in terms of primary line constants R, L, C, and G

$$Z_0 = \sqrt{\frac{R + j\omega L}{G + j\omega C}}$$

Where: R = resistance (ohms)/unit length
L = inductance (henry)/unit length
G = conductance (siemen)/unit length
C = conductance (farad)/unit length
$\omega = 2\pi f$
f = frequency of signal (Hz)

Propagation coefficient

$$\gamma = \sqrt{(R + j\omega L)(G + j\omega C)} \text{ /unit length}$$
$$\gamma = \alpha + j\beta\text{/unit length}$$

Where: α = attenuation constant [Neper (N)/unit length]
1 neper (N) = 8.686 decibels (db)
$\alpha = 0$ for lossless transmission lines
$\alpha \neq 0$ for lossy transmission lines
β = phase shift constant = $\frac{2\pi}{\lambda}$ radians/unit length
λ = wavelength of the signal

Reflection coefficient for lossless transmission line

$$\Gamma_L = \frac{V_R}{V_I}$$

$$\Gamma_L = \frac{Z_L - Z_o}{Z_L + Z_o}$$

Where: G_L = reflection coefficient at load
V_R = reflected voltage
V_I = incident voltage
Z_L = load impedance
Z_o = characteristic impedance of transmission line

For matched condition, i.e., $Z_L = Z_o$, $\Gamma_L = 0$
For open circuit condition, i.e., $Z_L = \infty$, $\Gamma_L = 1$
For short-circuit condition, i.e., $Z_L = 0$, $\Gamma_L = -1$
For lossy transmission lines, the reflection coefficient at a distance d from the load is equal to:

$$\Gamma_d = |\Gamma_L|\, e^{-2\alpha d}$$

(continues)

Voltage standing wave ratio (VSWR)

$$VSWR = \frac{V_{max}}{V_{min}}$$

$$VSWR = \frac{1 + |\Gamma_L|}{1 - |\Gamma_L|}$$

$$|\Gamma_L| = \frac{VSWR - 1}{VSWR + 1}$$

Where: V_{max} = maximum voltage of standing wave
V_{min} = minimum voltage of standing wave

The VSWR can have values in a range from unity to infinity. For a matched line, there is no standing wave, i.e.,

$$V_{max} = V_{min} \therefore \; VSWR = 1$$

ANTENNAS

Antenna size

$$\text{length of antenna } 1 \simeq \frac{\lambda}{2} \text{ or a multiple of } \frac{\lambda}{2}$$

Antenna gain
Directivity gain is given by:

$$g_d = \frac{\text{power density of antenna's radiation at distance } r}{\text{power density of an isotropic antenna at distance } r}$$

$$g_d = \frac{P}{P_t/4\pi r^2}$$

or

$$g_d = 4\pi r^2 \frac{P}{P_t}$$

Where: P_t = power radiated by an isotropic antenna

Directivity gain in decibels is given by:

$$[g_d]_{db} = G_d = 10 \log g_d$$

(continues)

Gain or power gain is given by:

$$g = \frac{P}{P_d/4\pi r^2}$$

Where: P_d = power delivered to antenna

The ratio of gain to directivity gain is a measure of the efficiency of the antenna:

$$\frac{g}{g_d} = \frac{P_t}{P_d}$$

Effective area of an antenna

$$A = \frac{\text{power received or available at antenna}}{\text{power density of the incident wave at antenna}}$$

$$A = \frac{P_r}{P}$$

$$A = \frac{\lambda^2}{4\pi} g_d(m^2)$$

Where: λ = wavelength of the incident wave
g_d = directivity gain of the receiving antenna

Effective isotropic radiated power (EIRP)

In a given direction:

$$EIRP = P_t g_t$$

$$EIRP(db) = 10 \log\left(\frac{P_t}{1\ \text{watt}}\right) + 10 \log g_t$$

Where: P_t = power delivered to antenna by the transmitter
g_t = directivity gain of the receiving antenna

RADIO WAVE PROPAGATION

Free space propagation

Received power:

$$P_r = P_t g_t g_r = \left(\frac{\lambda}{4\pi d}\right)^2$$

Where: P_r = power received (W)
P_t = transmitted power (W)
g_t = directivity gain of the transmitting antenna
g_r = directivity gain of the receiving antenna
d = distance between transmitting and receiving antennas (m)
λ = wavelength of radio signal (m)

(continues)

Free space transmission loss (FSTL)

$$FSTL = \left(\frac{4\pi d}{\lambda}\right)^2$$

Tropospheric propagation

The electric-field strength at the receiving antenna is given by:

$$E_r = 2E \sin\left(\frac{2\pi h_t h_r}{\lambda d}\right) V/m$$

$$E = \frac{\sqrt{30P_t g_t}}{d} V/m$$

Where: h_t = height of the transmitting antenna
h_r = height of the receiving antenna
d = distance between transmitting and receiving antennas

The maximum range for line-of-sight propagation is given by:

$$R_{max}(\text{miles}) = \sqrt{2h_t(\text{ft})} + \sqrt{2h_R(\text{ft})}$$

$$R_{max}(\text{km}) = \sqrt{17h_t(\text{m})} + \sqrt{17h_r(\text{m})}$$

Where: R_{max} = maximum range of propagation
h_t = transmitting antenna height
h_r = receiving antenna height

Ionospheric propagation

The critical frequency (the highest frequency of a signal transmitted vertically, i.e., the incident angle between the direction of wave propagation and the normal to the ionospheric boundary is zero) that gets reflected by an ionospheric layer is given by:

$$f_c = 9\sqrt{N_{max}} \text{ for a } \theta_i = 0$$

Where: N_{max} = maximum electron density (electrons/m^3)
θ_i = incident angle between direction of wave propagation and the normal to the ionospheric boundary

The maximum usable frequency [MUF] (the highest frequency of a signal transmitted at an angle) that is reflected back by an ionospheric layer is given by:

$$MUF = f_c \sec \theta_i$$

Due to irregularities in ionospheric behavior, a frequency called optimum working frequency (OWF), which is 15 percent less than the MUF, is used as the operating frequency for ionospheric transmissions. The OWF is given by:

$$OWF = 0.\eta 5(9\sqrt{N_{max}} \sec \theta;)$$

DIGITAL COMMUNICATIONS

Relationship between number of bits and possible binary combinations or signal levels

$$2^n = X$$

Where: n = number of bits
X = number of possible binary combinations or signal levels

Baud rate or signaling rate

$$\text{baud rate} = \frac{1}{T} \text{ baud}$$

Where: T = time duration of the smallest signaling element

Information rate or data rate

$$\text{Information rate} = \text{baud rate} \times (\text{number of bits/signaling level})$$
$$= \frac{1}{T} \times n \text{ bps}$$

Nyquist's theorem

$$C = 2W \text{ bps}$$

Where: C = transmission speed of channel or channel capacity
W = bandwidth of the noisless channel

Nyquist's sampling theorem

$$f_s \geq 2 f_a$$

Where: f_s = sampling frequency
f_a = highest frequency of the analog signal

Shannon's law

$$C = \text{channel capacity} = W \log_2 (1 + SNR) \text{ bps}$$

Where: W = channel's bandwidth (Hz)
SNR = channel's signal-to-noise ratio

COMPANDING

μ-law

The μ-law, normalized for a coding range of ± 1, is given by:

$$Y(s) = \text{sgn}(x)\,[\ln(1+\mu|x|)/\ln(1+\mu)]$$

for x in range ± 1.

In this expression, 255 is used for μ, thus it is called μ-255 law.

A-law

For the absolute value of x, $|X|$, falling beyond 1/A. The A-law is given by:

$$Y(s) = \text{sgn}(x)\,[(1+\log A|x|)/(1 + \log A)]$$

For the absolute value of x, $|X|$, falling between 0 and 1/A. The A-law is given by:

$$Y(s) = \text{sgn}(x)[(\log A|x|)/(1 + \log A)]$$

In this expression, 87.6 is used for A, thus it is called A87.6 law.

FIBER OPTICS COMMUNICATIONS

Numerical aperture of an optical fiber

$$NA = \frac{\sqrt{n_1^2 - n_2^2}}{n_0}$$

$$NA = \sqrt{n_1^2 - n_2^2} \text{ for } n_0 = 1$$

Where: n_1 = refractive index of the fiber core
n_2 = refractive index of the fiber cladding
n_0 = refractive index of the launching medium (usually air)

Maximum acceptance angle

For meridional rays:

$$\theta_{0\max} = \sin^{-1}\frac{\sqrt{n_1^2 - n_2^2}}{n_0}$$

For skew rays:

$$\theta_{0\,Sk\max} = \sin^{-1}\left(\frac{\frac{\sqrt{n_1^2 - n_2^2}}{n_0}}{\cos\left(\frac{\alpha}{2}\right)}\right)$$

Where: α = angle by which skew rays change direction

(continues)

Cutoff wavelength for an optical fiber

$$\lambda_0 = \frac{2\pi r}{X_{np}}\sqrt{(n_1^2 - n_2^2)}$$

Where: r = radius of the fiber core
X_{np} = a cutoff parameter (Bessel function)

The total number of modes in an optical fiber is proportional to this cutoff parameter.

Dispersion

$$\Delta t_{total} = \sqrt{\Delta t_{IM}^2 + \Delta t_{CR}^2 + \Delta t_{WV}^2} \text{ ns} - \text{km}$$

Where: Δt_{total} = total dispersion time
Δt_{IM} = intermodal dispersion (intermodal dispersion for monomode fiber is zero)
Δt_{CR} = chromatic or material dispersion
Δt_{WV} = waveguide dispersion

Maximum transmission rate

$$C_{max} \simeq \frac{1}{t_R} = \frac{1}{t_I + \Delta t_{total}} \text{ bps}$$

Where: t_R = output pulse duration
t_I = input pulse duration

Three-db modulation bandwidth for optical emitters (LEDs, laser diodes)

$$f_{3db} = \frac{1}{2\pi\tau} = \frac{0.35}{t_r}$$

Where: τ = charge carrier lifetime
t_r = rise time of the output signal pulse at the optical source

Three-db modulation bandwidth for optical detectors (PIN diode, APD, etc.)

$$f_{3db} = \frac{1}{2\pi C_J R_L} = \frac{0.35}{t_{rd}}$$

Where: R_L = load resistance
C_J = junction capacitance
t_{rd} = rise time of the output signal pulse at the detector

Total channel loss

$$C_L = \text{splice loss} + \text{cable loss} + \text{connector loss}$$

(continues)

Condition for system viability

system power margin ≥ total channel loss + safety margin

$$P_L - P_M \geq C_L + M_S$$

Where: P_L = power launched into the optical fiber
P_M = minimum power required at the optical detector
M_S = safety margin

Optical Detectors: Figures of Merit

Responsivity

Voltage responsivity (V/W)

$$R = \frac{V_S}{P} = \frac{V_S}{(H \times A_d)}$$

Where: H = irradiance in W/cm^2
A_d = sensitive area in cm^2
P = incident radiation power

Current responsivity (A/W)

$$R = \frac{\eta q}{hv} = \frac{\eta \lambda_0 (\text{in } \mu m)}{1.242}$$

Noise equivalent power (W)

$$NEP = \frac{HA_dV_n}{V_S} = \frac{V_n}{R}$$

Where: Vn = noise voltage at output

Response time (s)

$$\tau = \frac{1}{2\pi f_{3db}}$$

Detectivity (W^{-1})

$$D = \frac{1}{NEP}$$

Quantum efficiency

$$\eta = \frac{\text{\# of electrons collected}}{\text{\# of incident photons}}$$

(continues)

Photocurrent gain

$$G = \frac{I_P}{I_{ph}} = \frac{\tau}{\tau_T}$$

$$R = \frac{I_{out}}{P_{in}} \quad \text{or} \quad \frac{V_0}{P_{in}}$$

Where: I_p = photocurrent flowing between electrodes
I_{ph} = photocurrent
τ = carrier life time
τ_T = carrier transit-time

AVAILABILITY AND RELIABILITY OF COMPONENTS/SYSTEMS

Availability

The availability of a device is the measure of the functionality of its various components when in use.

$$A_d = \frac{MTBF}{MTTR + MTBF}$$

Where: A_d = availability of a device
MTBF = mean time before failure (hours)
MTTR = mean time to repair (hours)

Availability for a system

$$A_s = A_1 \times A_2 \times A_3 \times \ldots\ldots \times A_n$$

Where: $A_1, A_2, A_3 \ldots\ldots A_n$
are the individual availabilities of devices defining a system.

Reliability

The reliability of a device is the probability that it will not fail and continue to function over a specific time period.

$$R_d = e^{-\left(\frac{1}{MTBF} \times t\right)}$$

Where: R_d = reliability of a device
t = time period (in hours) for which reliability is to be defined

Reliability of a system

$$R_s = R_1 \times R_2 \times R_3 \times \ldots\ldots \times R_n$$

Where: $R_1, R_2, R_3 \ldots\ldots R_n$
are the individual reliabilities of devices defining a system.

(continues)

Reliability of a system with devices connected in parallel

$R_p = 1 - (1 - R_1)(1 - R_2) \ldots\ldots (1 - R_n)$

Where: $R_1, R_2 \ldots\ldots R_n$ are individual reliabilities of devices connected in parallel.

SYSTEM EFFECTIVENESS

System effectiveness is a measure of how well the system meets user demands.

system effectiveness = system availability × system reliability

$$E_s = A_s \times R_s$$

TRAFFIC MEASUREMENT

Determining the probability of a call not being completed

In telecommunication networks, erlang B and Poisson are the two most commonly used formulas to determine the probability of a call not being completed. Erlang B formula is given by:

$$P = (A^n/n!) / \sum_0^n (A^n/n!)$$

Where: P = Probability of loss (0–1)
A = Traffic intensity (erlangs)
n = Number of paths or trunks

The erlang B formula is based on the assumption that the number of sources is infinite and that lost calls are cleared, whereas the Poisson formula assumes that the number of sources are infinite but the lost calls either re-enter the system or are held in the system for a period of time not exceeding the average holding time of all calls. The Poisson formula is given by:

$$P = e^{-A} \sum_{n+1}^{\infty} (A^n/n!)$$

Where: P = Probability of blocking (0–1)
A = Traffic intensity (erlangs)
n = Base of natural logarithm

(continues)

Determining the traffic intensity

Traffic intensity (in erlangs) can be determined by the following formula:

$$A = \lambda t_m$$

Where: A = Traffic intensity (erlangs)
λ = Arrival rate of calls (average number of calls per unit of time)
t_m = Holding time (average call duration)

FOURIER SERIES

The Fourier series of a periodic function is given by:

$$x(t) = \frac{a}{2} + \sum_{n=1}^{\infty} (a_n \cos 2\pi nft + b_n \sin 2\pi ft)$$

Where: $a_n = \frac{z}{T} \int_{-T/2}^{T/2} x(t) \cos 2\pi ft \, dt$

$b_n = \frac{2}{T} \int_{-T/2}^{T/2} x(t) \sin 2\pi ft \, dt$

f = fundamental frequency
T = time period of the signal

FOURIER TRANSFORM

A Fourier transform decomposes a time domain function into its frequency components. The Fourier transform of a time domain function $x(t)$ is given by:

$$x(f) = \int_{-\infty}^{\infty} x(t) \, e^{-i2\pi ft} \, dt$$

Where: $x(f)$ = frequency domain representation of the signal
$x(t)$ = time domain representation of the signal
f = frequency of the signal

INVERSE FOURIER TRANSFORM

The inverse Fourier transform converts frequency domain representation, obtained by the Fourier transform, back to the time domain representation. The inverse Fourier transform is represented by:

$$x(t) = \int_{-\infty}^{\infty} X(f) \, e^{-i2\pi ft} \, df$$

MODULE 34

Integrated Services Digital Network (ISDN)

The ITU-TS (CCITT) has defined ISDN as a "network evolved from the telephone integrated digital network (IDN) that provides end-to-end connections to support a wide variety of services, including voice and nonvoice services to which users have access by a limited set of standard multipurpose customer interface."

Table 34-1 presents the evolution of ISDN. Table 34-2 compares integrated access (ISDN) with non-integrated access. Table 34-3 lists interface standards for ISDN. Table 34-4 lists the characteristics of ISDN channels. Table 34-5 presents the CCITT's reference model for ISDN. Table 34-6 presents potential applications for ISDN, which is also called narrowband ISDN (N-ISDN).

TABLE 34-1 Evolution of ISDN

	LOCAL LOOP	CENTRAL OFFICE	INTEROFFICE TRANSMISSIONS	TOLL / TANDEM OFFICE	INTEROFFICE TRANSMISSIONS	CENTRAL OFFICE	LOCAL LOOP	
1960s	ANALOG	ANALOG SWITCHING	DIGITAL	ANALOG SWITCHING	DIGITAL	ANALOG SWITCHING	ANALOG	INTRODUCTION OF : (1) DIGITAL INTEROFFICE TRANSMISSION AND (2) STORED PROGRAM CONTROLLED ELECTRONIC SWITCHING
1970s	ANALOG	ANALOG SWITCHING	DIGITAL	DIGITAL SWITCHING	DIGITAL	ANALOG SWITCHING	ANALOG	INTRODUCTION OF DIGITAL TOLL TANDEM SWITCHING
EARLY 1980s	ANALOG/ DIGITAL	ANALOG DIGITAL SWITCHING	DIGITAL	DIGITAL SWITCHING	DIGITAL	ANALOG/ DIGITAL SWITCHING	ANALOG/ DIGITAL	INTRODUCTION OF DIGITAL LOCAL LOOP AND CENTRAL OFFICE
LATE 1980s	DIGITAL	DIGITAL SWITCHING	DIGITAL	DIGITAL SWITCHING	DIGITAL	DIGITAL SWITCHING	DIGITAL	ACCOMPLISHMENT OF END-TO-END DIGITAL NETWORK

TABLE 34-2 Integrated access vs. non-integrated access

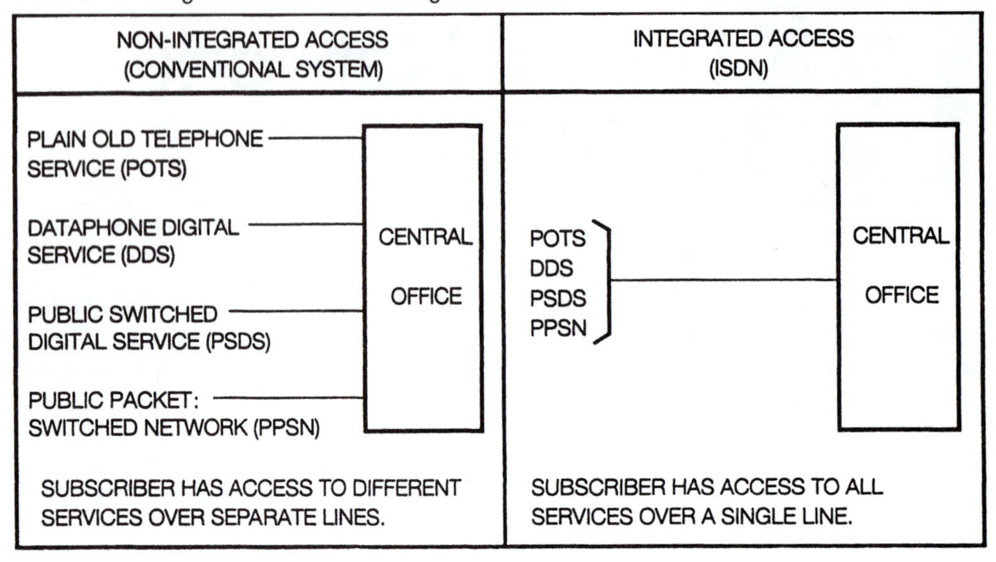

TABLE 34-3 ITU (CCITT) Interface standards for ISDN

INTERFACE	DESCRIPTION
Basic access interface (BAI) or Basic rate interface (BRI)	Data rate: 144 kbps Two 64-kbps B-channels One 16-kbps D-channel
Primary access interface (PAI) or Primary rate interface (PRI)	Data rate: 1.544 Mbps (equal to the bandwidth of T-1 line) Twenty-three 64-kbps B-channels One 64-kbps D-channel It can be reconfigured, e.g. $3H_0 + 1D$ H_{11}

B = The Bearer channel. It handles circuit-switched voice and circuit- or packet-switched data at 64-kbps.

D = The Delta channel. It is packet switched and handles out-of-band signaling for the B-channel. When not signaling for the B-channel, it can be used for low-speed data transmission.

TABLE 34-4 ISDN channel characteristics

ISDN CHANNEL	BIT RATE	INTERFACE
D	16 kbps 64 kbps	Basic Rate Interface (BRI) Primary Rate Interface (PRI)
B	64 kbps	Basic Rate Interface (BRI) Primary Rate Interface (PRI)
H_0	384 kbps	Primary Rate Interface (PRI)
H_{11}	1536 kbps	Primary Rate Interface (PRI)
H_{12}	1920 kbps	Primary Rate Interface (PRI)

TABLE 34-5 CCITT'S reference model for ISDN

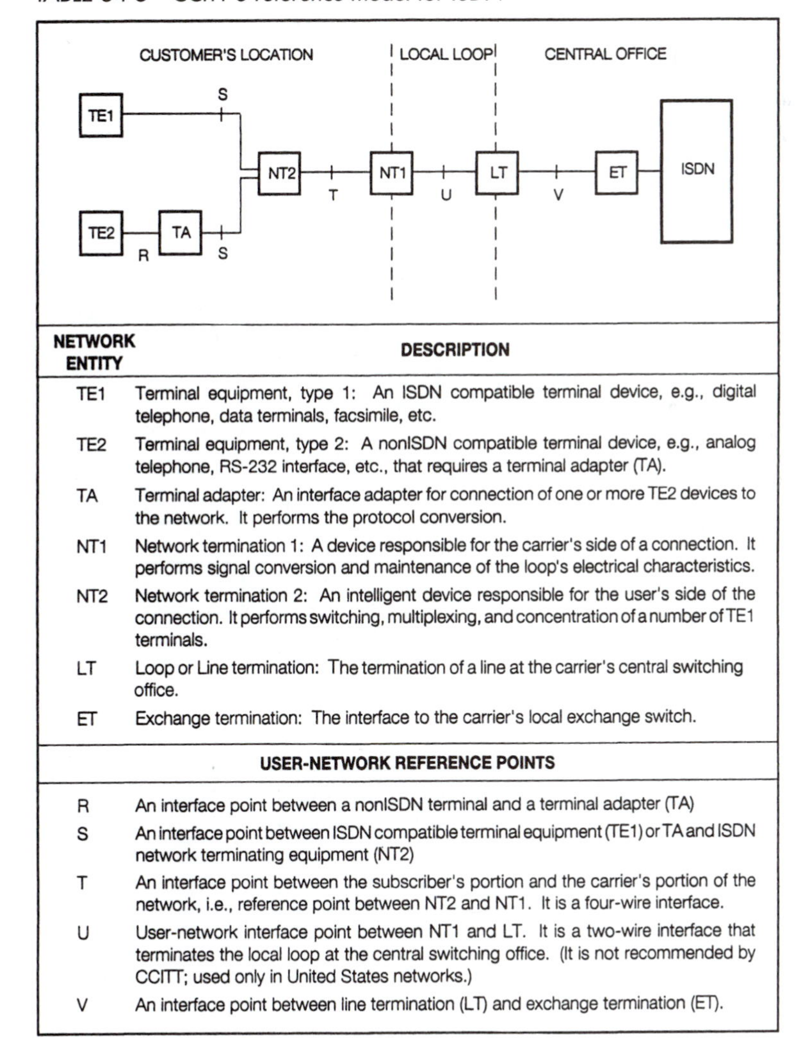

NETWORK ENTITY	DESCRIPTION
TE1	Terminal equipment, type 1: An ISDN compatible terminal device, e.g., digital telephone, data terminals, facsimile, etc.
TE2	Terminal equipment, type 2: A nonISDN compatible terminal device, e.g., analog telephone, RS-232 interface, etc., that requires a terminal adapter (TA).
TA	Terminal adapter: An interface adapter for connection of one or more TE2 devices to the network. It performs the protocol conversion.
NT1	Network termination 1: A device responsible for the carrier's side of a connection. It performs signal conversion and maintenance of the loop's electrical characteristics.
NT2	Network termination 2: An intelligent device responsible for the user's side of the connection. It performs switching, multiplexing, and concentration of a number of TE1 terminals.
LT	Loop or Line termination: The termination of a line at the carrier's central switching office.
ET	Exchange termination: The interface to the carrier's local exchange switch.

USER-NETWORK REFERENCE POINTS

R	An interface point between a nonISDN terminal and a terminal adapter (TA)
S	An interface point between ISDN compatible terminal equipment (TE1) or TA and ISDN network terminating equipment (NT2)
T	An interface point between the subscriber's portion and the carrier's portion of the network, i.e., reference point between NT2 and NT1. It is a four-wire interface.
U	User-network interface point between NT1 and LT. It is a two-wire interface that terminates the local loop at the central switching office. (It is not recommended by CCITT; used only in United States networks.)
V	An interface point between line termination (LT) and exchange termination (ET).

TABLE 34-6 Potential applications for ISDN

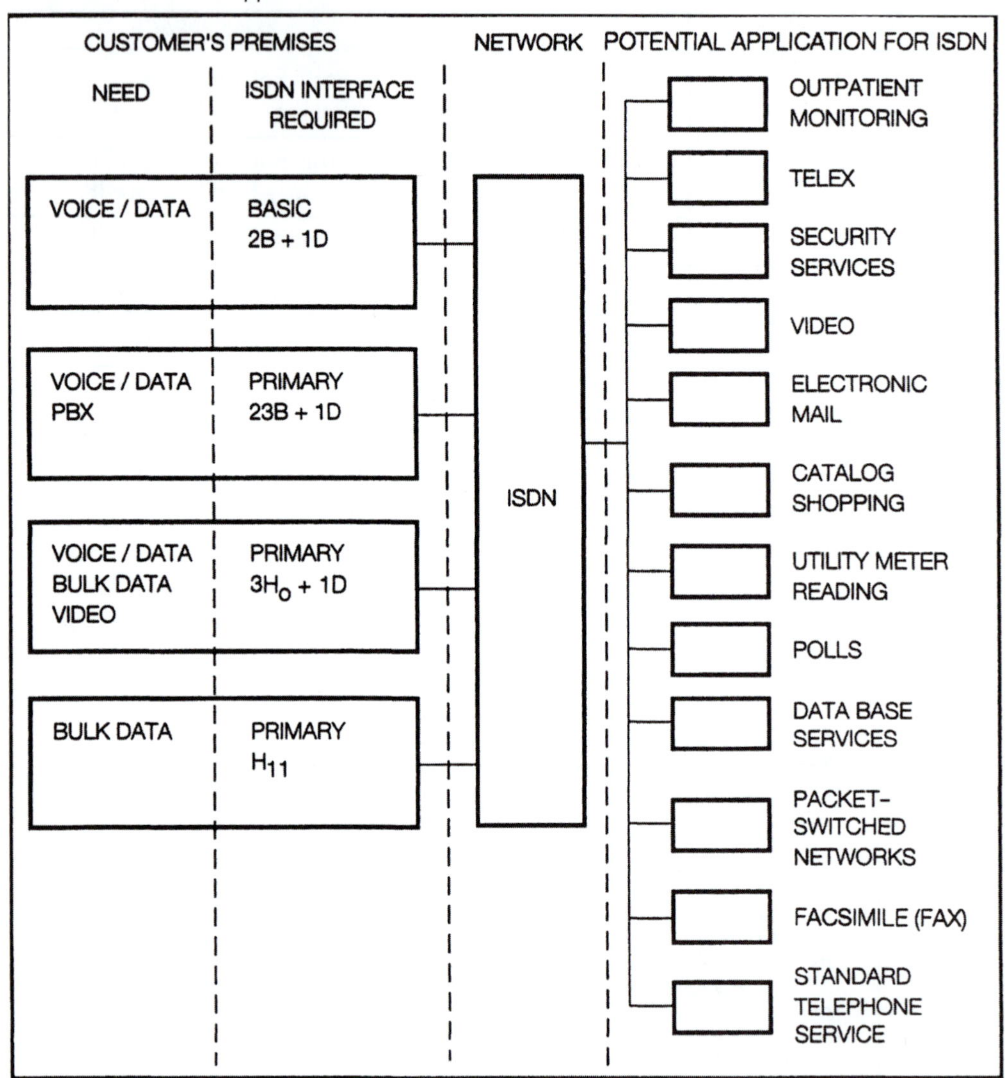

B-ISDN

Broadband integrated services digital network (ISDN) refers to digital service that uses broadband technology asynchronous transfer mode (ATM) to offer high-speed data transfer to support interactive services (e.g., video conferencing, video mail) and distribution services (e.g., television broadcast).

MODULE

35 Network Architecture

Network architecture is defined as an overall plan or a blueprint that governs the design of the hardware and software components of a data communication system. The following is a brief description of the two major network architectures: the OSI reference model and SNA.

The OSI Reference Model: The Open Systems Interconnect (OSI) reference model, generally known as the OSI model, was developed in 1977 by a committee of the International Standards for Organization (ISO) in response to the international need for an open set of communications standards that could:

1. provide an architectural reference point for developing standardized procedures,
2. allow full integration of communication networks at all levels, and
3. provide compatibility between multivendor products and services.

The OSI model acts as an international standard protocol hierarchy guideline:

1. It is not a protocol.
2. It is an architectural model.
3. It can be used to establish protocols.

The OSI model includes seven distinct software layers that define protocols and interfaces needed to support an open system (Figure 35-1). Table 35-1 lists and defines each of the seven layers.

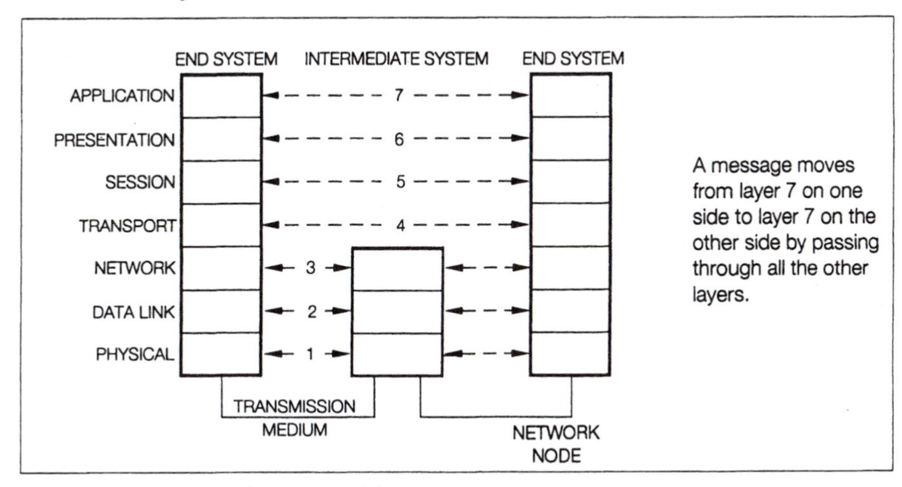

FIGURE 35-1 OSI reference model

TABLE 35-1 OSI reference model

LAYER NUMBER	NAME	FUNCTION	DATA UNITS	EXAMPLES	HARDWARE
7	Application	This layer is responsible for the actual end-user applications, such as file transfer control, distributed data base control, and applications specific to the network.	Messages	HTTP Telnet	Gateways
6	Presentation	Presents data in a format that is compatible with the receiving system. It can perform functions of syntax transformation for character sets, text strings, data display formats, data types, encoding and decoding, data compaction, etc.	Messages	EBCDIC ASCII JPEP TIFF	Gateways
5	Session	Coordinates interaction between end-application processes. It performs functions such as start and stop tasks, dialogue control (half-duplex, full-duplex, duration, etc.), exchange of data during a session, error checking, recovery, etc.	Messages	OS	Gateways
4	Transport	The software in this layer provides end-to-end data integrity and quality of service. It incorporates flow control, end-to-end error detection and recovery, assembles and disassembles data packets for layer 3, etc.	Segments	TCP	Gateways
3	Network	The software in this layer is responsible for establishing, maintaining, and terminating the information route from source to the destination.	Packets	IP	Routers
2	Data Link	This layer is responsible for the transmission of units of information between two or more network components via a specific physical linkage. It adds synchronization information and error-checking algorithms to the data to ensure an error-free transmission.	Frames	HDLC Ethernet Token Ring	Bridges/ switches
1	Physical	This layer is responsible for the transmission of actual data in the form of electrical pulses over a physical link. It consists of cables, connectors, and communicating devices.	Bits	EIA/TIA – 232, V.28, V.33	Repeaters

Systems Network Architecture (SNA): SNA was introduced in 1974 by IBM. SNA is a network architecture intended for IBM machines to be connected in the form of a network. Prior to the development of SNA, IBM had several hundred communications products with a large number of access methods and protocols. The SNA objective was to eliminate this disorder and provide a logical framework for distributed processing.

Like OSI, SNA also follows a seven-layered architecture (Figure 35-2). The two are similar but not identical. Table 35-2 lists and defines each of the seven layers.

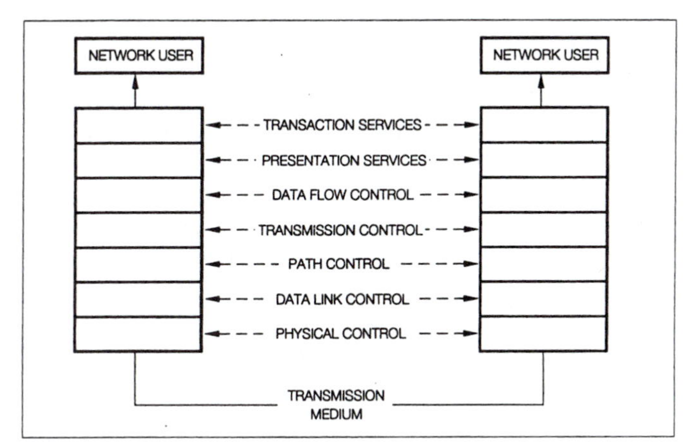

FIGURE 35-2 SNA software layers

TABLE 35-2 Functions of SNA layers

LAYER NUMBER	NAME	FUNCTION
7	Transaction services	The highest layer of SNA provides end-user interface to IBM's distributed transaction architecture.
6	Presentation services	This layer formats data for the end-user and coordinates the sharing of resources.
5	Data flow control	It is concerned with the overall integrity of data flow between the end points of a session.
4	Transmission control	This layer monitors the status of sessions between users. It controls the data flow, exchange, sequence, size, and encryption/decryption.
3	Path control	This layer is responsible for data routing through the network and data control.
2	Data link control	It is responsible for transmitting data between adjacent nodes. It is similar to the OSI model's data link layer.
1	Physical control	It connects adjacent nodes physically and electrically. It is similar to the OSI model's physical layer.

MODULE 36

Digital Subscriber Line (DSL): An Overview

Digital subscriber line (DSL) is a technology that allows a high-speed data connection over ordinary copper wires. With a DSL line, a user could receive data at rates up to 6.1 Mbps. Typically, a DSL connection provides speeds of 384 kbps–1.544 Mbps for downstream, or receiver, connectivity, and 128 kbps–768 kbps for upstream, or transmitting, connectivity.

A number of DSL variations are available to the consumer. DSL was started with the development of high-bit-rate digital subscriber line (HDSL) in the early 1990s in order to achieve data transmission rates equivalent to the T-1 (U.S. 1.544 Mbps)/E-1(Europe 2.04 Mbps) lines.

DSL uses advanced modulation techniques such as multitone and discrete multitone (DMT) and carrierless amplitude phase (CAP) modulation to achieve a very high data rate over twisted wires. Table 36-1 lists the characteristics of commonly used xDSL configurations.

TABLE 36-1 Characteristics of xDSL configurations

xDSL TYPE	DATA RATE	DISTANCE/WIRE TYPE	WIRE TYPE	TYPICAL APPLICATION
High bit-rate Digital Subscriber Line (HDSL)	1.544 Mbps duplex on two wire pairs, 2.048 Mbps duplex on three twisted-pair lines.	12,000	AWG 24	Used for Internet access (Provides low-cost replacement for T-1 and fractional T-1 lines)
Symmetric Digital Subscriber Line (SDSL)	n × 64 kbps–2 Mbps on a single duplex line.	12,000	AWG 24	Used for Internet access (Provides low-cost replacement for T-1 and fractional T-1 lines)

(continues)

TABLE 36-1 Characteristics of xDSL configurations (continued)

xDSL TYPE	DATA RATE	DISTANCE/WIRE TYPE	WIRE TYPE	TYPICAL APPLICATION
Asymmetric Digital Subscriber Line (ADSL)			AWG 24	Used for Internet access
ADSL-1	Downstream (1.5–2.0 Mbps)	18,000 feet		
ADSL-3	Downstream (6.144 Mbps) A bidirectional channel up to 640 kbps	12,000 feet		
Very High Speed DSL (VDSL)	Downstream (12.9–52.8 Mbps) Upstream (1.5–2.3 Mbps)	12.96 Mbps @ 4,500 feet 25.82 Mbps @ 3,000 feet 51.84 Mbps @ 1000 feet	AWG 24	ATM/SONET

MODULE 37

IP Addresses: An Introduction

37-1 IP ADDRESS

A unique 32-bit address used by TCP/IP protocol to identify a device connected to the Internet. It contains two fields: a network and a host. The size of the network field determines the maximum number of networks, and the size of the host field determines the maximum number of hosts per network. This addressing scheme is called IP version 4 or IPv4 (Figure 37-1).

Network Identifier	Host Identifier

FIGURE 37-1 IPv4 addressing scheme

37-2 CLASSES OF IP ADDRESSES

IP addresses are categorized into five classes (class A–class E). Class A, B, and C are commonly used to identify the network and the host, and Class D is used for multicasting and Class E is reserved for future use. Table 37-1 presents a summary of the characteristics of the five classes of IP addresses. The maximum number of available hosts can be calculated by using the following formula:

$$2^n - 2$$

Where n is the number of host bits and accounts for 2 octets with all zeros (identifies the network) and all ones (identifies the broadcast address).

Figure 37-2 presents the formats for class A–E IP addresses.

TABLE 37-1 Characteristics of IP addresses

CLASS	NUMBER OF BITS FOR NETWORK	NUMBER OF BITS FOR HOST	NUMBER OF POSSIBLE NETWORKS	NUMBER OF POSSIBLE HOSTS PER NETWORK	POSSIBLE RANGE OF ADDRESSES
A	8	24	126 (#0 & #127 are reserved)	16,777,217	0.0.0.0–127.255.255.255
B	16	16	16,384	65,536	128.0.0.0–191.255.255.255
C	24	8	2,097,152	254 (#0 & #255 are reserved)	192.0.0.0–223.255.255.255
D	See Figure 37-2				224.0.0.0–239.255.255.255
E	See Figure 37-2				240.0.0.0–247.255.255.255

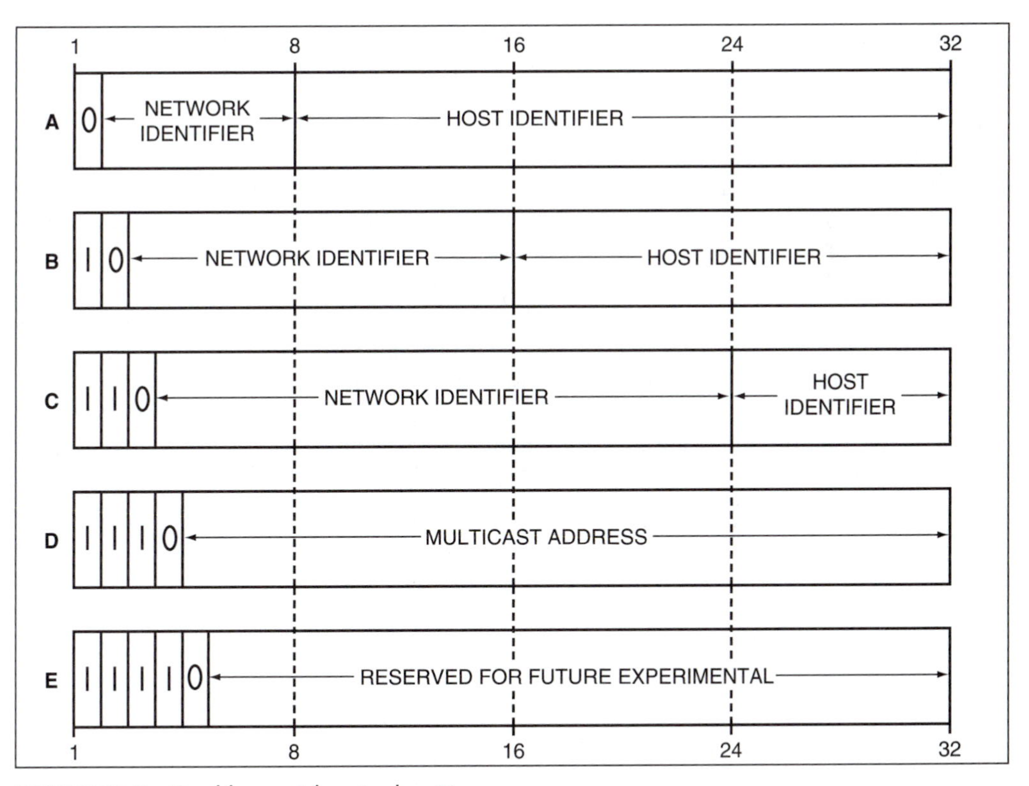

FIGURE 37-2 IP addresses (class A–class E)

Due to rapid growth of Internet and increased demands on the IP routing tables in Internet routers, a new IP addressig scheme called "Classless Interdomain Routing (CIDR)" was developed. CDIR improves IP address space utilization and routing scalability in the Internet. In contrast to the class-based addressing (class A, B, C and so on), the CDIR uses an arbitary length for the network address. CDIR address follows the format of w.x.y.z/n, where n indicates the number of bits in the network mask, i.e, the network portion of address.

■ 37-3 DOTTED DECIMAL NOTATION

The actual IP address is a binary bit stream format that is easily processed by telecommunication devices and computers. But in order to make it easier for human users to read and write the 32-bit binary pattern, the IP address is expressed as four groups of decimal numbers. This format is known as "dotted decimal notation." Table 37-2 presents the concept of dotted decimal notation.

TABLE 37-2 Dotted Decimal Notation

BIT POSITION	1 8	9 16	17 24	25 32
IP Address	10000011	00001010	00110111	00000111
Dotted Decimal Notation	131	10	55	7

■ 37-4 IP ADDRESS VERSION 6 (IPv6)

The IPv4 addressing scheme, developed in the 1980s, yields around 4 billion possible addresses. The exponential growth of the Internet and emergence of the Internet-compatible devices, such as cell phones and PDAs, diminish the availability of IPv4 addresses. To rectify this shortcoming of IPv4, a new IPv6 scheme was developed.

IPv6, also known as "Next Generation Internet Protocol" or IPng, was recommended by the Internet Engineering Task Force in 1994 through RFC 1752.

IPv6 uses 128 bits and therefore allows for a larger number of addresses (2^{128}) and subnets (10^{15}), and adds many improvements to IPv4 such as routing and network autoreconfiguation.

IPv6 can be installed as a software upgrade in Internet devices and is interoperable with IPv4.

MODULE

38 ITU-TS Recommendations

■ ITU-TS (CCITT)

The International Telecommunication Union (ITU), formerly the International Telegraph Union (established 1865), became an agency of the United Nations (UN) after its establishment in 1947. In 1992, the ITU underwent reorganization to meet the challenges of the new era of globalization. As a result, the ITU was restructured into three sectors: Telecommunication Standardization (ITU-T) [formerly CCITT], Radiocommunication (ITU-R) [formerly CCIR], and Telecommunication Development (ITU-D). Presently 189 countries are members of the ITU.

ITU-TS produces standards (Recommendations) on technical, operating, and tariff issues. The ITU-TS's standardization work is carried out by Technical Study (SGs) [http://www.itu.int/ITU-T/studygroups/]. Although ITU-TS recommendations are not intended to be mandatory, they have the effect of law in many countries of the world.

This module presents a brief introduction to the ITU-TS's recommendations. Table 38-1 lists a series of ITU-TS Recommendations. Table 35-2 lists some examples of the ITU-TS Recommendations. More information on ITU-TS Recommendations can be obtained from: http://www.itu.int/ITU-T/.

TABLE 38-1 ITU-TS (CCITT) Recommendations

ITU-T SERIES	DESCRIPTION
A	Organization of the work of the ITU-T
B	Means of expression: definitions, symbols, classification
C	General telecommunications statistics
D	General tariff principles
E	Telephone operation, quality of service, and tariffs
F	Non-telephone telecommunication services
G	Transmission systems and media, digital systems and networks
H	Audiovisual and multimedia systems
I	Integrated services digital network (ISDN)
J	Sound program, television transmission, and cable networks
K	Protection against interference
L	Protection of cable sheaths and poles

(continues)

TABLE 38-1 ITU-TS (CCITT) Recommendations (continued)

ITU-T SERIES	DESCRIPTION
M	Maintenance for telephone, telegraphs, and data transmission
N	Maintenance of sound program and television transmission
O	Specifications of measuring equipment
P	Quality of telephone transmission: local network and telephone installation
Q	Telephone signaling and switching
R	Telegraph transmission
S	Alphabetical telegraph and data terminal equipment
T	Apparatus and transmission for facsimile telegraphy
U	Telegraph switching
V	Data communications using the telephone system
X	Data transmission over public networks
Y	Global information infrastructure and Internet Protocol issues
Z	Software aspects for telecommunication systems
Source: http://www.itu.int/ ITU-T /publications/	

TABLE 38-2 ITU-T Recommendation examples

ITU-T RECOMMENDATION	DESCRIPTION
A.12	Collaboration with the International Electrotechnical Commission on the subject of definitions.
B.3	Use of the International Systems of Units.
C.1	Yearbook of Common Carrier Telecommunication Statistics.
D.4	Costs and values of services rendered as factors in the fixing of rates.
E.161	Numbering and dialing procedures for international service.
F.67	Charging and accounting in the international telex service.
G.733	Characteristics of primary PCM multiplex equipmet operating at 1544 kbps.
H.53	Transmission of wide-spectrum signals over wide band super group links.
I.211	B-ISDN service aspects.
I.630	ATM protection switching.
J.61/62	Specifications for long-distance television transmission.
K.11	Use of gas-filled protectors and fuses.
L.7	Application of joint cathodic protection.
M.880	International phototelegraph transmission.
N.15	Maximum permissible power during an international sound program transmission.

(continues)

TABLE 38-2 ITU-T Recommendation examples (continued)

ITU-T RECOMMENDATION	DESCRIPTION
0.71	Specifications for an impulsive noise-measuring instrument for telephone-type circuits.
P.32	Devices for recording messages or telephone conversations.
Q.13	International routing plan.
R.83	Changes of level and interruptions in voice-frequency telegraphy channels.
S.6	Characteristics of answer-back units.
T.1	Standardization of phototelegraph apparatus.
U.2	Standardization of dials and dial-pulse generators for the international telex service.
V.1	Equivalence between binary notation symbols and the significant conditions of a two-condition code.
V.2	Power-level standards for data transmission over telephone lines.
V.3	International alphabet number 5.
V.4	General structure of signals of international telegraph alphabet number 5 for data transmission.
V.7	Definitions of terms related to data communications over the telephone network.
V.10	Electrical characteristics for unbalanced double-current interchange circuits for general use with integrated circuit equipment in the field of data communications. (See X.26.)
V.11	Electrical characteristics for balanced double-current interchange circuits for general use with integrated circuit equipment in the field of data communications. (See X.27.)
V.13	Answer-back simulator.
V.15	Use of acoustic coupling for data communications.
V.19	Modems for parallel data transmission using telephone signaling frequencies.
V.21	300-bits-per-second (bps), full-duplex modem, standardized for use in the general switched telephone network.
V.22	1200-bps full-duplex modem standardized for use in the general switched telephone network.
V.23	600/1200-bps modem standardized for use in the general switched telephone network.
V.24	List of definitions for interchange circuits between data terminal equipment and data circuit-terminating equipment.

(continues)

TABLE 38-2 ITU-T Recommendation examples (continued)

ITU-T RECOMMENDATION	DESCRIPTION
V.25	Automatic calling and/or answering equipment on the general switched telephone network, including disabling of echo suppressors on manually established calls.
V.26	2400-bps modem standardized for use on four-wire leased circuits.
V.26bis	1200/2400-bps modem standardized for use in the general switched telephone network.
V.27	4800-bps modem with manual equalizer standardized for use on leased telephone-type circuits.
V.27bis	2400/4800-bps modem with automatic equalizer standardized for use on leased telephone-type circuits.
V.27ter	2400/4800-bps modem with standardized for use in the general switched telephone network.
V.28	Electrical characteristics for unbalanced double-current interchange circuits.
V.29	9600-bps modem standardized for use on point-to-point leased telephone-type circuits.
V.36	Modems for synchronous transmission using 60–108 kHz group band circuits.
V.50	Standard limits for transmission quality of data transmission.
V.54	Loop test devices for modems.
V.55	Specifications for an impulsive noise-measuring instrument for telephone-type circuits.
V.100	Interconnection between PDNs (public data networks) and PSTN (public switched telephone network).
V.110	Support of DTE (data terminal equipment) with V-series-type interface by an ISDN (integrated digital services network).
X.1	Designation of international user classes of service in public data networks.
X.2	Designation of international user facilities in public data networks.
X.3	Packet assembly/disassembly facility (PAD) in a PDN (public data network).
X.4	General structure of the signals of the international alphabet number 5 code for data transmission over a PDN (public data network).
X.15	Definition of terms related to public data networks (PDN).

(continues)

TABLE 38-2 ITU-T Recommendation examples (continued)

ITU-T RECOMMENDATION	DESCRIPTION
X.20	Interface between data terminal equipment (DTE) and data circuit-terminating equipment (DCE) for start-stop transmission on the public data network (PDN).
X.20bis	Use on the PDN (public data network) of DTE (data terminal equipment) that is designed for interfacing to asynchronous V-series modems.
X.21	General purpose interface between data terminal equipment (DTE) and data circuit-terminating equipment (DCE) for asynchronous operation on public data networks.
X.21bis	Use on a PDN (public data network) of DTE (data terminal equipment) that is designed for interfacing to synchronous V-series modems.
X.22	Multiplex DTE/DCE interface for user classes 3 to 6.
X.24	List of definitions for interchange circuits between data terminal equipment (DTE) and data circuit-terminating equipment (DCE) on a PDN (public data network).
X.25	Interface between data terminal equipment (DTE) and data circuit-terminating equipment (DCE) for terminal operating in the packet mode and connected to a public data network (PDN).
X.26	Electrical characteristics for unbalanced double-current interchange circuits for general use with integrated circuit equipment in the field of data communications. (Similar to V.10)
X.27	Electrical characteristics for balanced double-current interchange circuits for general use with integrated equipment in the field of data communications. (Similar to V.11)
X.28	DTE/DCE interface for start-stop-mode data terminal equipment accessing the packet assembly/disassembly facility (PAD) in a PDN (public data network) in the same country.
X.29	Procedures for the exchange of control information and user data between a packet assembly/disassembly (PAD) facility and a packet mode DTE or another PAD.
X.75	Terminal and transit call-control procedures and data-transfer mechanisms on international circuits between packet-switched data networks.
X.95	Networks parameters in public data networks.

(continues)

TABLE 38-2 ITU-T Recommendation examples (continued)

ITU-T RECOMMENDATION	DESCRIPTION
X.150	Data terminal equipment (DTE)/data terminal-circuit-equipment (DCE) test loops in public data networks.
X.200	OSI (open systems interconnect) reference model. (See Module 35.)
X.225	Session protocol specifications for open systems interconnection (OSI) for ITU-TS application.
X.250	Formal description techniques for data communication protocols.
X.400	Standards for international electronic mail networking.
X3.1	Synchronous signaling rates for data transmission.
X3.15	Bit sequencing of the American Standard Code for Information Interchange (ASCII) in serial-by-bit transmission.
X3.16	Character structure and character parity sense for serial-by-bit data communications in American Standard Code for Information Interchange (ASCII).
X3.36	Synchronous, high-speed data signaling rates between data terminal equipment (DTE) and data circuit-terminating equipment (DCE).
X3.41	Code extension techniques for use with the seven-bit coded-character set of American Standard Code for Information Interchange (ASCII).
X3.44	Determination of the performance of data communication devices.
X3.57	Message heading formats for information interchange using American Standard Code for Information Interchange (ASCII) for data communication system control.
X3.92	Algorithm for data encryption.
Z.101	General explanation of specification and description language.

MODULE 39

TIA/EIA Standards: 232, 449, 485, 644, and ITU-TS Equivalents

The Electronics Industry Alliance (EIA) and the Telecommunications Industry Association (TIA) are industry trade organizations that develop standards for data interfaces used in data communications. Figure 39-1 illustrates the classification of data communication standards, and Table 39-1 presents a summary of commonly used TIA/EIA and other standards for DTE/DCE interfaces.

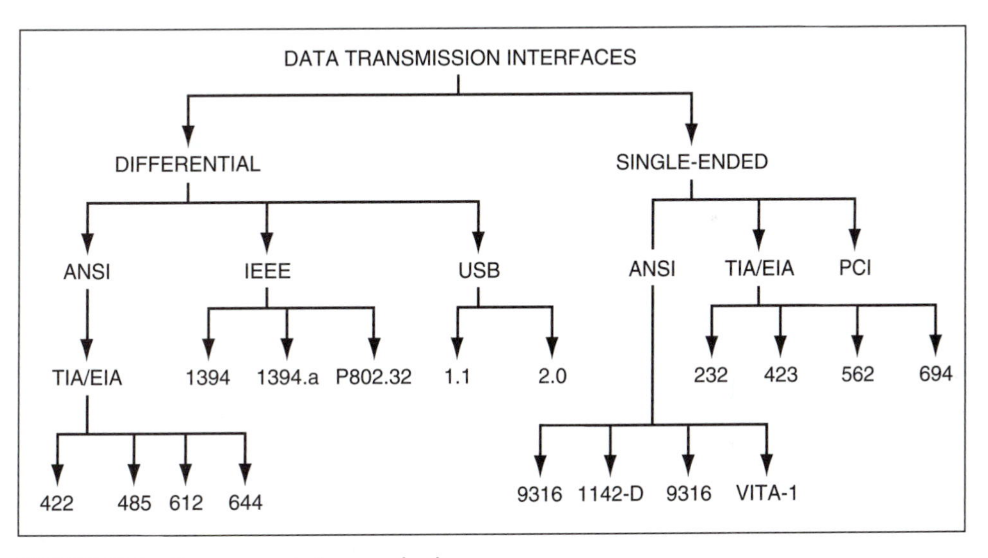

FIGURE 39-1 Data transmission standards

TABLE 39-1 Commonly used TIA/EIA standards

BALANCED INTERFACES	UNBALANCED INTERFACES	SIGNAL QUALITY STANDARDS
TIA/EIA-422-B	TIA/EIA-232 (A-F)	EIA-334-A
TIA/EIA-485-A	TIA/EIA-423-B	EIA-363
TIA/EIA-612	TIA/EIA-562	EIA-404-A
TIA/EIA-644	TIA/EIA-694	

TIA/EIA-232 is a set of specifications that apply to the transfer of data between DTE (data terminal equipment) and DCE (data communications equipment) (Figure 39-2). The first version of RS-232 was issued in 1960, and was revised in 1963 to revision A. In 1965, revision B was issued. Revision C, issued in 1969, is the most commonly used interface in the United States in spite of its speed and distance limitations. Revision D was issued in 1987. (Versions E (1991) and F (1997) represent the fifth and sixth revisions.) TIA/EIA-232-D specifies a 25-pin connector for which four types of data lines are defined: data signals, control signals, timing signals, and grounds. Of the 25 signal lines, two are grounds, four are data signals, three are timing signals, and twelve are control signals. In a given application, an equipment does not have to use all available lines. For example, a full-duplex operation between DTE and DCE can be established using only three lines (Figure 39-3). A TIA/EIA-232-D interface can also be used as a null modem to connect two DTEs with each other over a short distance (Figure 39-4). The CCITT's counterparts to RS-232 are the V.24 and V.28 recommendations. V.24 specifies the signal lead definitions and V.28 specifies the electrical characteristics. The ITU-TS and TIA/EIA recommendations have similar electrical specifications but different mechanical specifications.

For additional information about TIA/EIA standards visit:
http://www.eia.org
http://www.tiaonline.org

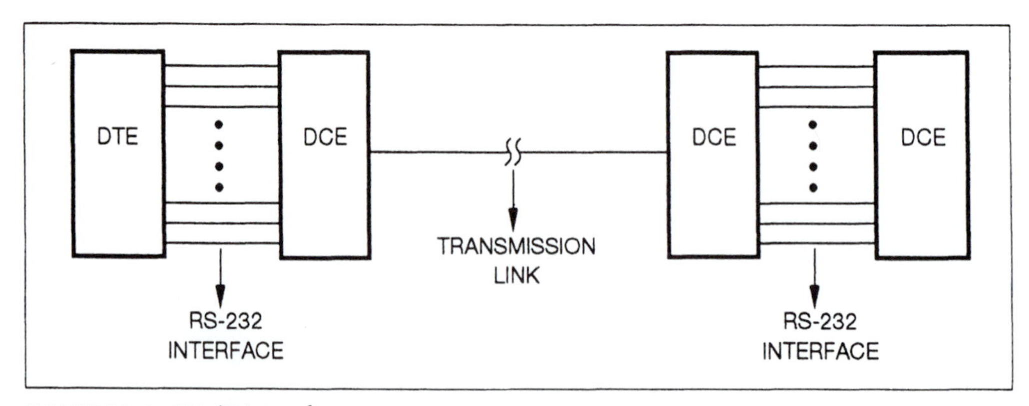

FIGURE 39-2 TIA/EIA interface

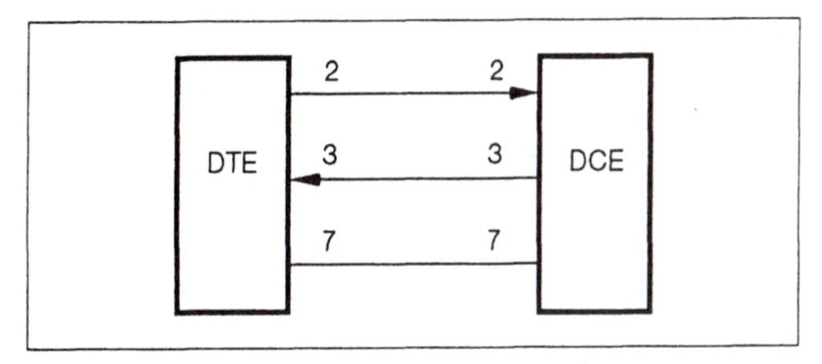

FIGURE 39-3 Basic DTE-DCE interface using RS-232-D

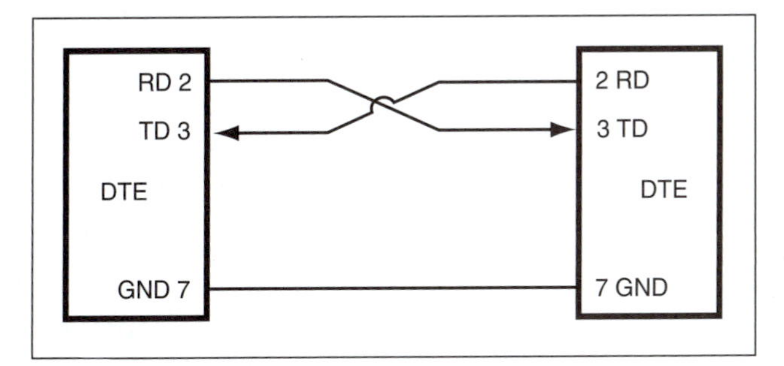

FIGURE 39-4 RS-232-D interface as null modem

In 1975, the EIA introduced RS-449 (TIA/EIA-449), which was designed in conjunction with RS-422 (TIA/EIA-422) and RS-423 (TIA/EIA-423), to overcome the limitations of RS-232 (TIA/EIA-232). RS-449 retains all functional capabilities of RS-232 and introduces ten additional interchange circuits to enhance interface capabilities. RS-422 (TIA/EIA-422) is a balanced electrical interface and RS-423 (TIA/EIA-423) is an unbalanced interface. Although ITU-TS (CCITT) has not developed any recommendation similar to TIA/EIA-449, it has recommended specifications V.10 (also X.2.6) and V.11 (X.27), which are similar to TIA/EIA-423 and TIA/EIA-442 respectively. V.10 (also X.26) deals with the electrical characteristics of unbalanced interchange circuits and V.11 (also X.27) deals with the balanced interchange circuits.

Table 39-2 compares the mechanical specifications of TIA/EIA-232 and TIA/EIA-449. Table 39-3 compares the electrical specifications of TIA/EIA-232-C, TIA/EIA-449, and ITU-TS (CCITT) V.24. Table 39-4 lists the specifications for TIA/EIA-232-D. Table 39-5 compares the characteristics of TIA/EIA-232, TIA/EIA-422, TIA/EIA-423, TIA/EIA-485, and TIA/EIA-644.

TABLE 39-2 Mechanical specifications of TIA/EIA-232 and TIA/EIA-449

PIN #	DESCRIPTION	
	TIA/EIA-232-D	TIA/EIA-449
1	Protective ground	Shield
2	Transmitted data	Signal rate indicator
3	Received data	Spare
4	Request to send	Send data
5	Clear to send	Send timing
6	DCE ready	Receive data
7	Signal ground	Request to send
8	Data carrier detect	Receive timing
9	Reserved for testing	Clear to send
10	Reserved for testing	Local loopback
11	Unassigned	Data mode
12	Secondary received line signal detector/data signal	Terminal ready
13	Secondary clear to send	Receiver ready
14	Secondary transmitted data	Remote loopback
15	Transmitter signal element timing	In-coming call
16	Secondary received data	Signal rate select/select frequency
17	Receiver signal element timing	Terminal timing
18	Local loopback	Test mode
19	Secondary request to send	Signal ground
20	DTE ready	Receive Common
21	Remote loopback signal quality detect	Spare
22	Ring indicator	Send data
23	Data signal rate selector	Send timing
24	Transmit signal element timing	Receive data
25	Test mode	Request to send
26	NA	Receive timing
27	NA	Clear to send
28	NA	Terminal in service
29	NA	Data mode
30	NA	Terminal ready
31	NA	Receiver ready
32	NA	Select standby
33	NA	Signal quality
34	NA	New signal
35	NA	Terminal timing
36	NA	Standby indicator
37	NA	Send common

TABLE 39-3 Signal comparison RS-232-D, RS-449, and CCITT standards

TIA/EIA-232-C		TIA/EIA-449		ITU-TS (CCITT) V.24
Circuit Mnemonic	Description	Circuit Mnemonic	Description	Circuit Mnemonic
AB	Signal ground	SG	Signal ground	102
		SC	Send common	102a
		RC	Receive common	102b
CE	Ring indicator	IC	In-coming call	125
DC	Data terminal ready	TR	Terminal ready	108/2
CC	Data set ready	DM	Data mode	107
		IS	Terminal in service	
BA	Transmitter data	SD	Send data	103
BB	Received data	RD	Receive data	104
DA	Transmitter signal Element timing (DTE source)	TT	Terminal timing	113
DB	Transmitter signal Element timing (DCE source)	ST	Send timing	114
DD	Receive signal Element timing	RT	Receive timing	115
CA	Request to send	RS	Request to send	105
CB	Clear to send	CE	Clear to send	106
CF	Received line Signal detector	RR	Receiver ready	109
CG	Signal quality detector	SQ	Signal quality	110
		NS	New signal	
CH	Data signal Rate selector (DTE source)	SF	Select frequency	126
		SR	Signaling rate selector	111
CI	Data signal rate selector (DCE source)	SI	Signaling rate indicator	112
SBA	Secondary transmitted data	SSD	Secondary send data	118
SBB	Secondary receive data	SRD	Secondary receive data	119
SCA	Secondary request to send	SRS	Secondary request to send	120
SCB	Secondary clear to send	SCS	Secondary clear to send	121
SCF	Secondary received line signal detect	SRR	Secondary receiver ready	122
		LL	Local loop back	141
		RL	Remote loop back	140
		TM	Test mode	142
		SS	Select standby	116
		SB	Standby indicator	117

TABLE 39-4 TIA/EIA-232-D specifications

PIN NUMBER	CIRCUIT MNEMONIC	DESCRIPTION	SIGNAL TYPE	CCITT CIRCUIT NUMBER
1	--	Protective ground	--	--
2	BA	Transmitted data	Data	103
3	BB	Received data	Data	104
4	CA	Request to send	Control	105
5	CB	Clear to send	Control	106
6	CC	DCE ready	Control	107
7	AB	Signal ground	Ground	102
8	CF	Received line signal detector	Control	109
9	--	Reserved for testing	--	--
10	--	Reserved for testing	--	--
11	--	Unassigned	--	--
12	SCF/CI	Secondary received line signal detector/data signal	Control	122/112
13	SCB	Secondary clear to send	Control	121
14	SBA	Secondary transmitted data	Data	118
15	DB	Transmitter signal element timing (DCE source)	Timing	115
16	SBB	Secondary received data	Data	119
17	DD	Receiver signal element timing (DCE source)	Timing	114
18*	LL	Local loop back	Control	141
19	SCA	Secondary request to send	Control	120
20	CD	DTE ready	Control	108.2
21+	RL/CG	Remote loop back/signal quality detector	Control	140/110
22	CE	Ring indicator	Control	125
23	CH/CI	Data signal rate selector (DTE/DCE source)	Control	111/112
24	DA	Transmit signal element timing (DTE source)	Timing	113
25*	TM	Test mode	Control	142

* Pins 18 and 25 are unassigned in TIA/EIA-232-C

+ Pin 21 is signal quality detector in TIA/EIA-232-C

TABLE 39-5 Comparison of TIA/EIA-232, TIA/EIA-422, TIA/EIA-423, TIA/EIA-485, and TIA/EIA-644 standards

CHARACTERISTIC	TIA/EIA-232	TIA/EIA-422	TIA/EIA-423	TIA/EIA-485	TIA/EIA-644
Form of operation	Single-ended (unbalanced)	Differential (balanced)	Single-ended (unbalanced)	Differential (balanced)	Differential (balanced)
Cable length (maximum)	50 feet	4000 feet	4000 feet	4000 feet	100 feet
Data rate (maximum)	20 kbps	10 Mbps	100 kbps	50 Mbps	655 Mbps
Number of receivers	1	10	10	32	32
Minimum driver output levels	± 5V	± 2V	± 3.6V	± 1.5V	± 0.247V
Receiver sensitivity	± 3V	± 200 mV	± 200 mV	± 200 mV	± 100 mV

MODULE 40 Codes for Information/Data Transmission

TABLE 40-1 Comparison of information/data codes

CODE	# OF BITS	PARITY BIT	APPLICATION
Morse	none	No	Telegraph
BCD	4/decimal digit	No	Storage of numeric information
Baudot	5	No	Teletype machines
EBCDIC	8	No	Data transmission
ASCII	7	Yes	Data transmission
UNICODE	16		A 16-bit code set developed as a universal set by Unicode consortium (http://www.unicode.org) to represent characters of any language used in the world, such as Arabic, Chinese, Greek, Hebrew, Hindi, Japanese, Urdu, etc.

TABLE 40-2 Binary coded decimal

DECIMAL NUMBER	BINARY EQUIVALENT	BCD CODE			
$10^n \ldots 10^4 10^3 10^2 10^1 10^0$	$2^n \ldots 2^{10} 2^9 2^8 2^7 2^6 2^5 2^4 2^3 2^2 2^1 2^0$	$2^3 2^2 2^1 2^0$	$2^3 2^2 2^1 2^0$	$2^3 2^2 2^1 2^0$	$2^3 2^2 2^1 2^0$
0	0				1010
1	1				0001
2	10				0010
3	11				0011
4	100				0100
5	101				0101
6	110				0110
7	111				0111
8	1000				1000
9	1001				1001
10	1010			0001	1010
11	1011			0001	0001
12	1100			0001	0010
13	1101			0001	0011
14	1110			0001	0100
15	1111			0001	0101
16	10000			0001	0110
20	10100			0010	1010
63	111111			0110	0011
64	1000000			0110	0100
255	11111111		0010	0101	0101
256	100000000		0010	0101	0110
511	111111111		0101	0001	0001
512	1000000000		0101	0001	0010
1023	1111111111	0001	0000	0010	0011
1024	10000000000	0001	0000	0010	0100

TABLE 40-3 Radio telephone Morse code

RADIO TELEPHONE MORSE CODE

A	• –
B	– • • •
C	– • – •
D	– • •
E	•
F	• • – •
G	– – •
H	• • • •
I	• •
J	• – – –
K	– • –
L	• – • •
M	– –
N	– •
O	– – –
P	• – – •
Q	– – • –
R	• – •
S	• • •
T	–
U	• • –
V	• • • –
W	• – –
X	– • • –
Y	– • – –
Z	– – • •
1	• – – – –
2	• • – – –
3	• • • – –
4	• • • • –
5	• • • • •
6	– • • • •
7	– – • • •
8	– – – • •
9	– – – – •
0	– – – – –
.	• – • – • –
,	– – • • – –
?	• • – – • •

• DOT
– DASH

TABLE 40-4 Baudot code

Character Case		$b_5\ b_4\ b_3\ b_2\ b_1$
Lower	Upper	
A	—	0 0 0 1 1
B	?	1 1 0 0 1
C	;	0 1 1 1 0
D	$	0 1 0 0 1
E	3	0 0 0 0 1
F	!	0 1 1 0 1
G	&	1 1 0 1 0
H	#	1 0 1 0 0
I	8	0 0 1 1 0
J	Bell	0 1 0 1 1
K	(	0 1 1 1 1
L	)	1 0 0 1 0
M	.	1 1 1 0 0
N	,	0 1 1 0 0
O	9	1 1 0 0 0
P	0	1 0 1 1 0
Q	1	1 0 1 1 1
R	4	0 1 0 1 0
S	'	0 0 1 0 1
T	5	1 0 0 0 0
U	7	0 0 1 1 1
V	;	1 1 1 1 0
W	2	1 0 0 1 1
X	/	1 1 1 0 1
Y	6	1 0 1 0 1
Z	"	1 0 0 0 1
Letters (Shift) ↓		1 1 1 1 1
Figures (Shift) ↑		1 1 0 1 1
Space (SP) =		0 0 1 0 0
Carriage Return <		0 1 0 0 0
Line Feed ≡		0 0 0 1 0
Blank		0 0 0 0 0

TABLE 40-5 EBCDIC

bits→Low $b_1b_2b_3b_4$ ↓ High $b_8b_7b_6b_5$		0000	0001	0010	0011	0100	0101	0110	0111	1000	1001	1010	1011	1100	1101	1110	1111
	1st HEX Digit→ 2nd HEX Digit↓	0	1	2	3	4	5	6	7	8	9	A	B	C	D	E	F
0000	0	NUL	SOH	STX	ETX	PF	HT	LC	DEL		RLF	SMM	VT	FF	CR	SO	SI
0001	1	DLE	DC1	DC2	DC3	RES	NL	BS	IL	CAN	EM	CC		ITS	IGS	IRS	IUS
0010	2	DS	SOS	FS		BYP	LF	EOB/ ETB	ESC/ PRE			3M			ENQ	ACK	BEL
0011	3			SYN		PN	RS	UC	EOT					DC4	NAK		SUB
0100	4	SP										¢	•	<	(	+	|
0101	5	&										!	$	*	)	;	¬
0110	6	-	/								/		,	%	—	>	?
0111	7											:	#	@	'	=	"
1000	8		a	b	c	d	e	f	g	h	i						
1001	9		j	k	l	m	n	o	p	q	r						
1010	A		~	s	t	u	v	w	x	y	z						
1011	B																
1100	C		A	B	C	D	E	F	G	H	I						
1101	D		J	K	L	M	N	O	P	Q	R						
1110	E			S	T	U	V	W	X	Y	Z						
1111	F	0	1	2	3	4	5	6	7	8	9						

ACK	Acknowledgement	BS	Backspace
BEL	Bell	BYP	Bypass

(continues)

TABLE 40-5 EBCDIC (continued)

CAN	Cancel	LC	Lower Case
CC	Unit backspace	LF	Line feed
CR	Carriage return	NAK	Negative acknowledgement
DC1	Device control 1	NL	New line
DC2	Device control 2	NUL	Null
DC3	Device control 3	PF	Punch off
DC4	Device control 4	PN	Punch on
DEL	Delete	PRE	Prefix
DLE	Data link escape	RES	Restore
DS	Digit select	RS	Record separaor
EM	End of medium	SI	Shift in
ENQ	Enquiry	S M	Start message
EOB	End of block	SMM	Repeat
EOT	End of transmission	SO	Shift out
ETX	End of text	SOH	Start of header
FF	Form feed	SOS	Start of significance
FS	File separator	SP	Space
HT	Horizontal tab	STX	Start of text
IFS	Interchange file separator	SUB	Substituute
IGS	Interchange group separator	SYN	Synchronization idle
IL	Idle	UC	Upper case
IRS	Interchange record separator	VT	Vertical tab
IUS	Interchange unit separator		

TABLE 40-6 ASCII

Low $b_3b_2b_1b_0$	HIGH $b_7{}^*b_6b_5b_4$ / 2nd Hex Digit → / 1st Hex Digit ↓	0000 / 0	0001 / 1	0010 / 2	0011 / 3	0100 / 4	0101 / 5	0110 / 6	0111 / 7
0 0 0 0	0	NUL	DLE	SP	0	@	P	'	p
0 0 0 1	1	SOH	DC1	!	1	A	Q	a	q
0 0 1 0	2	STX	DC2	"	2	B	R	b	r
0 0 1 1	3	ETX	DC3	#	3	C	S	c	s
0 1 0 0	4	EOT	DC4	$	4	D	T	d	t
0 1 0 1	5	ENQ	NAK	%	5	E	U	e	u
0 1 1 0	6	ACK	SYN	&	6	F	V	f	v
0 1 1 1	7	BEL	ETB	'	7	G	W	g	w
1 0 0 0	8	BS	CAN	(	8	H	X	h	x
1 0 0 1	9	HT	EM	)	9	I	Y	i	y
1 0 1 0	10	LF	SUB	*	:	J	Z	j	z
1 0 1 1	11	VT	ESC	+	;	K	[	k	{
1 1 0 0	12	FF	FS	,	<	L	\	l	|
1 1 0 1	13	CR	GS	-	=	M	]	m	}
1 1 1 0	14	SO	RS	.	>	N	^	n	~
1 1 1 1	15	SI	US	/	?	O	—	o	DEL

NUL	Null	DC3	Device control 3
SOH	Start of header	DC4	Device control 4
STX	Start of text	NAK	Negative acknowledgement
ETX	End of text	SYN	Synchronization idle
EOT	End of transmission	ETB	End of transmission block
ENQ	Enquiry	CAN	Cancel
ACK	Acknowledgement	EM	End of medium
BEL	Bell	SUB	Substitute
BS	Backspace	ESC	Escape
HT	Horizontal tab	FS	File separator
LF	Line feed	GS	Group separator
CR	Carriage return	RS	Record separator
SO	Shift out	US	Unit separator
SI	Shift in	DEL	Delete
DLE	Data link escape		
DC1	Device control 1		
DC2	Device control 2L		

* b_7 is always zero.

MODULE 41 International Television Systems and HDTV

INTERNATIONAL TELEVISION SYSTEMS AND HDTV

The television color standards used internationally are PAL, SECAM, and NTSC. The three systems are incompatible with each other. Table 41-1, Table 41-2, and Table 41-3 define these systems and list their users worldwide. Table 41-4 lists the HDTV formats, and Table 41-5 and Table 41-6 present the evolution of the HDTV systems and their characteristics.

TABLE 41-1 PAL

Phase alternate line is a color standard with 625 lines/frame, 25 frames/sec., 50 Hz field frequency, video BW of 5 to 6 MHz, 15.625 kHz line frequency, 7 or 8 MHz channel BW, negative video modulation, FM sound signal, and a 4.43 MHz color subcarrier. It reverses the phase of each 2nd line to correct color in the set.

USERS OF THE PAL SYSTEM

COUNTRY NAME	COUNTRY NAME	COUNTRY NAME
Azeros	Cameroon	Cyprus
Angola	Canary Islands	Czech Republic
Botswana	China (Peoples Republic)	East Timor
Bosnia/Herzegovina	Cooks Island	Easter Island
Cambodia (Kampuchea)	Croatia	Falkland Islands (Las Malvinas)

(continues)

TABLE 41-1 PAL (continued)

COUNTRY NAME	COUNTRY NAME	COUNTRY NAME
Germany	Luxembourg	Slovakia
Gibraltar	Macau	Slovenia
Greenland	Macedonia	Somalia
Grenada	Monaco	Sudan
Guinea	Namibia	Uruguay
Iceland	Palau	Vatican
Laos	Papua New Guinea	Yemen
Libya	Poland	Zambia
Liechtenstein	Romania	Zimbabwe
Lithuania	Saudi Arabia	

TABLE 41-2 SECAM

Sequential color and memory is a color standard with 625 lines/frame, 25 frames/sec., a 50 Hz field frequency, a 6 MHz video BW, 15.625 kHz line frequency, an 8 MHz channel BW, negative video modulation, FM sound signal, and a 4.43 MHz color subcarrier. SECAM sends hue in one line and saturation in the next. This makes the color almost error free.

USERS OF THE SECAM SYSTEM

COUNTRY NAME	COUNTRY NAME	COUNTRY NAME	COUNTRY NAME
Afars & Issas	Equatorial Guinea	Libya	Reunion
Albania	Estonia	Lithuania	Romania
Armenia	France	Liechtenstein	Russia
Azerbaijan	French Guiana	Luxembourg	Saudi Arabia
Benin	French Polynesia	Madagascar	Senegal
Bulgaria	Gabon	Mali	Syria
Burkina Faso	Georgia	Martinique	Tahiti
Burundi	Guadeloupe	Mauritania	Tajikistan
Central African Republic	Greece	Mauritius	Togo
	Guinea	Moldova	Tunisia
Chad	Haiti	(Moldavia)	Turkmenistan
Congo	Hungary	Monaco	Ukraine
Cyprus	Iran	Mongolia	Upper Volta
Czechoslovakia	Iraq	Morocco	Uzbekistan
Djibouti	Ivory Coast	New Caledonia	Vietnam
East Germany	Korea, North	Niger	Zaire
Egypt	Lebanon	Poland	

TABLE 41-3 NTSC

The National Television System Committee (NTSC) is a color standard with 525 lines/frame, 30 frames/sec., 60 Hz field frequency, 4.2 MHz video BW, 15.750 kHz line frequency, 6 MHz channel BW, negative video modulation, FM sound signal, and 3.58 MHz color subcarrier. It was the first color standard. It was introduced in 1953.

USERS OF THE NTSC SYSTEM

COUNTRY NAME	COUNTRY NAME	COUNTRY NAME	COUNTRY NAME
Antarctica	Cayman Island	Guyana	Peru
Antigua	Chile	Honduras	Philippines
Aruba	Colombia	Jamaica	Samoa
Azeros	Costa Rica	Japan	Surinam
Bahamas	Cuba	Korea, South	Taiwan
Barbados	Diego Garcia	Marshall Island	Trinidad
Belize	Dominican Republic	Mexico	Tobago
Bermuda	Ecuador	Micronesia	United States
Bolivia	El Salvador	Midway Island	Venezuela
British Virgin Islands	Fiji	Netherlands Antiles	Vietnam
Burma	Galapagos Islands	Nicaragua	West Indes, St. Kitts
Canada	Guatemala	Panama	

■ HIGH-DEFINITION TELEVISION (HDTV)

High-definition television system (HDTV), a form of digital television (DTV), offers high resolution audio and video, compared to the analog systems (NTCS, PAL, and SECAM), by using multiple transmission format, digital compression, packetization, and modulation techniques.

In 1993, to develop the U.S. HDTV standard, seven organizations (AT&T, GI, MIT, Philips, Sarnoff, Thompson, and Zenith) formed a Grand Alliance. In 1995, the Grand Alliance proposed an HDTV standard that would enable television viewers at home to receive an HDTV picture composed of more than two million pixels (picture elements), with a high-quality spatial resolution (six times better than the NTSC image), and a panoramic horizontal-to-vertical aspect ratio of 16:9 (a big improvement over the traditional television picture aspect ratio of 4:3).

The Grand Alliance HDTV standard defined five system functions: format selection, video coding, audio coding, transport, and transmission. Digital representation and processing in an HDTV system supports more than one scanning format. The formats facilitate interoperability among different formats, and can be used for different video services and applications. Table 41-4 presents a summary of HDTV formats and their potential applications.

TABLE 41-4 HDTV formats

HDTV FORMAT	APPLICATION
720 × 1280 (progressive scanning) at 60 frames per second	Sports, graphics, commercials, etc.
720 × 1280 (progressive scanning) at 24 or 30 frames per second	Graphics, animation, complex film scenes, etc.
1080 × 1920 (interlaced scanning) at 30 frames per second	Images/video shot with interlaced-scan-camera
1080 × 1920 (progressive scanning) at 24 or 30 frames per second	Films/video who highest spatial resolution

Source: The U.S. HDTV standard The Grand, special report: digital TV, IEEE spectrum, April 1995, p. 40.

In 1996, the federal communications commission (FCC) adopted the advanced television systems committee (ATSC) digital television standard, which is based on MPEG-2 compression techniques proposed by the Grand Alliance. In 1997, the FCC allocated digital spectrum to broadcasters, and required them to broadcat digital television signals by 2006.

TABLE 41-5 Advanced television systems (high-definition television—HDTV)

HDTV SYSTEMS AND THEIR MAKERS	
SYSTEM NAME	PROPOSERS OF THE SYSTEM
# NHK Studio	NHK Japan
* ACTV	NBC, RCA, Samoff
* Bandwidth-efficient	MIT
* Fukinuki	Takahiko Fukinuki, Hitachi
* HD-NTSC	R.J. Iredale, Del Rey Gp.
* Receiver-compatible	MIT
* SuperNTSC	Faroudja Labs
* Yasumoto	Yoshio Yasumoto, Matsushita
+ AT&T Bell Labs	Theodore S. Rzeszewski
+ Glenn	N.Y. Institute of Technology
+ HdNTSCSignal	North American Phillips
- HDB-MAC	Scientific Atlanta
- HDMAC-60 Signal	North American Phillips
- HI-Vision w/ Muse	NHK, Japan
(#) Studio Systems (*) Single-Channel System (+) Dual-Channel Systems (-) Wide-Bandwidth Systems	

Source: (From Jurgen, Ronald K. "High Definition Television Update." IEEE Spectrum (April 1988). © IEEE. Reprinted with permission.)

TABLE 41-6 HDTV

CHARACTERISTICS OF HDTV SYSTEMS				
NAME	BW MHZ	LINES/FRAME FRAMES/SEC FIELDS/SEC	VER/HOR RESOLUTION LINES	ASPECT RATIO
# NHK Studio	50	1125/30/60	750/750	16:9
* ACTV	6	1050/30/60	480/410	5:3
* BW-Effic.	6	1200/60/60	720/1275	4:3
* Fukinuki	6	1050/60/60	400/425	4:3
* HD-NTSC	6	1125/60/60	450/450*	5:3
* Rec.-Comp.	6	1050/60/60	600/660	16:9
* SuperNTSC	6	1050/30/60	330/400	4:3
* Yasumoto	6	525/30/60	330/420	4:3
+ AT&T Bell	6+6	1050/30/60	480/600	5:3
+ Glenn	6+3	1125/30/60	800/800	16:9
+ HdNTSC SIG.	6+6	1050/60/60	480/495	16:9
- HDB-Mac	10.7	1050/30/60	450/450	4:3
- HDMAC-60	9.5	1050/60/60	480/495	16:9
- HI-Vision	10	1125/30/60	750/580	16:9

(#) Studio System
(*) Single-Channel Systems
(+) Dual-Channel Systems
(-) Wide-Bandwidth Systems
*Varies Between 450 and 700

Source: (From Jurgen, Ronald K. "High Definition Television Update." IEEE Spectrum (April 1988). © IEEE. Reprinted with permission.

MODULE 42

International Telephone Dialing Codes

An International call can be made using the following format:

International access code + country code + city code + local number access code

Table 42-1 lists the country and city codes.

TABLE 42-1 Country and city codes

COUNTRY/CITIES (INTERNATIONAL ACCESS CODE)	COUNTRY CODE	CITY CODE
Afghanistan (00)	93	
Albania (00)	355	
Algeria (00)	213	
American Samoa (00)	684	
Andorra (00)	376	
All points		
Angola	244	
Anguila (011)	1-264	
Antigua (011)		
(including Barbuda)	1-268	
Argentina (00)	54	
Bahia Blanca		91
Buenos Aires		1
Cordoba		51
Corrientes		783
La Plata		21
Mar Del Plata		23
Mendoza		61
Merlo		220
Posadas		752
Resistencia		722
Rio Cuatro		586
Rosario		41
San Juan		64
San Rafael		627
Santa Fe		42
Tandil		293
Armenia (00)	374	
Aruba (00)	297	
All points		8
Ascension Island (00)	247	
Australia (0011)	61	
Adelaide		8
Ballarat		53
Brisbane		7
Canberra		62
Darwin		89
Geelong		52
Gold Coast		75
Hobart		02
Launceston		03
Melbourne		3
Newcastle		49
Perth		9
Sydney		2
Toowoomba		76
Townsville		77
Wollongong		42
Austria (00)	43	
Bludenz		5552
Graz		316
Innsbruck		5222
Kitzbuhel		5356
Klagenfut		4222

(continues)

TABLE 42-1 Country and city codes (continued)

COUNTRY/CITIES (INTERNATIONAL ACCESS CODE)	COUNTRY CODE	CITY CODE
Krems An Der Donau		2732
Linz Donau		732
Neunkirchen Niederosterreich		2635
Salzburg		662
St. Polten		2742
Vienna		1 or 222
Villach		4242
Wels		7242
Wiener Neustadt		2622
Azerbaijan (00)	994	
Bahamas (011)	1-242	
Bahrain (00)	973	
Bangladesh (00)	880	
Barisal		431
Bogra		51
Chittagong		31
Comilla		81
Dhaka		2
Khulna		41
Maulabi Bazar		861
Mymensingh		91
Rajshahi		721
Sylhet		821
Barbados (011)	1-246	
Barbuda (011)	1-268	
Belarus (8~10)	375	
Belgium (00)	32	
Antwerp		3
Bruges		50
Brussels		2
Charleroi		71
Courtral		56
Ghent		91
Hasselt		11
La Louviere		64
Leuvenj		16
Libramont		61
Liege		41
Malines		15
Mons		65
Namur		81
Ostend		59
Verviers		87
Belize (00)	501	

COUNTRY/CITIES (INTERNATIONAL ACCESS CODE)	COUNTRY CODE	CITY CODE
Belize City		2
Belmopan		8
Benque Viego Del Carmen		93
Corozal Town		4
Dangviga		5
Independence		6
Orange Walk		3
Punta Gorda		7
San Ignacio		92
Benin (00)	229	
Bermuda (011)	1-441	
Bhutan (00)	975	
Bolivia (0010 (ENTEL) 0011 (AES COMMUNICATIONS BOLIVIA) 012 (TELEDATA) 0013 (BOLIVIATEL)	591	
Cochabamba		42
Cotoga		388
Guayafamerin		47
La Belgica		923
La Paz		2
Mineros		984
Montero		92
Oruro		52
Portachuelo		924
Saavedra		924
Santa Cruz		33
Trinidad		46
Warnes		923
Bosnia & Herzegovina (00)	387	
Botswana(00)	267	
Francistown		21
Gaborone		31
Jwaneng		38
Kanye		34
Lobatse		33
Mahalapye		41
Mochudi		37
Molepolole		32
Orapa		27
Palapye		42
Serowe		43

(continues)

TABLE 42-1 Country and city codes (continued)

COUNTRY/CITIES (INTERNATIONAL ACCESS CODE)	COUNTRY CODE	CITY CODE	COUNTRY/CITIES (INTERNATIONAL ACCESS CODE)	COUNTRY CODE	CITY CODE
Brazil (0014) (Brasil Telecom)	55		Chad (15)	235	
0015 (Telefonica)			Chalham Island (New Zealand (00)	64	
0021 (Embratel)			Chile (00)	56	
0023 (Intelig)			Chiquayante		41
0031 (Telemar)			Concepcion		41
Belem		91	Penco		41
Belo Horizonte		31	Recreo		31
Brasilia		61	San Bernardo		2
Curitiba		41	Santiago		2
Fortaleza		85	Talcahuano		41
Goiania		62	Valparaiso		32
Niteroi		21	Vina del Mar		32
Pelotas		532	China (PRC) (00)	86	
Porto Alegre		512	Beijing		1
Recife		81	Fuzhou		591
Rio De Janeiro		21	Ghuangzhou		20
Salvador		71	Shanghai		21
Santo Andre		11	Colombia		
Santos		132	005 Orbital	57	
Sao Paulo		11	007 ETB		
Vitoria		27	009 Telecom		
British Virgin			# 555 Bellsouth		
Islands (011)	1-284		# 999 Comcel		
Brunei	673		Armenia		60
Bandar Seri			Barranquilla		58
Begawan		2	Bogota		1
Kuala Belait		3	Bucaramanga		73
Mumong		3	Cali		23
Tutong		4	Cartagena		59
Bulgaria (00)	359		Cartago		66
Kardjali		361	Cucuta		70
Pazardjik		34	Giradot		832
Plovdiv		32	Ibague		82
Sofia		2	Manizales		69
Varna		52	Medellin		4
Burkina Faso (00)	226		Neiva		80
Burundi (00)	257		Palmira		31
Cambodia (001)	855		Pereira		61
Cameroon (00)	237		Santa Marta		56
Canada (similar to			Comoros (00)	269	
U.S.) [011]	1		Congo (00)	242	
Cape Verde Islands (0)	238		Congo (Zaire) (00)	243	
Cayman Islands (011)	1-345		Cook Island (00)	682	
Central African			Costa Rica (00)	506	
Republic (00)	236		Croatia (00)	385	

(continues)

TABLE 42-1 Country and city codes (continued)

COUNTRY/CITIES (INTERNATIONAL ACCESS CODE)	COUNTRY CODE	CITY CODE
Cuba (53)	119	
Cuba (Guantanamo Bay) 00	5399	
Curacao (00)	599	
Cyprus (00)	357	
Famagusta		536
Kyrenia		581
Kythrea		2313
Lapithos		8218
Larnaca		41
Lefka		57817
Lefkonico		3313
Limassol		51
Moni		5615
Morphou		71
Nicosia		2
Paphos		61
Platres		54
Polis		63
Rizokarpaso		3613
Yialousa		3513
Czech republic (00)	420	
Denmark (00)	45	
Copenhagen		1
(outlying suburbs)		2
Haderslev		4
Sonderborg		4
Diego Garcia (00)	246	
Djibouti (00)	253	
Dominica (011)	1-767	
Dominican Republic (011)	1-809	
East Timor (00)	670	
Easter Island (00)	56	
Ecuador (00)	593	
Ambato		2
Esmeraldas		2
Ibarra		2
Manta		4
Salinas		4
Santo Domingo		2
Tulcan		2
Egypt (00)	20	
Alexandria		3
Cairo		2
El Mahallah		43
Port Said		66
Shebin El Kom		48
El Salvador (00 144 + 00 Telefonica)	503	
Equatorial Guinea (00)	240	
Eritrea (00)	291	
Ethiopia (00)	251	
Addis Ababa		1
Asmara		4
Awassa		6
Harrar		5
Makale		4
Nazareth		2
Falkland Islands (Malvinas) (00)	500	
Fiji Islands (00)	679	
Finland (00)	358	
Epoo-Ebbo		15
Helsinki		0
Lahti		18
Vaasa		61
Vanda-Vantaa		0
France (00 France Telecom)	33	
40 (TELE 2)		
50 (OMNICOM)		
70 (LE 7 CEGETEL)		
90 (9 TELECOM)		
Bordeaux		56
Cannes		93
Chauvigny		49
Lourdes		62
Marseille		91
Paris		1
Toulouse		61
French Antilles (00)	596	
French Guiana (00)	594	
French Polynesia (00)	689	
Gabon Republic (00)	241	
Gambia (00)	220	
Georgia (8~10)	995	
Germany (00)	49	
Berlin		30
Bonn		228
Bremen		421
Cologne		221

(continues)

TABLE 42-1 Country and city codes (continued)

COUNTRY/CITIES (INTERNATIONAL ACCESS CODE)	COUNTRY CODE	CITY CODE
Dresden		51
Dusseldorf		211
Essen		201
Frankfurt		69
Hamburg		40
Heidelberg		6221
Karl-Marx-Stadt		71
Leipzig		41
Potsdam		33
Mannheim		621
Munich		89
Nurnberg		911
Stuttgart		711
Ghana (00)	233	
Gibraltar (00)	350	
Greece (00)	30	
Argos		751
Athens		1
Corinth		741
Thesaloniki		31
Tripolis		71
Greenland (00)	299	
Godthaab		2
Soendre Stroemfjord		11
Thule		50
Grenada (011)	1-473	
Guadeloupe (00)	590	
Guam (011)	671	
Guantanamo Bay (00)	5399	
Guatemala (00 130 + 00 Tefonica 147-00 Telgua)	502	
Guatemala City		2
All other cities		9
Guinea-Bissau (00)	245	
Guinea (00)	224	
Guyana (001)	592	
Georgetown		2
Linden		4
New Amsterdam		3
Rosignol		30
Whim		37
Haiti (00)	509	
Cap-Haitien		3
Cayes		5
Port-au-Prince		1
Honduras (00)	504	
Hong Kong (001-PCCW 0080-Hutchison 009-New World)	852	
Castle Peak		0
Fan Ling		0
Hong Kong		5
Hungary (00)	36	
Abasar		37
Budapest		1
Gyor		96
Szolnok		56
Iceland (00)	354	
Husavik		6
Rein		6
Reykjavik		1
Selfoss		9
Varma		1
India (00)	91	
Bhopal		755
Bombay		22
Calcutta		33
Madras		44
New Delhi		11
Surat		261
Indonesia (001, 008)	62	
Jakarta		21
Surabaya		31
Iran (00)	98	
Abadan		631
Arak		2621
Shiraz		71
Tehran		21
Iraq (00)	964	
Ireland (00, 048 for Northern Ireland)	353	
Cork		21
Donegal		73
Dublin		1
Kildare		45

(continues)

TABLE 42-1 Country and city codes (continued)

COUNTRY/CITIES (INTERNATIONAL ACCESS CODE)	COUNTRY CODE	CITY CODE	COUNTRY/CITIES (INTERNATIONAL ACCESS CODE)	COUNTRY CODE	CITY CODE
Limerick		6 or 61	Kisumu		35
			Nairobi		2
Tipperary		62	Thika		151
Waterford		51	Kiribati (00)	686	
Israel (00 012			Korea, North (00)	850	
Golden Lines	972		Korea, South		
013 Barak LTD			(001, 002, 00700)	82	
014 Bezev LTD)			Inchon		32
Beer Sheva		57	Osan		399
Jerusalem		2	Pusan		51
Nazareth		65	Seoul		2
Tel Aviv		3	Wonju		371
Tiberias		67	Kuwait (00)	965	
Italy (00)	39		Kyrgyz Republic (00)	996	
Bologna		51	Laos (00)	856	
Capri		81	Latvia (00)	371	
Florence		55	Lebanon (00)	961	
Genoa		10	Lesotho (00)	266	
Milan		2	Liberia (00)	231	
Naples		81	Libya (00)	218	
Pisa		50	Benghazi		61
Rome		6	Tripoli		21
Torino		11	Tripoli Int'l Airport		22
Venice		41	Liechtenstein (00)	423	
Verona		45	Lithuania (00)	370	
Ivory Coast (00)	225		Luxembourg (00)	352	
Jamaica (011)	1-876		Macao (00)	853	
Japan (001 KDD	81		Madagascar (00)	261	
010 Myline			Macedonia (00)	389	
0001 Cable &			Malawi (00)	265	
wireless ID			Domasi		531
041 Japan Telecom)			Thornwood		486
Hiroshima		82	Malaysia (00)	60	
Kawasaki		44	Alor Star		4
Nagasaki		958	Kuala Lumpur		3
Osaka		6	Maran		95
Tokyo		3	Maldives, Rep. of (00)	960	
Yokohama		45	Mali (00)	223	
Jordan (00)	962		Malta (00)	356	
Amman		6	Mauritius (00)	230	
Sult		5	Marshall Islands (011)	692	
Kazakhstan (8~10)	7		Mexico (00)	52	
Kenya (000)	254		Acapulco		748
Anmer		154	Cancun		988
Kabete		2	Chihuahua		14

(continues)

TABLE 42-1 Country and city codes (continued)

COUNTRY/CITIES (INTERNATIONAL ACCESS CODE)	COUNTRY CODE	CITY CODE
Guadalajara		36
Mexico City		5
Puebla		22
Puerta Vallarta		322
Tijuana		66
Veracruz		29
Micronesia (011)	691	
Monaco (00)	377	
All points		93
Montserrat (011)	1-664	
Moldova (00)	373	
Mongolia (001)	976	
Morocco, Kingdom of	212	
El Jadida (00)		34
Marrakech		4
Tanger		9
Mozambique (00)	258	
Myanmar (00)	95	
Nambia (00)	264	
Industria		61
Olympia		61
Windhoek		61
Windhoek Airport		626
Nepal (00)	977	
Netherlands (00)	31	
Amsterdam		20
Heemstede		23
Rotterdam		10
The Hague		70
Netherlands Antilles (011)	599	
Curacao		9
St. Maarten		5
Nevis	809	
New Caledonia (00)	687	
New Zealand (00)	64	
Auckland		9
Christchurch		3
Napier		70
Wellington		4
Nicaragua (00)	505	
Granda		55
Managua		2
San Marcos		43
Niger Republic (00)	227	
Nigeria (00)	234	
Norway (00)	47	
Drammen		3
Larvik		34
Oslo		2
Svalbard		80
Oman (00)	968	
Pakistan (00)	92	
Bahawalpur		621
Islamabad		51
Karachi		92
Lahore		42
Okara		442
Peshawar		521
Quetta		81
Sukkur		71
Palau (011)	680	
Palestine Settlement (00)	974	
Panama	507	
Cable & wireless 080 + 00 Tele carnia 053 + 00 Clarocom)		
Papua New Guinea (05)	675	
Paraguay (002)	595	
Asuncion		21
Concepcion		31
Hernandarias		63
San Antonia		27
Villeta		25
Peru (00)	51	
Callao		14
Ica		34
Lima		14
Philippines (00)	63	
Angeles		455
Clarkfield		4535
Manila		2
San Pablo		93
Tarlac City		452
Poland (0**0)	48	
Crakow		12
Gdansk		58
Lublin		81
Warsaw		22
Portugal (00)	351	

(continues)

TABLE 42-1 Country and city codes (continued)

COUNTRY/CITIES (INTERNATIONAL ACCESS CODE)	COUNTRY CODE	CITY CODE
Almada		1
Beja		84
Estoril		1
Lisbon		1
Santa Cruz		92
Velas		95
Qatar (00)	974	
Rawanda	250	
Romania (00)	40	
Arad		66
Bucharest		0
Sibiu		250
Russia (8~10)	7	
Rwandese Republic (00)	250	
St. Kitts (011)	1-869	24
St. Lucia (011)	1-758	
St. Pierre & Miquelon (00)	508	
St. Vincent & Grenadines (01)	1-784	
Saipan	670	
San Marino (00)	378	
All points		549
Saudi Arabia (00)	966	
Al Khobar		3
Damman		3
Jeddah		2
Makkah (Mecca)		2
Medina		4
Qatif		3
Senegal (00)	221	
Serbia and Montenego (99)	381	
Seychelles Republic (99)	381	
Sierra Leone (00)	233	
Singapore	65	
Singtel		001
MobileOne		002
Starhub IDP		008
Fax over IP		012
Singtel BudgetCall		013
Starhub I-Call		018
Singtel V019		019
South Africa (09)		
Cape Town		21
East London		431
Johannesburg		11
Port Elizabeth		41
Pretoria		12
Slovak Republic (00)	421	
Slovenia (00)	386	
Soloman Islands (00)	677	
Spain (00)	34	
Barcelona		3
Cadiz		56
Granada		58
Leon		87
Madrid		1
Seville		54
Valencia		6
Sri Lanka (00)	94	
Colombo (Central)		1
Galle		9
Kandy		8
Panadura		46
Sudan (00)	249	
Suriname (00)	597	
Swaziland (00)	268	
Sweden (00)	46	
Boras		33
Lund		46
Stockholm		8
Vasteras		21
Syria (00)	963	
Switzerland (00)	41	
Baden		56
Berne		31
Geneva		22
Interlaken		36
Lucerne		41
St. Gallen		71
Zurich		1
Taiwan (002)	886	
Chunan		36
Taipei		2
Tajikistan (8~10)	992	
Tanzania (000)	255	
Dar Es Salam		51

(continues)

TABLE 42-1 Country and city codes (continued)

COUNTRY/CITIES (INTERNATIONAL ACCESS CODE)	COUNTRY CODE	CITY CODE
Thailand (001)	66	
Bangkok		2
Lampang		54
Saraburi		36
Togo (00)	228	
Tonga Islands (00)	676	
Trinidad and Tobaco (011)	1-868	
Tunisia (00)	216	
Carthage		1
Khenis		3
Tunis		1
Turkey (00)	90	
Ankara		4
Istanbul		1
Izmir		51
Samsun		361
Turkmenistan (8~10)	993	
Turks and Caicos Islands (011)	1-649	
Tuvalu (00)	688	
Uganda (000)	256	
Entebbe		42
Kampala		41
Ukraine (8~10)	380	
United Arab Emirates (00)	971	
Abu Dhabi		2
Ajman		6
Al Ain		3
Dubai		4
Fujairah		70
Ras-al-Khaimah		77
Sharjah		6
Umm-al-Quwain		6
United Kingdom (00)	44	
Edinburgh		31
Glasgow		41
Gloucester		452
Liverpool		51
London		1
Manchester		61

COUNTRY/CITIES (INTERNATIONAL ACCESS CODE	COUNTRY CODE	CITY CODE
Nottingham		602
Southhampton		703
United States of America (011)	1	
US Virgin Islands (011)	1-340	
Uruguay (00)	598	
Atlantida		372
Florida		352
Montevideo		2
San Jose		342
San Jose De Carrasco		382
Uzbekistan (8~10)	998	
Vanuatu (00)	678	
Vatican City (00)	39, 379	
All Points		6
Venezuela (00)	58	
Barcelona		81
Caracas		2
San Cristobal		76
Valencia		41
Vietnam (00)	84	
Wake Island (00)	808	
Wallis and Futuna Islands (19)	681	
Western Samoa (00)	685	
Yemen Arab (00) Republic	967	
Amran		2
Dhamar		2
Sanaa		2
Taiz		4
Zabid		3
Zambia (00)	260	
Chingola		2
Lusaka		1
Ndola		2
Zanzibar (000)	255	
Zimbabwe (00)	263	
Harare		4
Mutare		20

For domestic area codes visit:
http://www.nanpa.com
http://www.bennetyee.org/ucsd-pages/area./htm

MODULE 43 Telecommunication Agencies/ Organizations

ACRONYM	ORGANIZATION/ AGENCY	WEBSITE	FUNCTION OF ORGANIZATION
ABNT	Associacao Brasileira de Normos Técnicas	http://www.abnt .org.br	National body for standards development in Brazil.
AFNOR	Association francaise de normalisation	http://www.afnor.fr	National clearinghouse for standards in France.
ANSI	American National Standards Institute	www.ansi.org	National clearinghouse for standards implemented in the United States.
	Commerce, Science & Transportation Committee	http://commerce .senate.gov/	Determines government policy for interstate and foreign communications. Has jurisdiction over the shuttle program and government satellite systems.
	Communications Subcommittee	http://commerce .senate/gov/ subcommittees	Determines information, communications, and telecommunications policy legislation, oversight, and funding authority over NTIA and the FCC.
ATSC	Advanced Television Systems Committee	http://www.atsc.org	Establishes technical standards for the high-definition television (HDTV)
BSI	British Standards Institution	htt;://www.bsi. org.uk	National body for standards development in United Kingdom.

(continues)

ACRONYM	ORGANIZATION/ AGENCY	WEBSITE	FUNCTION OF ORGANIZATION
CEN	Comité Européen de Normalisation	http://www.cenom.be/cenom/index.htm	One of three regional organizations for developing standards in the European Union (EU).
DIN	Deutsches Institut für Normung	http://www.din.de	A non-government agency involved with standards development in Germany.
EIA	Electronics Industries Alliance	http://www.eia.org/	A trade alliance that standardizes sizes, specifications, and terminology for electronic products. Developed the RS-232 interface specification.
ETSI	European Telecommunications Standards Institute	http://www.etsi.org	European Union's (EU) organization for establishing and coordinating standards for telecommunications, broadcasting, and information technology.
Eutelsat	European Telecommunications Satellite Organization	http://www.eutelsat.org/	Administrative agencies that design, develop, construct, operate, and maintain the space segment of the European telecommunication satellite system. Provides telephone, TV, business communications, and transponder lease services.
FCC	Federal Communications Commission	http://www.fcc.gov/	The FCC is responsible for regulating all interstate and foreign communications originating in the United States by means of radio, television, wire, cable, and satellite.
ICCC	International Council for Computer Communications	http://www.icccgovernors.org/	A council that promotes scientific research, development, and applications of computer communication.

(continues)

ACRONYM	ORGANIZATION/ AGENCY	WEBSITE	FUNCTION OF ORGANIZATION
IEEE	Institute of Electrical and Electronics Engineers	http://www.ieee.org/	A publishing and standards-making body responsible for many standards used in LANs, including the 802 series.
	International Economic Policy and Trade Subcommittee	http://www.iie.com/	Oversees telecommunications trade and export matters.
Intelsat	International Telecommunications Satellite Organization	http://www.intelsat.com	Responsible for the design, development, construction, and operation of the space segment of the global communications satellite system. *See* Module 46.
ISO	International Organization for Standardization	http://www.iso.org/	A body consisting of national standardization organizations from each member country responsible for the OSI model, encryption conventions, and other standards. *See also* Module 35.
ITU	International Telecommunication Union	http://www.itu.int/home/	The telecommunications agency of the United Nations established to provide standardized telecommunications on a worldwide basis. *See* Module 38.
ITU-TS (CCITT)	International Telecommunication Union – Telecommunication Sector		An international advisory sector of the ITU that recommends international standards for data communications. *See also* Module 38.
NCS	National Communications System	http://www.ncs.gov/	The arm of the General Services Administration that is responsible for U.S. government standards. Federal telecommunication standards are developed jointly with the National Bureau of Standards.

(continues)

ACRONYM	ORGANIZATION/ AGENCY	WEBSITE	FUNCTION OF ORGANIZATION
NTIA	National Telecommunications and Information Administration	http://www.ntia.doc.gov/	Part of the U.S. Department of Commerce. It helps develop federal telecommunications policy.
SAA	Standards Australia	http://standards.com.au	National body for standards development in Australia
SCC	Standards Council of Canada	http://www.scc.ca	National clearinghouse for standards in Canada.
SIRIM	Standards and Industrial Institute of Malaysia	http://sirim.my	National body for standards development in Malaysia.
TIA	Telecommunications Industries Alliance	http://www.tiaonline.org/	TIA is accredited by the American National Standards Institute (ANSI) to develop standards for telecommunications products.

MODULE 44 Telecommunications Journals and Magazines

NAME	WEBSITE/PUBLISHER
America's Network	http://www.americasnetwork.com
AES-Journal of the Audio Engineering Society	http://www.aes.org
Artificial Intelligence: an International Journal	http://www.elsevier.nl
Applied Optics	http://www.osa.org
AT&T Technology	http://www.att.com
AV Video	http://www.avvideo.com
Bell Labs Technical Journal	http://www.lucent.com/minds/techjournal/
Broadband Guide	http://www.broadband-guide.com
Business Communications Review	http://www.bcr.com
Byte Digest	http://www.byte.com
Cable Installation & Maintenance Magazine	http://www.cable-install.com
Cabling Business Magazine	http://www.cablingbusiness.com
Communications Engineer & Design	http://www.cedmagazine.com
Communications News	http://www.comnews.com
Communications Standards Review	http://www.csrstds.com
Connect Magazine	http://www.connectweb.co.uk
CQ Amateur Radio	http://www.cq-amateur-radio.com
Data Communications	http://www.data.com
Educause Review	http://www.educause.edu/er/

(continues)

NAME	WEBSITE/PUBLISHER
Electronic Design	http://www.elecdesign.com
Fiber and Integrated Optics	http://www.taylorandfrancis.com
Fiber Optics Online Community Newsletter	http://www.fiberopticsonline.com
Fiber Optics and Communications	http://www.igigroup.com/nl/pages/focomm.html
Fiber Optics News	http://telecomweb.com/reports/fon20/index.html
Fiberoptic Product News	http://www.fpnmag.com
Ham Radio	http://www.dxzone.com
IEEE Communications Magazine	http://www.comsoc.org/pubs/commag
IEEE Journal on Selected Areas in Communications	http://www.comsoc.org/livepubs/sac/index.html
IEEE/OSA Journal of Lightwave Technology	http://jlt.osa.org/journal/jlt/about.cfm
IEEE Transactions on Mobile Computing	http://www.comsoc.org/pubs/journals.html
IEEE Journal of Quantum Electronics	http://ieeexplore.ieee.org
IEEE Photonics Letters	http://ieeexplore.ieee.org
IEEE Lasers and Electro-Optics Society (LEOS) Newsletter	http://www.i-leos.org/
IEEE Journal of Selected Topics in Quantum Electronics	http://ieeexplore.ieee.org
IEEE Network	http://www.comsoc.org/pubs/net/
IEEE Spectrum	http://www.spectrum.ieee.org
IEEE Transactions on Acoustics, Speech, and Signal	http://www.ieee.org/xpl
IEEE Transactions on Communications	http://www.comsoc.org/pubs/jrnal/transcom.html

(continues)

NAME	WEBSITE/PUBLISHER
IEEE Transactions on Wireless Communications	http://www.comsoc.org/pubs/jrnal/twc.html
IEEE Internet Computing	http://www.computer.org/internet/
IEEE/ACM Transactions on Network-online	http://www.comsoc.org/pubs/jrnal/trasnet.html
Internet Research	http://www.emeraldinsight.com
International Insider	http://www.insider-online.com http://www.solvox.net/
International Journal of Electronics	http://www.tandf.co.uk/journals/titles/00207217.asp
International Journal of Satellite Communication	http://www3.interscience.wiley.com/cgi-bin/jhome/3524
International Telecommunication Union Operational Bulletin	http://www.itu.int/itudoc/itu-t/ob-lists/op-bull/index.html
Internet Telephone	http://www.tmcnet.com/voip
Information Week	http://www.cmp.com/pubinfo/?pubID=1
IPTC News	http://iptc.org
Laser Focus World	http://www.optoelectronics-world.com
Lightwave	http://www.light-wave.com
Microwave News	http://www.microwavenews.com
Network World	http://www.nwfusion.com
Optical Engineering	http://www.spie.org
Optics and Lasers in Engineering	http://www.elsevier.com/wps/find/journaldescription.cws_home/405906/description
Optics and Photonics News	http://www.osa.org
Outside Plant Magazine	http://www.ospmag.com
Pay Phone News	http://www.payphone-directory.org/news.html
PC Magazine	http://www.pcmagazine.org
Photonics Spectra	http://www.photonics.com

(continues)

NAME	WEBSITE/PUBLISHER
Popular Communications	http://www.cq-amateur-radio.com
Progress in Quantum Electronics: An International Review Journal	http://authors.elsevier.com/JournalDetail.html?PubID=410&Precis=DESC
QST: devoted entirely to amateur radio	http://www.arrl.org
QST	http://www.dxzone.com
Radio Electronics	http://www.radio-electronics.com
Satellite News	http://www.satnews.com
Telecom Digest	http://mirror.lcs.mit.edu
TeleGeography, Inc.	http://www.telegeography.com
Telecom Direct	http://www.telecomdirect.pwcglobal.com
Telecommunications Online	http://www.telecommagazine.com
Telecommunications Reports	http://www.tr.com
Telecom Update	http://www.angustel.ca
Telecom Web	http://www.telecomweb.com
Telephony	http://www.telephonyonline.com
Telephony World	http://www.telephonyworld.com
TeleSpan	http://www.telespan.com
VHF Communications	http://www.vhfcomm.co.uk
Video Business	http://www.videobusiness.com
Video Systems	http://www.videosystems.com
Voice+ Magazine	http://www.callvoice.com
WAP Insight	http://www.wapinsight.com
Washington Update	http://www.educause.edu
Wire & Cable Technology International	http://www.wiretech.com
Wired	http://www.wired.com
Year Book of Common Carrier Telecommunication Statistics	http://www.itu.int/ITU-D/ict/Pulications/yb

MODULE

45 FCC Frequency Allocations

TABLE 45-1 Frequency prefixes

EXPONENT	PREFIX	ABBREVIATION
10^3	*kilo*hertz	kHz
10^6	*mega*hertz	MHz
10^9	*giga*hertz	GHz
10^{12}	*tera*hertz	THz
10^{15}	*penta*hertz	PHz
10^{18}	*exa*hertz	EHz

TABLE 45-2 Electromagnetic spectrum: Frequency bands

FREQUENCY RANGE	WAVELENGTH RANGE	ABBREVIATION	CLASSIFICATION
30Hz–300Hz	10–1 Mm	ELF	Extremely low frequency
300Hz–3kHz	1–0.1 Mm	VF	Voice frequency
3kHz–30kHz	100–10 km	VLF	Very low frequency
30kHz–300kHz	10–1 km	LF	Low frequency
300kHz–3MHz	1–0.1 km	MF	Medium frequency
3MHz–30MHz	100–10 m	HF	High frequency
30MHz–300MHz	10–1 m	VHF	Very High frequency
0.3GHz–3GHz	100–10 cm	UHF	Ultrahigh frequency
3GHz–30GHz	10–1 cm	SHF	Super high frequency
30GHz–300GHz	10–1 mm	EHF	Extremely high frequency
0.3THz–4.29THz	100–0.7 μm	IR	Infrared
4.29THz–6.98THz	0.7–0.43 μm		Visible light
6.98THz–100THz	0.43–0.003 μm	UV	Ultraviolet
100PHz–1000EHz	300–0.003 Å		X rays

*TABLE 45-3 Medium frequency**

FREQUENCY (kHz)	ALLOCATIONS
300	Marine
400	Aviation
500	AM Radio Broadcast
600	(535 kHz – 1605 kHz)
700	Marine, Aviation, and
800	Land Mobile
900	Amateur Land Mobile
1000	
1100	
1200	
1300	
1400	
1500	
1600	
2000	
3000	

*For a detailed description of FCC frequency allocations visit http://www.fcc.gov

*TABLE 45-4 High frequency**

FREQUENCY (MHZ)	ALLOCATIONS
3	Marine
4	Marine, Aeronautical Mobile (3.4 MHz–3.5 MHz), Amateur (3.5 MHz–4.0 MHz)
5	Land Mobile
6	Marine Aviation
7	Amateur Aviation
8	Land Mobile
9	Marine
10	Marine Aviation Shortwave
13.36–13.41	Radio Astronomy
14.0–14.25	Amateur, Amateur-Satellite
16	Aviation Shortwave
18.06–18.16	Amateur, Amateur-Satellite
20	Aviation Shortwave
28–29.7	Amateur, Amateur-Satellite
30	Amateur Land Mobile, Shortwave Broadcast Radio (5.95 MHz–26.1MHz)

*For a detailed description of FCC frequency allocations visit http://www.fcc.gov

*TABLE 45-5 Very high frequency**

FREQUENCY (MHz)	ALLOCATIONS
30	Fiexd Land Mobile
	Government (33–38)
38–38.25	Radio Astronomy
40	Government
50	Government
60	TV Channels 2–4 Broadcast
70	Aviation R/C
80	Land Mobile
90	TV Channels 5–6 Broadcast
100	FM Broadcast (88–108)
137–137.02	Space Operation (Space-to-Earth)
160	Amateur Land Mobile
200	TV Channels 7–13 Broadcast (174–216)
300	Government Satellite

*For a detailed description of FCC frequency allocations visit http://www.fcc.gov

*TABLE 45-6 Ultrahigh frequency**

FREQUENCY (MHz)	ALLOCATIONS
300	Aviation
400	Government, Satellite
454–456	Fixed Land Mobile
	(Public Mobile, Maritime)
500	GMRS, Land Mobile
512–608	Broadcast Radio (TV)
600	TV Channels 14–83 Broadcast
700	
776–794	Wireless Communications
800	Land Mobile
900	Land Mobile
901–902	Personal Communications
902–928	ISM Equipment
930–931	Personal Communications
940–941	Personal Communications
960–1215	Aeronautical Radionavigation
1000	Fixed, Microwaves
1215–1240	Earth Exploration-Satellite
1240–1300	Earth Exploration-Satellite
1432–1435	Wireless Communications
1525–1535	Mobile-Satellite
1600	Aviation
1610–1660	Satellite Communications, Aviation
1670–1675	Wireless Communications

(continues)

TABLE 45-6 Ultrahigh frequency (continued)*

FREQUENCY (MHz)	ALLOCATIONS
1710–1755	Wireless Communications
1850–2000	RF Devices, Personal Communications
2000	Fixed Mobile
2000–2020	Satellite Communications
2020–2290	Space Operation (Space-to-Earth)
2345–2655	Wireless Communications, Aviation, Amateurs ISM equipment, Satellite Communication
2655–2900	Wireless Communications, Earth Exploration-Satellite
3000	Radar

*For a detailed description of FCC frequency allocations visit http://www.fcc.gov

TABLE 45-7 Frequency allocations for AM, FM, TV, Radar, Citizen, Amateur radio, and Personal Communications Services (PCS)

BAND	FREQUENCY RANGE
AM broadcast	535–1605 kHz
FM broadcast	88–108 MHz
VHF television	54–88 MHz and 174–216 MHz
UHF television	470–638 MHz
Radar	230 MHz to > 40 GHz
Citizen	26.965 MHz–27.40637 MHz
Amateur radio	1.8–148 MHz
Personal Communications Services (PCS)	18.5 GHz–19.9 GHz

TABLE 45-8 Radar bands

BAND NAME	FREQUENCY
P	230–1000 MHz
L	1–2 GHz
S	2–4 GHz
C	4–8 GHz
X	8–12.5 GHz
Ku	12.5–18 GHz
K	18–26.5 GHz
Ka	26.5–40 GHz
Millimeter	40 GHz–300 GHz
Submillimeter	300–3000 GHz

*TABLE 45-9 TV channel frequencies ****

CHANNEL NO.	FREQUENCY RANGE (MHz)	PICTURE CARRIER (MHz)	SOUND CARRIER (MHz)
2	54–60	55.25	59.75
3	60–66	61.25	65.75
4	66–72	67.25	71.75
5	76–82	77.25	81.75
6	82–88	83.25	87.75
7	174–180	175.25	179.75
8	180–186	181.25	185.75
9	186–192	187.25	191.75
10	192–198	193.25	197.75
11	198–204	199.25	203.75
12	204–210	205.25	209.75
13	210–216	211.25	215.75
14	470–476	471.25	475.75
15	476–482	477.25	481.75
16	482–488	483.25	487.75
17	488–494	489.25	493.75
18	494–500	495.25	499.75
19	500–506	501.25	505.75
20	506–512	507.25	511.75
21	512–518	513.25	517.75
22	518–524	519.25	523.75
23	524–530	525.25	529.75
24	530–536	531.25	535.75
25	536–542	537.25	541.75
26	542–548	543.25	547.75
27	548–554	549.25	553.75
28	554–560	555.25	559.75
29	560–566	561.25	565.75

(continues)

*TABLE 45-9 TV channel frequencies *** (continued)*

CHANNEL NO.	FREQUENCY RANGE (MHz)	PICTURE CARRIER (MHz)	SOUND CARRIER (MHz)
30	566–572	567.25	571.75
31	572–578	573.25	577.75
32	578–584	579.25	583.75
33	584–590	585.25	589.75
34	590–596	591.25	595.75
35	596–602	597.25	601.75
36	602–608	603.25	607.75
37*	608–614	609.25	613.75
38	614–620	615.25	619.75
39	620–626	621.25	625.75
40	626–632	627.25	631.75
41	632–638	633.25	637.75
42	638–644	639.25	643.75
43	644–650	645.25	649.75
44	650–656	651.25	655.75
45	656–662	657.25	661.75
46	662–668	663.25	667.75
47	668–674	669.25	673.75
48	674–680	675.25	679.75
49	680–686	681.25	685.75
50	686–692	687.25	691.75
51	692–698	693.25	697.75
52	698–704	699.25	703.75
53	704–710	705.25	709.75
54	710–716	711.25	715.75
55	716–722	717.25	721.75
56	722–728	723.25	727.25
57	728–734	729.25	733.75
58	734–740	735.25	739.75

(continues)

*TABLE 45-9 TV channel frequencies *** (continued)*

CHANNEL NO.	FREQUENCY RANGE (MHZ)	PICTURE CARRIER (MHZ)	SOUND CARRIER (MHZ)
59	740–746	741.25	745.75
60	746–752	747.25	751.75
61	752–758	753.25	757.75
62	758–764	759.25	763.75
63	764–770	765.25	769.75
64	770–776	771.25	775.75
65	776–782	777.25	781.75
66	782–788	783.25	787.75
67	788–794	789.25	793.73
68	794–800	795.25	799.75
69	800–806	801.25	805.75
70**	806–812	807.25	811.75
71**	812–818	813.25	817.75
72**	818–824	819.25	823.75
73**	824–830	825.25	829.75
74**	830–836	831.25	835.75
75**	836–842	837.25	841.75
76**	842–848	843.25	847.75
77**	848–854	849.25	853.75
78**	854–860	855.25	859.75
79**	860–866	861.25	865.75
80**	866–872	867.25	871.75
81**	872–878	873.25	877.75
82**	878–884	879.25	883.75
83**	884–890	885.25	889.75

* Channel 37 not available for TV transmission

** Channels 69 to 83 are also used in land mobile radio. For television, these channels are used for special services such as educational TV broadcasting.

*** When HDTV mandatory switch over occurs in 2006 many of the existing TV channels will not exist.

TABLE 45-10 CATV channel designations and carrier frequencies

CHANNEL	FREQUENCY RANGE (MHz)	PICTURE CARRIER (MHz)	SOUND CARRIER (MHz)
Low Band:			
2	54–60	55.25	59.75
3	60–66	61.25	65.75
4	66–72	67.25	71.75
5	76–82	77.25	81.75
6	82–88	83.25	87.75
Mid Band:			
A-2	108–114	109.25	113.75
A-1	114–120	115.25	119.75
A	120–126	121.25	125.75
B	126–132	127.25	131.75
C	132–138	133.25	137.75
D	138–144	139.25	143.75
E	144–150	145.25	149.75
F	150–156	151.25	155.75
G	156–162	157.25	161.75
H	162–168	163.25	167.75
I	168–174	169.25	173.75
High Band:			
7	174–180	175.25	179.75
8	180–186	181.25	185.75
9	186–192	187.25	191.75
10	192–198	193.25	197.75
11	198–204	199.25	203.75
12	204–210	205.25	209.75
13	210–216	211.25	215.75
Super Band:			
J	216–222	217.25	221.75
K	222–228	223.25	227.75
L	228–234	229.25	233.75
M	234–240	235.25	239.75
N	240–246	241.25	245.75
O	246–252	247.25	251.75
P	252–258	253.25	257.75
Q	258–264	259.25	263.75
R	264–270	265.25	269.75
S	270–276	271.25	275.75
T	276–282	277.25	281.75
U	282–288	283.25	287.75
V	288–294	289.25	293.75
W	294–300	295.25	299.75

(continues)

TABLE 45-10 CATV channel designations and carrier frequencies (continued)

CHANNEL	FREQUENCY RANGE (MHz)	PICTURE CARRIER (MHz)	SOUND CARRIER (MHz)
Hyper Band:			
AA	300–306	301.25	305.75
BB	306–312	307.25	311.75
CC	312–318	313.25	317.75
DD	318–324	319.25	323.75
EE	324–330	325.25	329.75
FF	330–336	331.25	335.75
GG	336–342	337.25	341.75
HH	342–348	343.25	347.75
II	348–354	349.25	353.75
JJ	354–360	355.25	359.75
KK	360–366	361.25	365.75
LL	366–372	367.25	371.75
MM	372–378	373.25	377.75
NN	378–384	379.25	383.75
OO	384–390	385.25	389.75
PP	390–396	391.25	395.75
QQ	396–402	397.25	401.75
RR	402–408	403.25	407.75
SS	408–414	409.25	413.75
TT	414–420	415.25	419.75
UU	420–426	421.25	425.75
VV	426–432	427.25	431.75
WW	432–438	433.25	437.75
XX	438–444	439.25	443.75
YY	444–450	446.25	449.75
ZZ	450–456	451.25	455.75
AAA	456–462	457.25	461.75
BBB	462–468	463.25	467.76

For additional information about FCC frequency allocations visit:
http://www.fcc.gov

MODULE 46

Satellite Communications: An Overview

■ 46-1 EVOLUTION OF COMMUNICATIONS SATELLITES

In the twenty-first century, satellite communications has become a ubiquitous technology for providing telecommunications services at the local, national, and international levels. The concept of geostationary orbit was conceived by the Russian writer Konstantin Tsiolkovsky in the early twentieth century. In the 1920s, Herman Oberth and Herman Potocnik (also known by the pseudonym "Herman Noordung") presented the idea of placing space stations in orbits with altitudes of 35,900 km, whose period matches the rotational period of the earth, thus making it appear stationary with respect to a fixed point on the equator. Arthur C. Clarke, in his article "Extra-terrestrial Relays: Can Rocket Stations Give World-wide Radio Coverage?" published in *Wireless World* (October 1945), presented the idea of global communication with the help of three geostationary satellites placed around the earth's equator. The first man-made satellite, *Sputnik 1*, was launched into a low earth orbit by the Soviet Union on October 4, 1957. The United States launched its first satellite, *Explorer 1*, on January 31, 1958. Today, satellites have become a common mode of transmission for providing fast, reliable, flexible, and scalable telecommunications services at the local, national, and global levels.

■ 46-2 TYPES OF SATELLITES

Satellites can be classified using the criteria of low earth orbit (LEO), medium earth orbit (MEO), and geosynchronous earth orbit (GEO). Table 46-1 compares the characteristics of LEO, MEO, and GEO satellites. Table 46-2 lists the characteristics of GEO satellites. Table 46-3 compares the characteristics of satellites with fiber optics technology, and Table 46-4 lists the characteristics of the Intelsat satellites. Table 46-5 presents a summary of useful websites for satellite communications.

TABLE 46-1 Comparison of LEO, MEO, and GEO satellites

CHARACTERISTICS			
ORBIT	GEO	MEO	LEO
Typical altitude (see Figure 46-1)	33,000 km	10,000 km	500 km
Power level	High	Low	Very low
Typical signal delay (one-way transmission between two points)	220 ms	66.7 ms	3.3 ms
Average life span	7–15 years	7–15 years	5 years
System examples	Intelsat (see section 42-3) (http://www.intelsat.com)	Global positioning system (GPS) Globalstar	Iridium

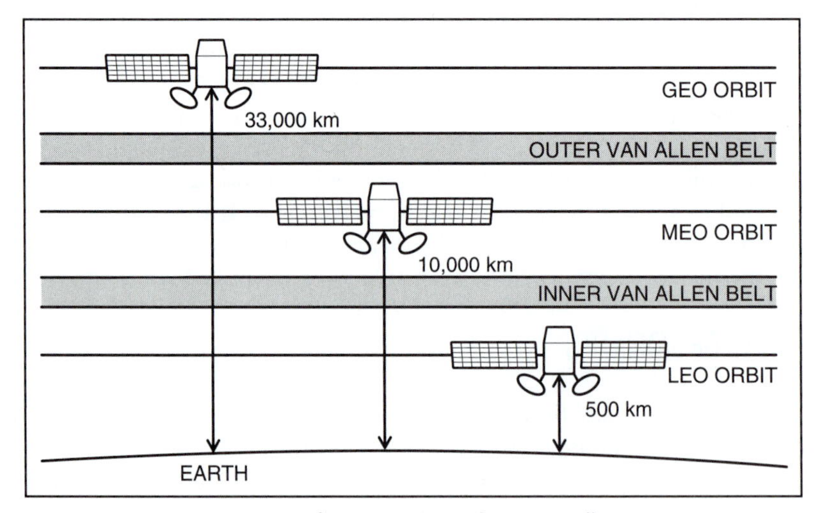

FIGURE 46-1 Comparison of LEO, MEO, and GEO satellites

TABLE 46-2 Characteristics of GEO satellites

ORBIT	Geosynchronous
ALTITUDE	33,000 km
TIME TO COMPLETE ONE REVOLUTION	23 hours 56 minutes and 4 seconds
SYSTEM COMPONENTS	Earth stations (Tx and Rx) Space segment (uplink, transponder, downlink); see Figure 46-2
MODES OF COMMUNICATION	Point-to-point Point-to-multipoint (see Figure 46-3)
SYSTEM BANDWIDTH	Depends on number of transponders (Typical bandwidth: 36 MHz, 54 MHz, and 72 MHz /transponder)
COST	100 million+
SIGNAL DELAY	Significant 220 ms (one-way transmission between two points) 440 ms (round-trip transmission between two points)
SIGNAL SECURITY	No, unless signal is scrambled or encrypted
INTERFERENCE	Sources: EMI/RFI, climatic conditions To prevent the interference of satellite transmission beams, satellites are placed in orbit with an angular separation of 4 degrees (or 1827 miles); see Figure 46-4
NUMBER OF SATELLITES REQUIRED TO PROVIDE GLOBAL COVERAGE	Three (see Figure 46-5)

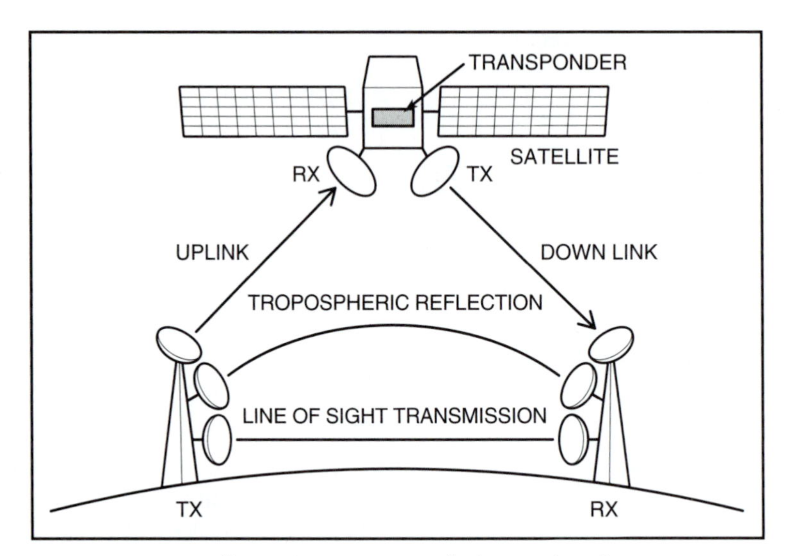

FIGURE 46-2 Satellite communication and other modes of VHF/UHF propagation

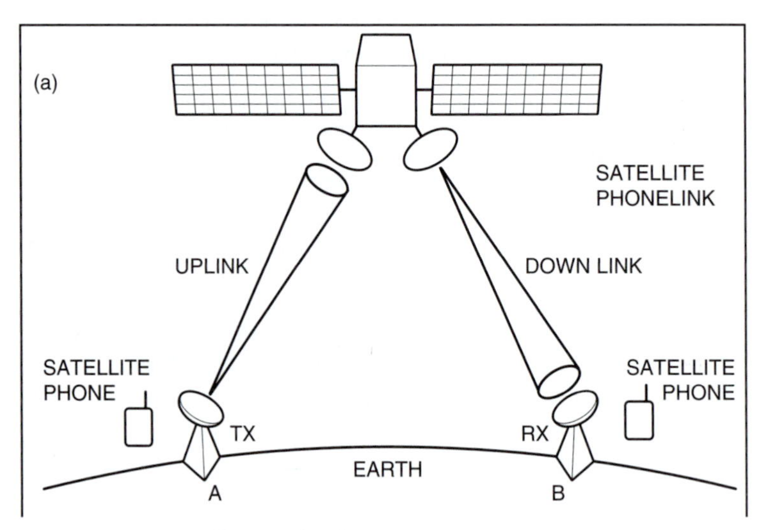

FIGURE 46-3 Satellite: mode of transmission (a) point-to-point; (b) point-to-multipoint (c) multipoint-to-point *(continues)*

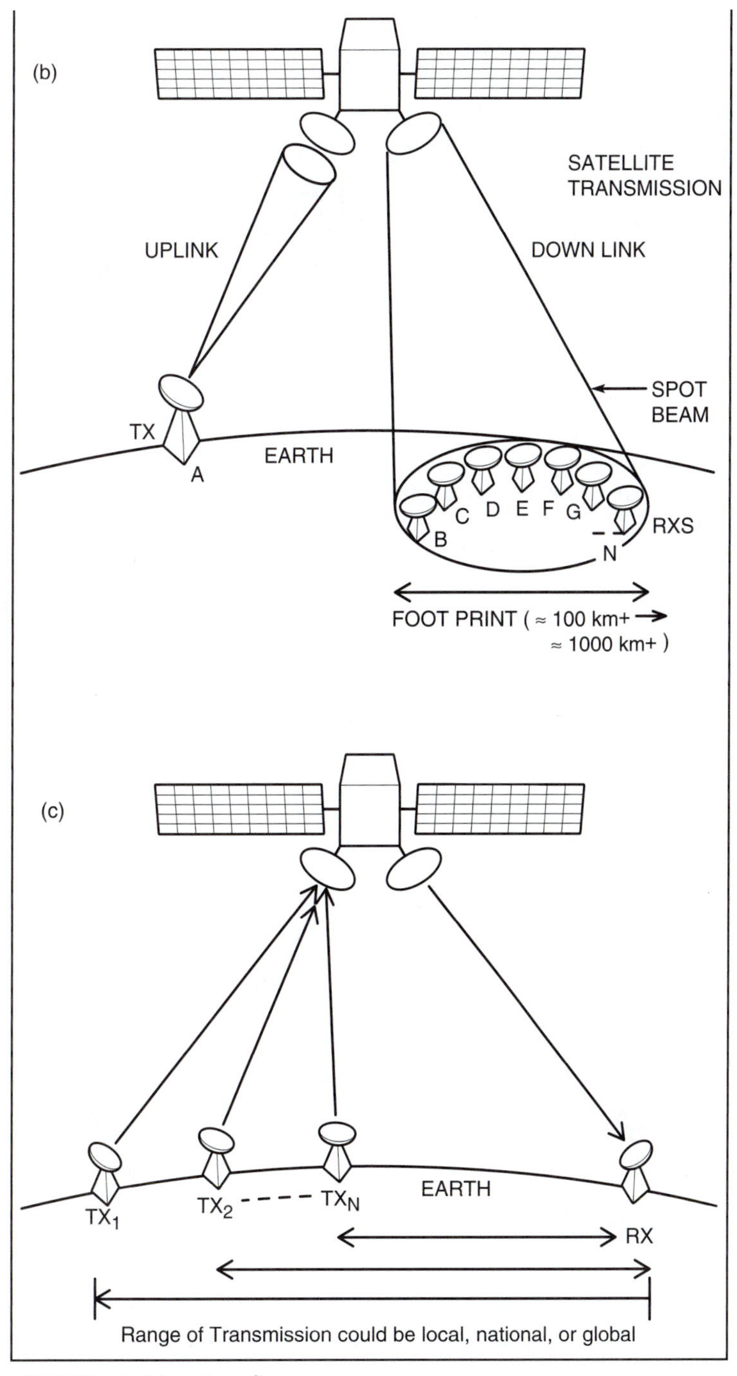

FIGURE 46-3 (continued)

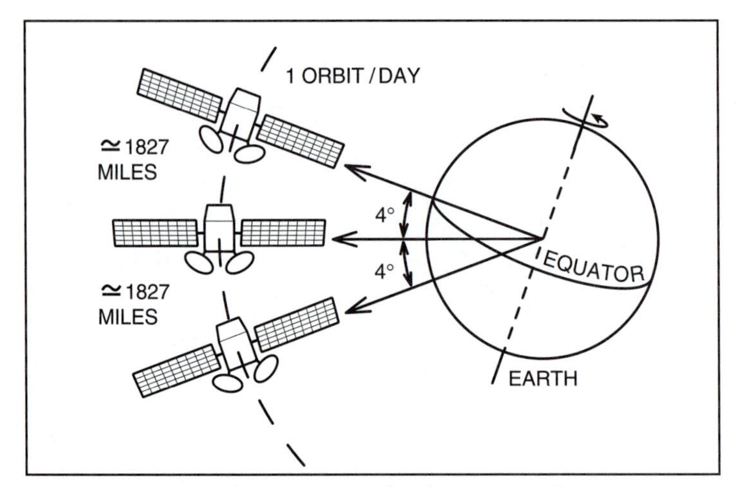

FIGURE 46-4 Angular separation of 4° between satellites prevents interference

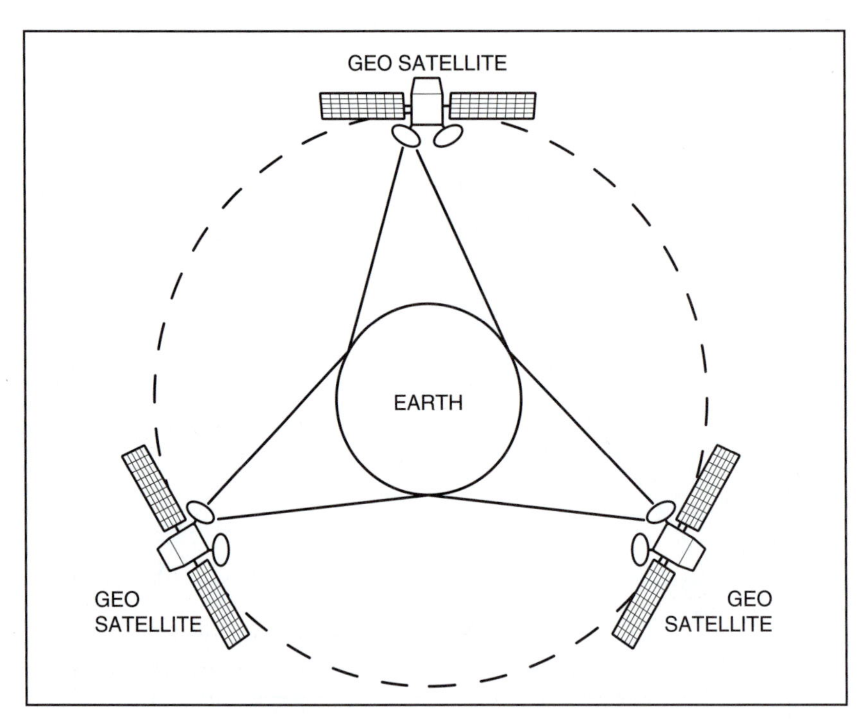

FIGURE 46-5 Three GEO satellites can provide global coverage (except the poles)

TABLE 46-3 Fiber optics technology versus satellite technology

CHARACTERISTICS	SATELLITE	OPTICAL FIBER
Typical system bandwidth	System bandwidth depends on the number of transponders Typical bandwidth/transponder = 36 MHz–72 MHz	Theoretical bandwidth = 100 THz Practical bandwidth = 10 GHz–100 GHz
Primary mode of communication/application	Point-to-multipoint	Point-to-point
System flexibility	Yes, additional Tx or RX units can be easily added or removed	No
Interference	Subject to EMI/RFI	No EMI/RFI
Signal delay	Significant (GEO)	Insignificant
Cost	High	Low/medium/high (depends on type of system: short-haul, long-haul, land-based or oceanic link)
Signal security	No, unless signal is secured	Yes, difficult to tap
Life	Short	Long

TABLE 46-4 Evolution of INTELSAT satellites

INTELSAT DESIG-NATION	YEAR OF FIRST LAUNCH	PRIME CON-TRACTOR	WIDTH DIMEN-SIONS (UNDE-PLOYED)	HEIGHT DIMEN-SIONS (UNDE-PLOYED)	LAUNCH VEHICLES	LIFETIME (YEARS)	BAND-WIDTH (MHZ)	FREQUENCY BANDS	CAPACITY (CIRCUITS)	NEW TECH-NOLOGIES INTRODUCED
INTELSAT I	1965	Hughes	0.7 m. 2.3 ft.	0.6 m. 2 ft.	Thor Delta	1.5	50	6/4 GHz (C)	240 or 1 TV	First commercial communications satellite
INTELSAT II	1967	Hughes	1.4 m. 4.6 ft.	0.7 m. 2.3 ft.	Thor Delta	3	130	6/4 GHz (C)	240 or 1 TV	Multipoint communications capability between earth stations
INTELSAT III	1968	TRW	1.4 m. 4.6 ft.	1.0 m. 3.3 ft.	Thor Delta	5	300	6/4 GHz (C)	1500 and 4 TV	Equipped with mechanically despun antenna. Expanded multipoint communications simultaneously: telephone, telegraph, television, high-speed data, and facsimile. TV service provided without interrupting telephone service
INTELSAT IV	1971	Hughes	2.4 m. 7.9 ft.	5.3 m. 17.4 ft.	Atlas Centaur	7	500	6/4 GHz (C)	4000 and 2 TV	Two steerable spot-beam transmit antennas.
INTELSAT IV-A	1975	Hughes	2.4 m. 7.9 ft.	6.8 m. 22.3 ft.	Atlas Centaur	7	800	6/4 GHz (C)	6000 and 2 TV	Design permitted the simultaneous use of the same frequencies through the use of directional antennas for both reception and

(continues)

TABLE 46-4 *Evolution of INTELSAT satellites (continued)*

INTELSAT DESIGNATION	YEAR OF FIRST LAUNCH	PRIME CONTRACTOR	WIDTH DIMENSIONS (UNDEPLOYED)	HEIGHT DIMENSIONS (UNDEPLOYED)	LAUNCH VEHICLES	LIFETIME (YEARS)	BANDWIDTH (MHZ)	FREQUENCY BANDS	CAPACITY (CIRCUITS)	NEW TECHNOLOGIES INTRODUCED
										transmission, yielding equivalent of nearly twice the number of transponders.
INTELSAT V	1980	Ford Aerospace	2.0 m. 6.6 ft.	6.4 m. 21 ft.	Atlas Centaur or Ariane 1,2	7	2,144	6/4 GHz (C) 14/11 GHz (Ku)	12,000 and 2 TV	Both C- & Ku-bands. The 6/4 frequencies are used 4 times, by orthogonally polarized east and west hemispheric and zone beams. Also carried L-band maritime communications subsystmes.
INTELSAT V-A	1985	Ford Aerospace	2.0 m. 6.6 ft.	6.4 m. 21 ft.	Atlas Centaur or Ariane 1,2	7	2,250	6/4 GHz (C) 14/11 GHz (Ku) 14/12 GHz (Ku)	15,000 and 2 TV	3 crosspolarized spot beams at 6/4 GHz for domestic leased services and use of nickel hydrogen batteries. The last three satellites in the V-A series are capable of operating both in 14/11 and 14/12 GHz. Also had extended spot beam coverage.

(continues)

TABLE 46-4 *Evolution of INTELSAT satellites (continued)*

INTELSAT DESIG-NATION	YEAR OF FIRST LAUNCH	PRIME CON-TRACTOR	WIDTH DIMEN-SIONS (UNDE-PLOYED)	HEIGHT DIMEN-SIONS (UNDE-PLOYED)	LAUNCH VEHICLES	LIFETIME (YEARS)	BAND-WIDTH (MHZ)	FREQUENCY BANDS	CAPACITY (CIRCUITS)	NEW TECH-NOLOGIES INTRODUCED
INTELSAT VI	1989	Hughes	3.6 m. 11.8 ft.	5.3 m. 17.4 ft.	Ariane 4 or Titan	13	3,300	6/4 GHz (C) 14/11 GHz (Ku)	24,000 and 3 TV (up to 120,000 with DCME)	Introduction of SS/TDMA. The 6/4 GHz frequencies are reused six times. The 14/11 GHz frequencies are reused twice. Higher power in the 14/11GHz promotes access by smaller earth stations.
INTELSAT VII	1992	Ford Aero-space	2.7 m. 8.9 ft.	4.2 m. 13.8 ft.	Ariane-space or General Dynamics	15	2,432	6/4 GHz (C) 14/11GHz (Ku) 14/12GHz (Ku)	18,000 and 3 TV (up to 90,000 with DCME)	Independently steerable C- & Ku-band spot beams. Four times frequency reuse at C-band. Two times frequency reuse at Ku-band. Solid state power amplifiers (SSPAs) at C-band and linearized traveling wave tube amplifiers (TWTAs) provide improved transmission.

For characteristics/features of Intelsat IX series (901–907) and Intelsat 10-02 visit Intelsat's website: http://www.intelsat.com/resources/satellites.aspx

■ 46-3 EXAMPLE OF SATELLITE SERVICE PROVIDER: INTELSAT

INTELSAT started as a nonprofit cooperative of 114 nations. INTELSAT owns and operates a global system of communications satellites that provide international telecommunications services. It also provides domestic telecommunications services to more than 30 nations. The INTELSAT network contains satellites in geosynchronous orbit over the Atlantic Ocean (AO), Indian Ocean (IO), and Pacific Ocean (PO) regions (see Table 46-5). These satellites, coupled with the more than 700 antennas in participating nations, make INTELSAT the major provider of transoceanic telephone, data, and television service.

Since its creation in August of 1964, INTELSAT's satellites have brought history-making events to televisions and radios all around the world. More than one-half of all international telephone calls and nearly all transoceanic television are carried by the INTELSAT system. Among the daily users that rely on INTELSAT's global network are the international financial networks, multinational corporations, international news services, and radio and television broadcasters. (For the current status of INTELSAT visit http://www.intelsat.com.)

TABLE 46-5 Useful websites for satellite communications

WEBSITE	DESCRIPTION
http://www.intelsat.com	Official website of INTELSAT, provider of global satellite services for voice and data.
http://www.Satsig.net/sslist.htm	List of satellites in geostationary orbit
http://www.amsat.org	The official website of Radio Amateur Satellite Corporation
http://www.noaa.gov/satellites.html	Homepage for the National Oceanic and Atmospheric Administration's operational environment satellite system

(continues)

TABLE 46-5 Useful websites for satellite communications (continued)

WEBSITE	DESCRIPTION
http://www.allaboutsatcom.com	Provides a list of companies that offer services in satellite communications
http://www.inmarsat.com/home.aspx	Homepage of Inmarsat, which owns and operates a global satellite network
http://www.hq.nasa.gov/office/pao/History/satcomhistory.html	Provides a brief history of satellite communications
http://www.itso.int/	Homepage of the International Telecommunications Satellite Organization
http://www.globexplorer.com/	Provides a library of aerial/satellite images and maps
http://www.colorado.edu/geography/gcraft/notes/gps/gps_f.html	Provides an overview of the Global Positioning System (GPS)

MODULE 47

Wireless Technologies and Standards: A Summary

"It is dangerous to put limits on wireless."
—Guglielmo Marconi (1932)

A century ago, Marconi transmitted the first radio signals. Fifty years later, AT&T's Bell Laboratories invented cellular technology. Today, the cellular industry is one of the world's fastest growing major industries, with cellular technology playing a pivotal role in how people communicate around the world. In 1983, fewer than 10 countries had cellular systems. By 1993, there were approximately 140 countries with cellular service. In 1993, North America's cellular market grew 43 percent to 16 million and Western Europe's grew 47 percent to 8.8 million. By contrast, Eastern Europe grew 129 percent, while Malaysia's grew by 64 percent, Argentina's by 220 percent, and China's by 300 percent. Research analysts forecast that the worldwide wireless penetration will reach 21 percent of the world's population by 2006, or a total of 1.3 billion subscribers.

Table 47-1 compares generations of the wireless systems, Table 47-2 presents a summary of wireless access (air interface) technologies, Table 47-3 compares the characteristics of the major wireless standards, and Table 47-4 presents a summary of useful websites for wireless communications.

TABLE 47-1 Comparison of wireless generations

WIRELESS SYSTEM	AIR INTERFACE TECHNOLOGY	DESCRIPTION/COMMENTS
First Generation	FDMA	Coverage could be spotty
Second Generation (2G)	TDMA and CDMA	• Better voice quality • Circuit-switched data • Low data rates
Second Generation (2G+)	TDMA and CDMA	• Packet-data capability • Higher data rates (up to 144 kbps)
Third Generation (3G)	TDMA and CDMA	• Higher data rates (144 kbps+) • Uses WCDMA/UMTS/CDMA 2000/UWC

TABLE 47-2 Comparison of wireless access (air interface) technologies

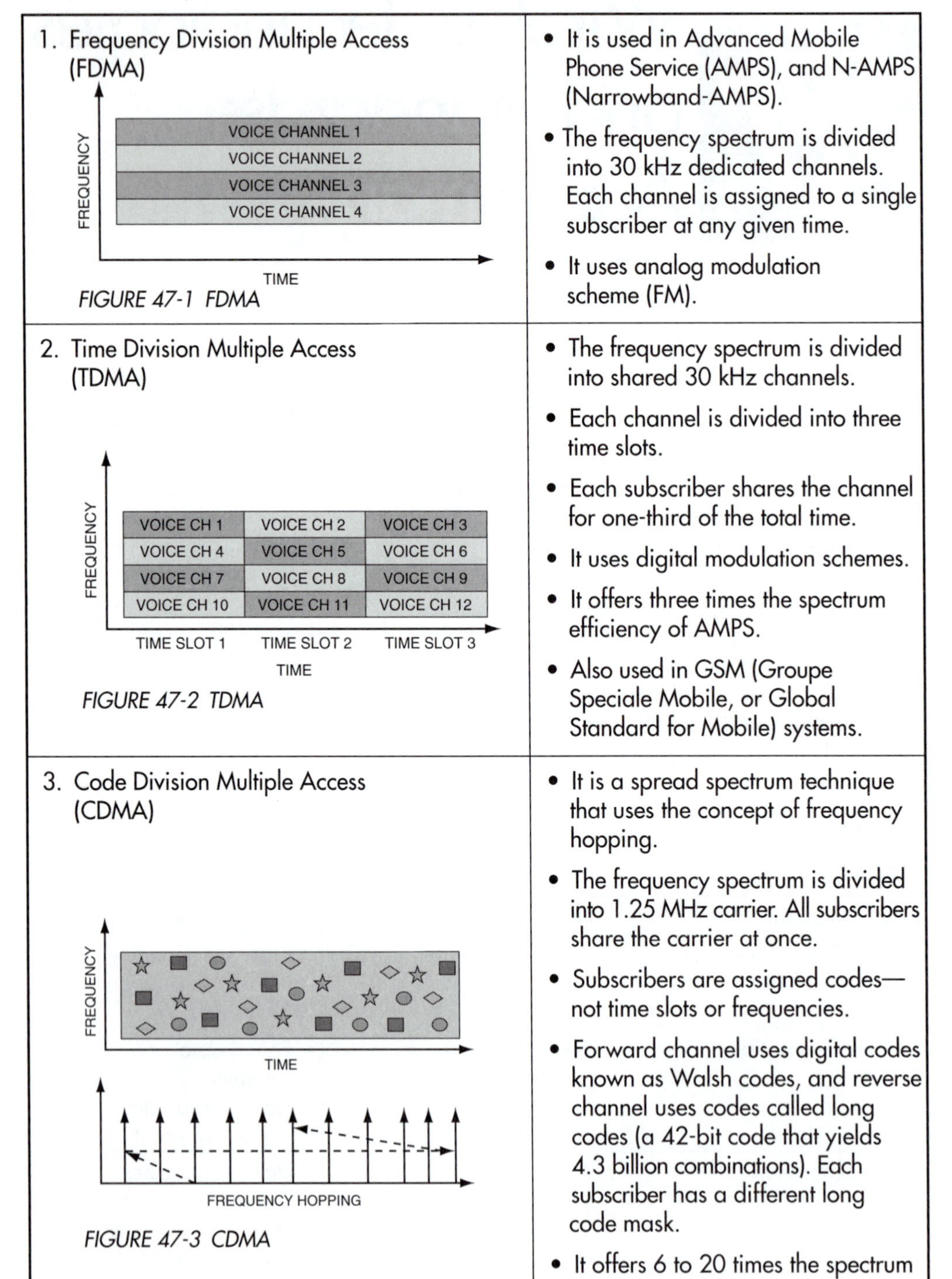

1. Frequency Division Multiple Access (FDMA) *FIGURE 47-1 FDMA*	• It is used in Advanced Mobile Phone Service (AMPS), and N-AMPS (Narrowband-AMPS). • The frequency spectrum is divided into 30 kHz dedicated channels. Each channel is assigned to a single subscriber at any given time. • It uses analog modulation scheme (FM).
2. Time Division Multiple Access (TDMA) *FIGURE 47-2 TDMA*	• The frequency spectrum is divided into shared 30 kHz channels. • Each channel is divided into three time slots. • Each subscriber shares the channel for one-third of the total time. • It uses digital modulation schemes. • It offers three times the spectrum efficiency of AMPS. • Also used in GSM (Groupe Speciale Mobile, or Global Standard for Mobile) systems.
3. Code Division Multiple Access (CDMA) *FIGURE 47-3 CDMA*	• It is a spread spectrum technique that uses the concept of frequency hopping. • The frequency spectrum is divided into 1.25 MHz carrier. All subscribers share the carrier at once. • Subscribers are assigned codes—not time slots or frequencies. • Forward channel uses digital codes known as Walsh codes, and reverse channel uses codes called long codes (a 42-bit code that yields 4.3 billion combinations). Each subscriber has a different long code mask. • It offers 6 to 20 times the spectrum efficiency of AMPS.

TABLE 47-3 Comparison of major wireless standards

	AMPS	N-AMPS	TDMA (IS-136)	CDMA (IS-95)	GSM
Nature of Transmission	Continuous	Continuous	Burst	Continuous or Burst	Burst
Type of Transmission	Analog	Analog	Digital	Digital	Digital
RF Bandwidth	Narrow	Narrow	Medium/ Wide	Very Wide	Medium/ Wide
Modulation Scheme	Frequency Modulation (FM)	Frequency Modulation (FM)	Phase Modulation (DQPSK)	• Direct Sequence Code Modulation • QPSK • BPSK	Phase Modulation (GMSK)
Air Interface/Access Technology	FDMA	FDMA	TDMA	CDMA	TDMA
Channel Bandwidth	30 kHz	10 kHz	30 kHz	1.25 MHz	200 kHz

TABLE 47-4 Useful websites for wireless communications

WEBSITE	DESCRIPTION
http://www.gwec.org	Global Wireless Education Consortium
http://www.ctia.org	Cellular Telecommunications & Internet Association (CTIA)
http://www.wcai.com/	Wireless Communications Association International (WCA)
http://wireless.fcc.gov/	Wireless Telecommunications Bureau
http://www.centurion.com	Supplier of antenna and power products
http://www.comsoc.org/	IEEE communications society
http://www.cellular.co.za/celltech.htm	Comparison of cellular technologies
http://www.wave-guide.org /archives/ waveguide_3/cellular-history.html	Brief history of cellular technology
http://www.wca.org/	Wireless Communications Alliance (WCA)

MODULE 48 Computer Networks: An Overview

Computer networks can be classified by size (distance between nodes), configuration (topology), and ownership (Figure 48-1). Table 48-1 lists the definitions of a LAN, MAN, WAN, Enterprise network, and Global network. Table 48-2 lists the components of a LAN. Table 48-3 presents various network topologies. Table 48-4 lists LAN access methods. Table 48-5 presents a summary of inter-networking devices and backbone technology. Module 0 lists IEEE 802 standards for LANs. Table 48-6 lists 802.11 standards, a group of specifications developed by the Institute of Electrical and Electronics Engineers (IEEE) for wireless local area networks (WLANs). The 802.11 specifications, also known as Wi-Fi, define an over-the-air-interface over a short range between a wireless access point and a client or base station.

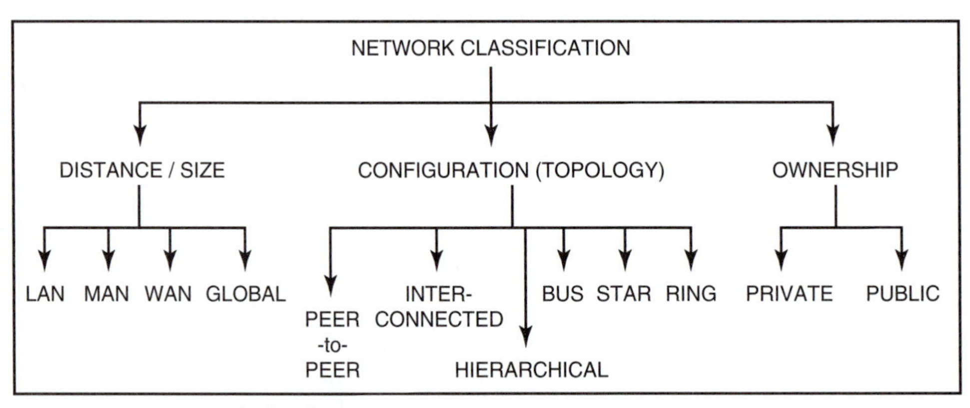

FIGURE 48-1 Network classification

TABLE 48-1 Types of networks

NETWORK TYPE	DESCRIPTION
Local Area Network (LAN)	A blending of computer hardware, software, and transmission media to form a network over short distances. A LAN is usually contained within a building or a campus.
Metropolitan Area Network (MAN)	A computer network that covers a larger area than a LAN. Typically a MAN covers a metropolitan area.
Wide Area Network (WAN)	A network that is larger than a MAN. It interconnects LANs and MANs that may be separated by long distances.
Enterprise Network	A network that interconnects LANs belonging to a single organization.
Global Network	A large-scale network that provides universal communications, e.g., the Internet.

TABLE 48-2 Components of a LAN

Software	Network operating system (e.g., Novel Netware, Windows 2003, Windows XP, Windows 2000, Linux)
Transmission medium	Twisted pair • Unshielded twisted pair (UTP) • Shielded twisted pair (STP) Coaxial cable • Thin • Thick Optical fiber • Monomode/single mode step-index • Multimode step-index • Multimode graded-index
Nodes	Computers, laptops, file server, print server
Linking hardware	Network interface card (NIC), cables
Connectors	RJ-45, T-Connector, BNC connector, DB-25, DB-15, etc.
Security hardware/software	Firewall

TABLE 48-3 Network Topologies

TOPOLOGY	ILLUSTRATION
Peer-to-Peer	 Node 1 — Node 2 *FIGURE 48-2 Peer-to-peer topology*
Interconnected • Fully interconnected • Partially interconnected	(a) 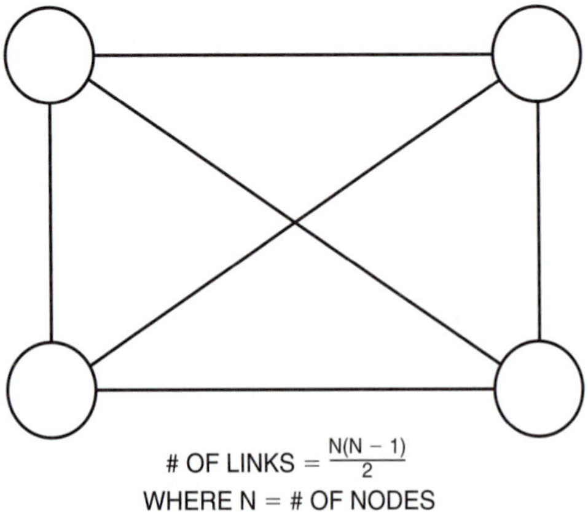 # OF LINKS = $\frac{N(N-1)}{2}$ WHERE N = # OF NODES *FIGURE 48-3 (a) Fully interconnected topology* (b) 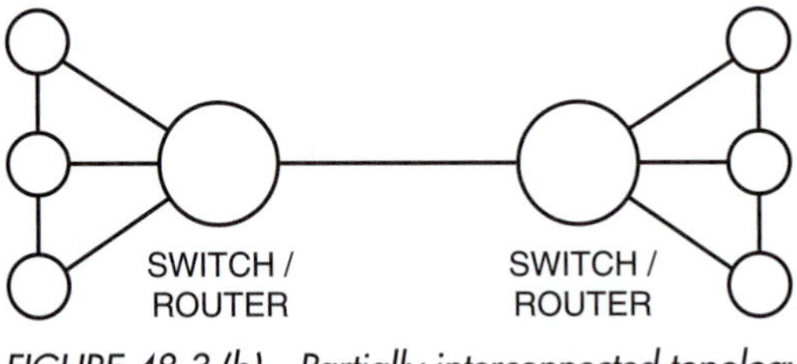 SWITCH / ROUTER SWITCH / ROUTER *FIGURE 48-3 (b) Partially interconnected topology*
Hierarchical	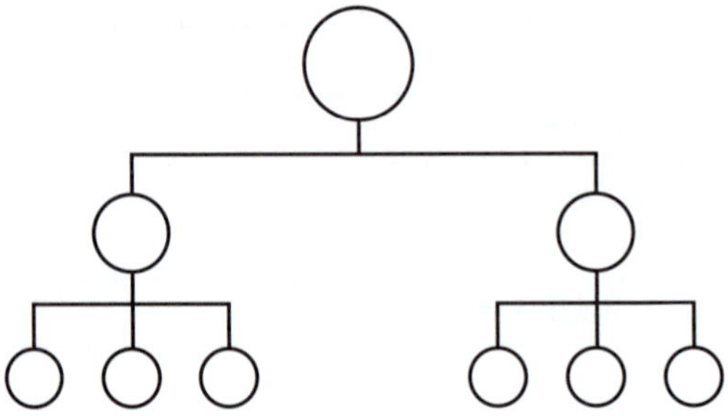 *FIGURE 48-4 Hierarchical topology*

(continues)

TABLE 48-3 Network Topology (continued)

TOPOLOGY	ILLUSTRATION
Bus	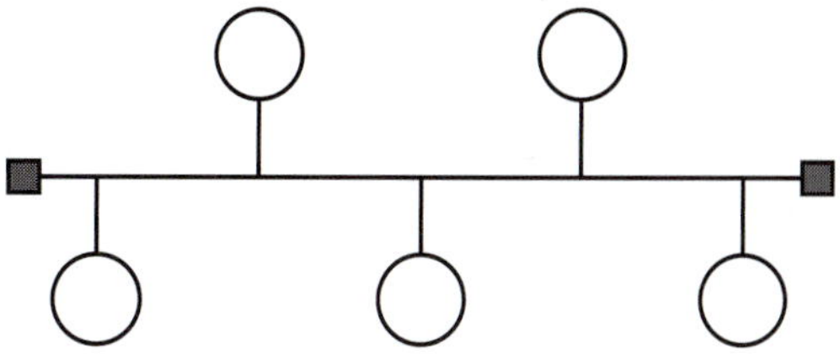 *FIGURE 48-5 Bus topology*
Star	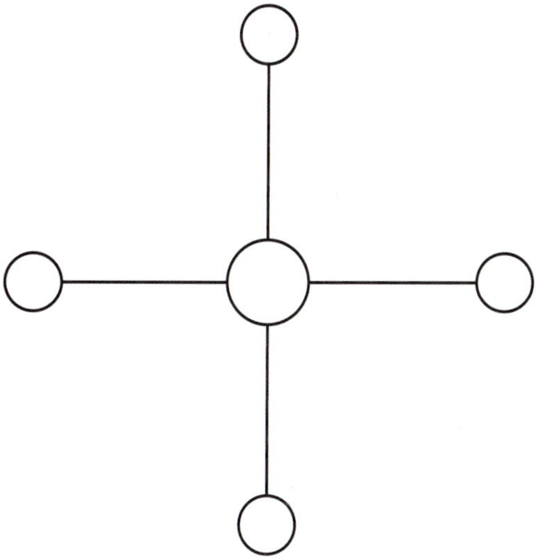 *FIGURE 48-6 Star topology*
Ring	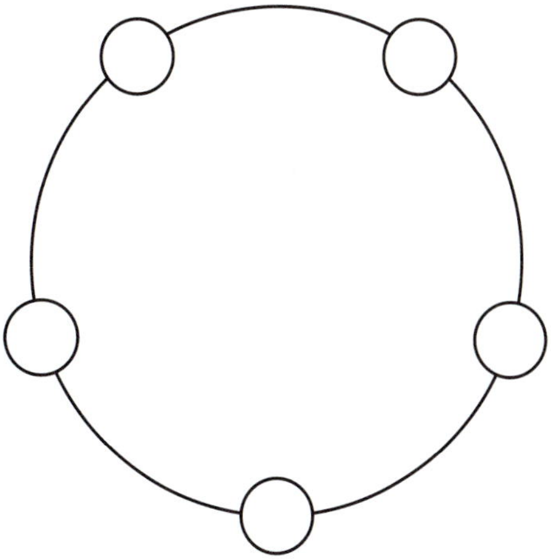 *FIGURE 48-7 Ring topology*
For comparison of topologies, see Table 20–1.	

TABLE 48-4 LAN access methods

ACCESS METHOD	DESCRIPTION
Token Passing	A special bit pattern (token) is distributed sequentially to all nodes. Once a token is received, the user gets the right to tansmit the data. After the transmission is completed, token is passed on to the next user in queue.
Contention	Carrier Sense Multiple Access/Collision Detection (CSMA/CD); and Carrier Sense Multiple Access/Collision Avoidance (CSMA/CA)
Polling	A primary node acts as a media access administrator and queries all nodes in a predetermined manner if they have any data to transmit.

TABLE 48-5 Inter-networking devices and backbone technologies

DEVICE	DESCRIPTION
Repeater	Extends the range of transmission between two network segments (works at the physical layer of OSI model).
Bridge	Connects two segments of similar networks; does not perform any protocol conversion (operates at the data link layer of OSI model).
Router	Connects logically separate networks (subnets); operates at network layer of OSI model.
Brouter	Combines the functions of a bridge and a router.
Gateway	Connects to dissimilar networks by translating the protocols (operates at Transport, Session, Presentation, and Application layers of OSI model).
Hub	• Connects cables from multiple computers to form a single network connection. • A *passive hub* just repeats the signal, whereas an active hub (multi-port repeaters) extends the range of signal transmission. • An *intelligent hub* also performs remote management, and diagnostic functions about the network.

(continues)

TABLE 48-5 Inter-networking devices and backbone technologies (continued)

DEVICE	DESCRIPTION
Switch	• Filters network traffic, i.e., forwards frames to a specific port if the destination is known; and floods network traffic, i.e., forwards frames to all ports except the one on which the frames were received, if the destination address is not known. • Switches have emerged as the replacements for bridges. They offer enhanced throughput performance, increased port density, lower per-port cost, compared to bridges. Switches operate at the data link layer of the OSI model.
Backbone Technologies	• T-1 • DSL • ISDN • FDDI • ATM • Frame Relay • SONET

TABLE 48-6 IEEE 802.11 specifications

IEEE GROUP/SPECIFICATION	DESCRIPTION
802.11	The original wireless local area network (WLAN) standard, supports data rates of 1–2 Mbps.
802.11a	A high-speed wireless local area network (WLAN) standard, specifies wireless data transmission in 5 GHz band, and supports data rates of 54 Mbps.
802.11b	A WLAN specification for 2.4 GHz band, supports data rate of 11 Mbps.
802.11d	A WLAN standard for international roaming. A device is automatically reconfigured to meet local RF requirements.
802.11e	A WLAN standard that addresses the quality of service (QoS).
802.11f	A WLAN specification that defines roaming between multi-vendor access points and distribution systems.
802.11g	A WLAN specification that defines an additional modulation technique for 2.4 GHz, and supports data rate of 54 Mbps.

(continues)

TABLE 48-6 IEEE 802.11 specifications (continued)

IEEE GROUP/SPECIFICATION	DESCRIPTION
802.11h	A WLAN specification that defines the spectrum management of the 5 GHz band.
802.11i	This group is working to develop new WLAN standards to fix the security flaws for authentication and encryption protocols.
802.11n	This group tried to create a global standard in the 5 GHz band, but the work was discontinued due to inoperability of hyper LAN and 802.11a.
For additional information about IEEE 802. Standards visit: http://www.ieee802.org http://grouper.ieee.org/groups/	

MODULE 49

Telecommunications Technology Trends and Statistics

The exponential growth of telecommunications technologies (satellite communications, cellular communications, fiber optics systems, the Internet, and so forth) during the past two decades has impacted people around the globe at personal and national levels. The rapid pace of technological change continues to transform cultures and societies around the globe. Analysts predict that telecommunications technologies will continue to grow at a rapid pace in the coming years. Figure 49-1 illustrates the exponential increase in the number of transistors on a single CPU. Figure 49-2 shows the world Internet usage as a penetration percent of world population. Figure 49-3 shows the growth of host computers worldwide. Figure 49-4 shows the top 10 nations by number of Internet users. Figure 49-5 shows the number of telephone lines. Figure 49-6 shows the number of cellular phone subscribers around the world, and Figure 49-7 illustrates the trend for world semiconductor sales.

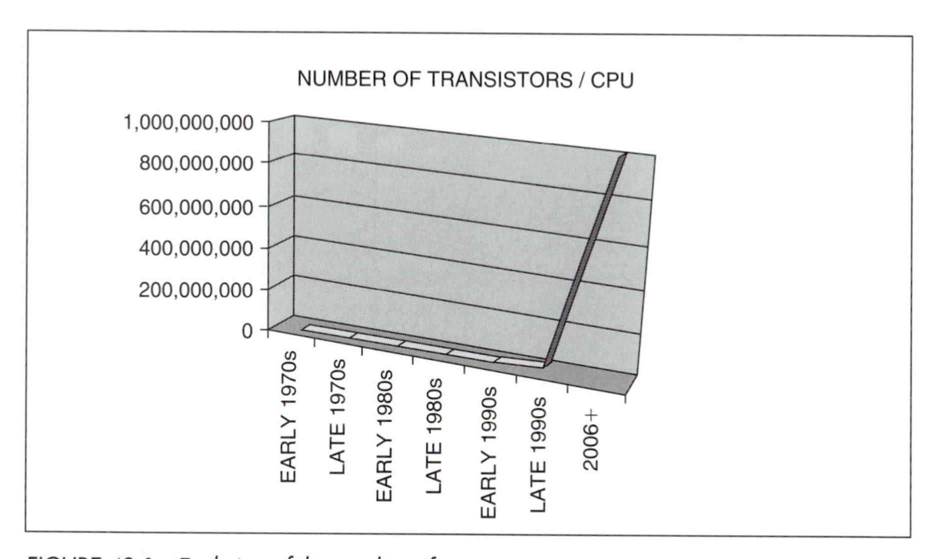

FIGURE 49-1 Evolution of the number of transistors on a CPU

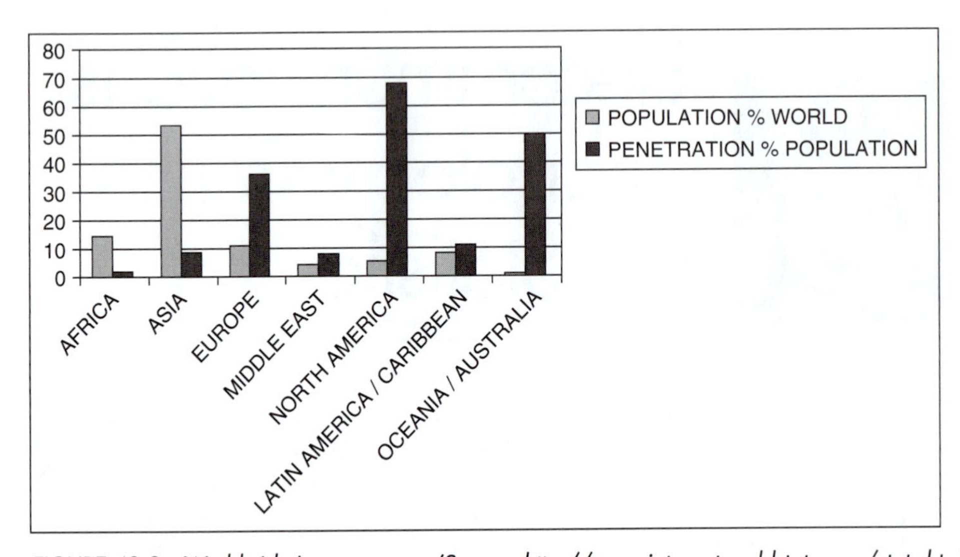

FIGURE 49-2 Worldwide Internet usage (Source: http://www.internetworldstats.com/stats.htm)

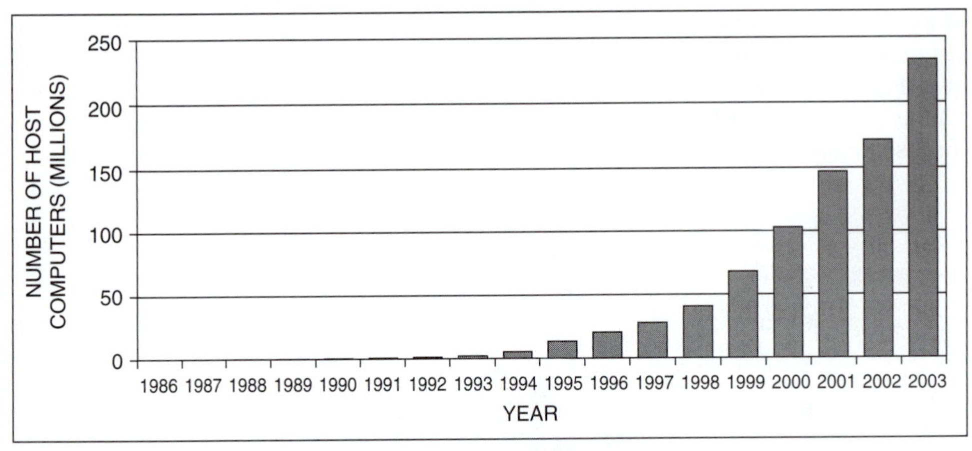

FIGURE 49-3 Number of host computers worldwide 1986–2003
(Source: http://www.internetworldstats.com/stats.htm)

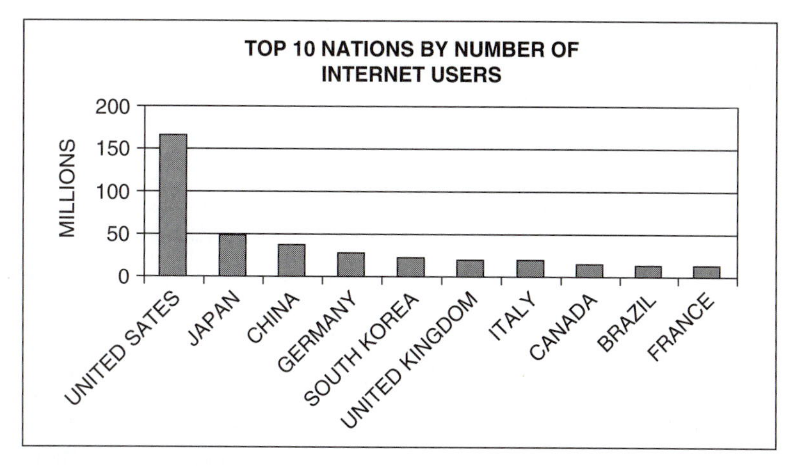

FIGURE 49-4 Top ten nations by number of Internet users (Source: Vital Signs 2002, *Worldwatch Institute, Washington, D.C.)*

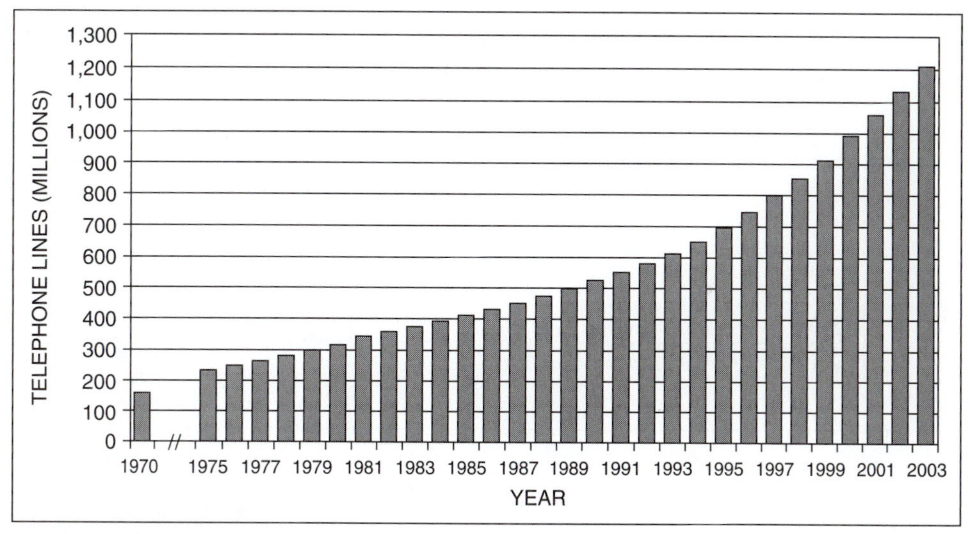

FIGURE 49-5 Telephone lines worldwide (Source: ITU and Vital Signs 2002, *Worldwatch Institute, Washington, D.C.)*

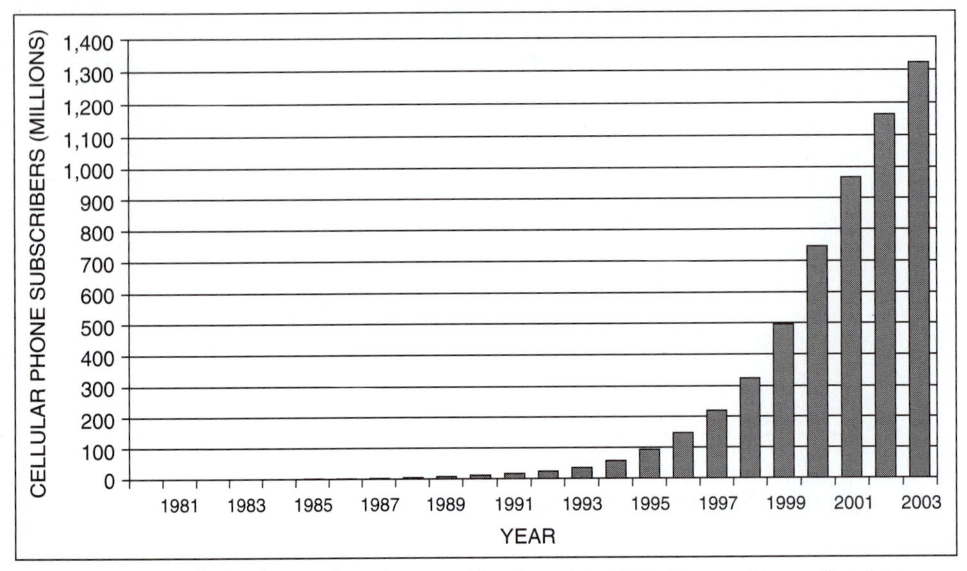

FIGURE 49-6 Cellular phone subscribers worldwide 1980–2003 (Source: ITU and Vital Signs 2002, *Worldwatch Institute, Washington, D.C.)*

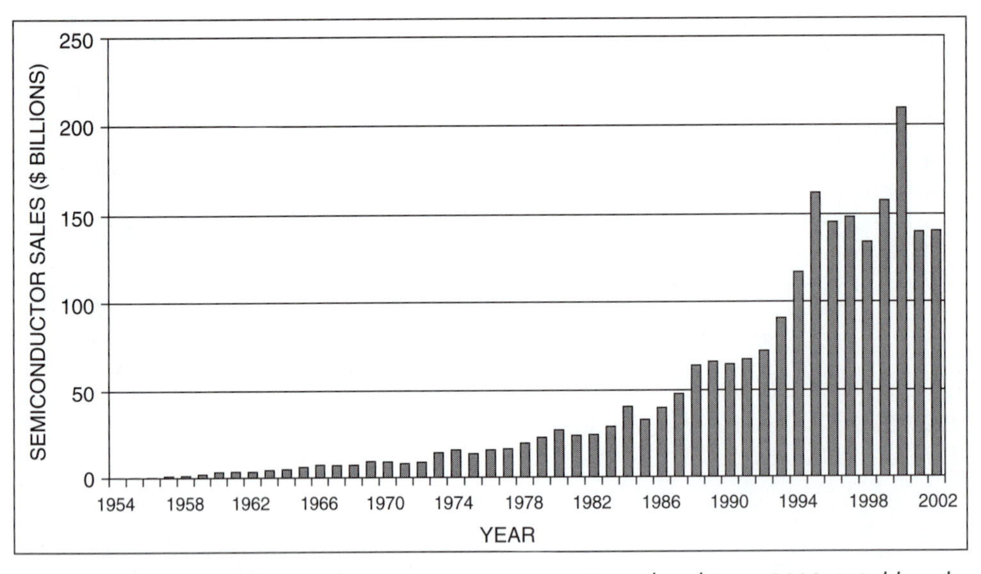

FIGURE 49-7 Semiconductor sales 1954–2002 (Source: ITU and Vital Signs 2002, *Worldwatch Institute, Washington, D.C.)*

MODULE 50 Telecommunications Acronyms

1xDO	3G extension of IS-95B (data only)
1XDV	3G extension of IS-95B (shared data and voice)
1xEV	3G extension of IS-95B (data with circuit-switched voice)
1xRTT	3G extension of IS-95B (1 RF channel)
2G	Second Generation wireless technology
2.5G	Wireless systems between second and third generation
3G	Third Generation wireless technology
3GPP	3G Partnership Project for wideband CDMA standards based on backward compatibility with GSM and IS-136/PDC
3GPP	3G Partnership Project for CDMA2000 based on backward compatibility with IS-95
3xRTT	3G extension of IS-95 (3 RF channels)
4GL	Fourth Generation Language
AAA	Authentication, Authorization, and Accounting
AAL	Application Adaptation Layer
AAR	Automatic Alternative Routing
ABC	Atanasoff-Berry Computer
ABM	Asynchronous Balanced Mode
ABR	Available Bit Rate
ABS	Average Busy Season
ABT	Abort Timer or Answer Back Tone
AC	Access Control
AC	Acoustic Coupler
ACA	Automatic Circuit Assurance
ACBH	Average Consistent Busy Hour
ACC	Automatic Callback Calling
ACD	Automatic Call Distributor
ACF	Advanced Communications Facility
ACF	Advanced Communications Function
ACF/NCP	Advanced Communication Function/ Network Control Program
ACF/VTAM	Advanced Communication Function/Virtual Terminal Access Method
ACK	Affirmative Acknowledgement
ACK0	Positive Acknowledgement
ACK1	Positive Acknowledgement
ACL	Access Control List
ACL	Asynchronous Connectionless Link
ACP	Access Control Points
ACP	Action Control Points
ACR	Abandon Call and Retry
ACR	Actual Cell Rate
ACSI	Abstract Communication Service Interface
ACTGA	Attendant Control of Trunk Group Access
ACTIUS	Association of Computer Telephone Integration Users and Suppliers
ACU	Automatic Calling Unit
A/D	Analog/Digital
ADC	Analysis Date Concentrator
ADCCP	Advanced Data Communication Control Procedure
ADM	Adaptive Delta Modulation
ADM	Add Drop Multiplexer
ADMD	Administrative Management Domain
ADP	Automatic Data Processing
ADPCM	Adaptive Differential Pulse Code Modulation
ADSL	Asymmetric Digital Subscriber Line
ADSP	Apple DataStream Protocol
AEC	Adaptive Echo Cancellation
AES	Advanced Encryption Protocol
AFC	Automatic Frequency Control
AFP	Appletalk File Protocol
AGBH	Average Group Busy Hour
AGC	Automatic Gain Control
AHT	Average Holding Times
AI	Artificial Intelligence
AIM	Asynchoronous Interface Module
AIOD	Automatic Identification of Outward Dialing
ALGOL	Algorithmic Language
ALOHANet	University of Hawaii Local Area Network
ALU	Arthimetic Logical Unit
AM	Amplitude Modulation
AMA	Automatic Message Accounting
AMA	Automated Message Account
AMI	Alternative Mark Inversion
AMM	Agent Management Module

AMPS Advanced Mobile Phone System
AMT Alternative Minimum Tax
ANBH Average Network Busy Hour
AND Automatic Network Dialing
ANI Automatic Number Identification
ANM Advanced Network Management
ANN Auditing Network Needs
ANSI American National Standards Institute
AP Applications Processors
APD Avalanche Photodiode
API Applications Programming Interfaces
APL A Programming Language
APLT Advanced Private Line Termination
APR Address Resolution Protocol
APPC Advanced Program-to-Program Communications
APPC/PC A version of APPC developed by IBM to run on PC-based token ring networks
APPN Advanced Peer to Peer Networking
APON Asynchronous transfer mode Passive Optical Network
ARC Attached Resource Computer
ARL Attendant Release Loop
ARO After Receipt of Order
ARP Address Resolution Protocol
ARPA Advanced Research Projects Agency
ARPANET Advanced Research Projects Agency Network
ARQ Auto Request for Transmission
ARR Automatic Repeat Request
ARS Automatic Route Selection
ARU Audio Response Unit
AS Autonomous System
ASCII American Standard Code for Information Interchange
ASIC Application-Specific Integrated Circuit
ASP Application Service Provider
ASR Automatic Send/Receive
ASYNC Asynchronous Transmission
ATAS Analog Test Access System
ATB All Trunks Busy
ATD The Association of Telecommunications Dealers
ATM Asynchronous Transfer Mode
ATT Applied Transmission Technologies
AT&T American Telephone and Telegraph Company
ATTIS AT&T Information Systems
AUI Attachment Unit Interface
AVD Alternative Voice/Data
AVN Automated Voice Network
B-ISDN Broadband ISDN
BASIC Beginners All Symbolic Instruction Code
BASR Buffered Automatic Send/Receive
BAUD A Measure of Signaling Rate
BAUDOT Teleprinter Code (Fifth Level)
BBH Bouncing Busy Hour
BCC Block Check Character
BCD Binary Coded Decimal
BCS Business Communications Systems
BCW Burst Code-Word
BDLC Burroughs Data Link Control
BDN Bell Data Network
BER Bit Error Rate
BERSIM Bit Error Rate Simulator
BERT Bit Error Rate Test
BEZS Bandwidth Efficient Zero Suppressing
BFO Beat Frequency Oscillator
BGP Border Gateway Protocol
BIND Berkeley Internet Name Domains
BISYNC Bisynchronous Transmission
BIT Binary Digit
BIU Base Station Interface Unit
BIU Bus Interface Unit
BLER Block Error Rate
BLERT Block Error Rate Test
BLF Busy Lamp Field
BLSR Bidirectional Line-Switched Ring
BLU Basic Link Unit
BMOS Bytex Matrix Operating System
BMS Basic Mapping Support
BNA Burroughs Network Architecture
BOC Bell Operating Company
BOI Basic Operators Interface
BPNRZ Bipolar Non-Return-to-Zero
BPRZ Bipolar Return-to-Zero
bps bits per second
Bps Bytes per second
BPSK Binary Phase Shift Keying
BPSS Bell Packet Switching System
BRA Basic Rate Access
BRAN Broadband Radio Access Network
BRI Basic Rate Interface
BSC Base Station Controller
BSC Binary Synchronous Communications
BSI British Standards Institute
BSRF Basic System Reference Frequency
BTAM Basic Telecommunications Access Method (IBM)
BTR Bit Transfer Rate
BTS Base Transceiver Station
C1-C5 Levels of Line Conditioning
CA Certificate Authority
CACS Customer Administration Communication System
CAD/CAM Computer-Aided Design/ Computer-Aided Manufacturing
CAI Common Air Interface
CALC Customer Access Line Charge
CAMA Centralized Automatic Message Accounting
CAP Customer Administration Panel

CAS Centralized Attendant System
CAS Communications Applications Specifications
CASM Common Application Service Model
CAT Category of twisted pair cable
CAT 1 Category 1 UTP cable
CAT 2 Category 2 UTP cable
CAT 3 Category 3 UTP cable
CAT 4 Category 4 UTP cable
CAT 5 Category 5 STP cable
CAT 6 Category 6 STP cable
CAT 7 Category 7 STP cable
CATT China Academy of Telecommunication Technology
CATV Cable Television (Community Antenna Television)
CB Citizen's Band
CBS Cell Broadcast Service
CBT Computer-Based Terminal
CBX Computerized Branch Exchange
CCDN Corporate Consolidated Data Network (IBM)
CCIR Committee Consultive International de Radiocommunications (former name for ITU-R)
CCIS Common Channel Interoffice Signaling
CCITT Committee Consultative International Telegraphique Telephonique (former name of ITU-T)
CCK Complimentary Code Keying
CCR Customer Controlled Reconfiguration
CCS Centum Call Seconds
CCSA Common Control Switching Arrangement
CCTV Closed-Circuit Television
CCU Communications Control Unit
CD Collision Detector
CD Carrier Detect
CDCCP Control Data Communications Control Procedure
CDMA Code Division Multiple Access
CDPD Cellular Digital Packet Data
CDR Call Detail Recording
CDV Cell Delay Variation
CEI Comparably Efficient Interconnection
CER Cell Error Rate
CFAC Call Forwarding All Calls
CFS Cryptographic File System
CGA Color Graphic Adapter
CGI Common Gateway Interface
CHT Call Holding Time
CHT Constant Holding Time
CICS Customer Information Control Systems
CICSPARS CICS Performance Analysis Reporting System
CIDR Classless Inter-Domain Routing
CIMAP Circuit Installation Maintenance Access Package
CIO Chief Information Officer
CIR Carrier-to-Interference Ratio
CIR Committed Information Rate
CIU Communications Interface Unit
CLB Common Logic Board
CM Call Manager
CMOS Complementary Metal-Oxide Semiconductor
CMS Communications Management Services
CO Central Office
COAX Coaxial Cable
COBOL Common Business-Oriented Language
COCOT Customer-Owned Coin-Operated Telephone
CODASYL Conference of Data Systems Language
COM Computer Output Microfilm
COS Call Originate Status
COT Central Office Trunks
CPE Customer Premises Equipment
CPH Characters Per Hour
CPI Computer to PABX Interface
CPM Cost Per Minute
CPODA Contention Priority-Oriented Demand Assignment
CPR Customer Premise Equipment
CPS Cycles Per Second
CPU Central Processing Unit
CR Carriage Return
CRC Cyclic Redundancy Check
CRQ Call Request
CRT Cathode Ray Tube
CSA Canadian Standards Association
CSMA/CA Carrier Sense Multiple Access/ Collision Avoidance
CSMA/CD Carrier Sense Multiple Access/ Collision Detection
CSN Carrier Service Mode
CSR Customer Service Record
CSU Channel Service Unit
CSU Central Switching Unit
CTAK Cipher Text Auto Key
CT2 Cordless Telephone-2
CTI Computer Telephone Integration
CTS Clear To Send
CTS Communications Technology Satellite
CVSD Continuous Variable Slope Delta Modulation
CVTC Conversational Voice Technologies Corporation
CWDM Course Wavelength Division

	Multiplexing
DA	Data Available
DAA	Data Access Arrangement
DACS	Digital Access and Cross-Connect Systems
DAD	Duplicate Address Detection
DAK	Data Acknowledge
DAL	Data Access Line
DAMA	Data Assigned Multiple Access
DAP	Data Access Protocol
DARPA	Defense Advanced Research Projects Agency
DASD	Direct Access Storage Device
DATAPAC	Canadian Packet Switching Network
DATEL II	RCA Global Communication Data Service in Conjunction with Telenet
DATEX-P	German Packet Switching Network
DB	Decibel
DBi	DB gain with respect to an isotropic antenna
DBMS	Data Base Management System
DBS	Data Base Service
DBS	Direct Broadcast via Satellite
DBTG	Data Base Task Group
DBX	Digital Branch Exchange
DCA	Defense Communication Agency
DCA	Digital Communications Association
DCA	Document Content Architecture
DCAA	Dual Call Auto Answer
DCE	Data Communications Equipment
DCF	Distributive Computing Facility
DCIU	Data Communications Interface Unit
DCL	Digital Command Language
DCPSK	Differentially Coherent Phase Shift Keying
DCS	Data Circuit Switches
DCS	Distributed Communication Systems
DCS	Distributed Computing System
DCS1800	Digital Communication System 1800
DCTE	Data Circuit-Terminating Equipment
DDCMP	Digital Data Communications Message Protocol
DDD	Direct Distance Dialing
DDE	Dynamic Data Exchange
DDI	Direct Dialing In
DDL	Data Definition Language
DDM	Distributed Data Management Architecture
DDN	Defense Data Network
DDP	Distributive Data Processing
DDR	Dial-on-Demand Routing
DDS	Data-Phone Digital Service
DDSs	Distributed Denial-of-Service attacks
DDX-2	Japanese Packet Switching Network
DECNET	Network of Digital Equipment Corp.
DEF	Direct Equipment Failure
DES	Data Encryption Standard
DFEP	Diagnostic Front End Processor
DFL	Distributed Feedback Laser
DFS	Distributed File Systems
DHCP	Dynamic Host Configuration Protocol
DIA	Document Interchange Architecture
DID	Direct Inward Dialing
DIMM	Dual In-line Memory Module
DIP	Dual In-Line Package
DISA	Direct Inward System Access
DISOSS	Distributed Office Support System
DLC	Data Link Control
DLCF	Data Link Control Field
DLE	Data Link Escape
DLM	Data Line Monitor
DLO	Data Line Occupied
DLS	Data Link Switching
DM	Delta Modulation
DMA	Direct Memory Access
DMI	Desktop Management Interface
DMI	Digital Multiplexed Interface
DML	Data Manipulation Language
DNA	Digital Network Architecture
DNC	Dynamic Network Controller
DNDS	Distributed Network Design System
DNIS	Dialed Number Identification Service
DNP	Distributed Network Processing
DNR	Data Network Routing
DNS	Domain Name Service or Domain Name Server
DOD	Direct Outward Dialing
DOMSAT	Domestic Satellite Service
DoS	Denial of Service
DOS	Disk Operating System
DOV	Data Over Voice
DP	Dial Port or Data Processing or Dial Pulse
DPC	Data Processing Center
DPL	Dedicated Private Line
DPM	Defects Per Million
DPO	Dial Pulse Originating
DPSK	Differential Phase Shift Keying
DPT	Dial Pulse Terminating
DPT	Dynamic Packet Transport
DQDB	Distributed Queue Dual Bus
DRAM	Dynamic Random Access Memory
DRN	Data Routing Network
DRS	Data Rate Selector
DS	Digital Signal
DSA	Dedicated Switched Access
DSB	Double Sideband
DSBFC	Double Sideband Full Carrier
DSBSC	Double Sideband Suppressed Carrier
DSC	Direct Satellite Communications
DSDS	Dataphone Switched Digital Service
DSE	Digital Switching Exchange
DSE	Distributed System Environment

DSL	Digital Subscriber Line
DSLAM	DSL Access Multiplexer
DSLO	Distributed System License Option
DSM	Distributed Switching Matrix
DSP	Digital Speech Processing
DSR	Data Set Ready
DSS	Direct Station Selection
DSS	Digital Signature Standard
DSSS	Direct Sequence Spread Spectrum
DSU	Data Service Unit
DT	Detection Threshold
DTAS	Digital Test Access System
DTE	Data Terminal Equipment
DTMF	Dual Tone Multifrequency Signaling
DTR	Data Terminal Ready
DTS	Data Transfer System
DTS	Digital Transmission Systems
DWDM	Dense Wavelength Division Multiplexing
EAP	Extensible Authentication Protocol
EAROM	Electrical Alterable Read Only Memory
EAS	Extended Area Service
EAX	Electronic Automatic Exchange
EB	Erlang B
EBCDIC	Extended Binary Coded Decimal Interchange Code
ECDNS	Educational Content Delivery Networks
ECL	Emitter Coupled Logic
ECMA	European Countries Manufacturers Association
ECOS	Extended Communications Operating System
ECS	European Fixed-Service Satellite System
EDF	Execution Diagnostic Facility
EDI	Electronic Data Interchange
EDP	Electronic Data Processing
EEB	Extended Erlang B
EFT	Electronic Funds Transfer
EGA	Enhanced Graphic Adapter
EGP	Exterior Gateway Protocol
EIA	Electronics Industries Association
EIA/TIA	Electronics Industries Association/Telecommunications Industries Association
EIES	Electronic Information Exchange System
EINOS	Enhanced Interactive Network Optimization System
EIRP	Effective Isotropic Radiated Power
EKTS	Electronic Key Telephone System
ELEPL	Equal Level Echo Path Loss
EMF	Electromagnetic Force
EMI	Electromagnetic Interference
EMP	Electromagnetic Pulse
ENQ	Enquiry
EOA	End of Address
EOM	End of Message
EON	End of Number
EOT	End of Transmission
EPABX	Electronic Private Automatic Branch Exchange
EPON	Ethernet Passive Optical Network
EPROM	Eraseable Read Only Memory
EPSCS	Enhanced-Private Switched Communications Service
EQEEB	Equivalent Queue Extended Erlang B
ERL	Echo Return Loss
ESA	European Space Agency
ESC	Enhanced Signaling Link
ESF	Extended Superframe Format
ESN	Electronic Switched Network
ESP	Enhanced Service Providers
ESS	Electronic Switching System
ETB	End of Transmission Block
ETN	Electronic Tandem Network
ETSI	European Telecommunications Standards Institute
ETX	End of Text
EV	Extreme Value
EVE	Extreme Value Engineering
EXOS	Excelan Open System
FAX	Facsimile
FCC	Federal Communications Commission
FCIP	Fiber Channel over IP
FCS	Frame Check Sequence
FDDI	Fiber Distributed Data Interface
FDHM	Full Duration Half Maximum
FDM	Frequency-Division Multiplexer
FDMA	Frequency-Division Multiple Access
FDX	Full Duplex
FEC	Forward Error Correction
FED-STD	Federal Standard
FEP	Front End Processor
FF	Form Feed
FGND	Frame Ground
FHD	Fixed-Head Disk
FHSS	Frequency Hopping Spread Spectrum
FIFO	First In First Out
FIGS	Figures Shift
FIM	Fiber Interface Module
FIPS	Federal Information Processing Standards
FIR	Finite Impulse Response
FISUS	Fill In Signal Units
FM	Frequency Modulation
FMS	File Management System
FNC	Federal Networking Council
FOC	Fiber Optics Communications
FOIRL	Fiber Optic Inter-Repeater Link
FOM	Fiber Optic Modem

FORTRAN	Formula Translation
FOT	Frequency of Optimum Traffic
FOTS	Fiber Optic Transmission System
FQDN	Fully Qualified Domain Name
FRS	Flexible Route Selection
FSK	Frequency Shift Keying
FSL	Free Space Loss
FTAM	File Transfer Access and Management
FTP	File Transfer Protocol
FTTC	Fiber To The Curb
FTTH	Fiber To The Home
FTTL	Fiber To The Loop
FTTN	Fiber To The Node
FX	Foreign Exchange
GFI	General Format Identification
GFSK	Gaussian Frequency Shift Keying
GIAC	Global Information Assurance Certification
GMSK	Gaussian Minimum Shift Keying
GPD	General Purpose Discipline
GRE	Generic Route Encapsulation
GRIN	Graded Indicies
GSM	Global System for Mobile communications
GTE	General Telephone and Electronics
GUI	Graphical User Interface
GW	User Through Gateways
HASP	Houston Automatic Spooling Priority
HC	Hyperchannel
HCS	Hard Clad Silica
HD	High Day
HDLC	High-level Data Link Control
HDTV	High Definition Television
HDX	Half Duplex
HEHO	Head-In Hop-Off
HEMP	High-altitude Electromagnetic Pulses
HEMS	High-level Entity Monitoring System
HFC	Hybrid Fiber Coax
HID/LOD	High-Density/Low-Density Tariff
HLMS	Heterogeneous LAN Management Specification
HPFS	High-Performance File System
HPM	High-Power Microwave
HN	Host to Network
HNDS	Hybrid Network Design System
HSDC	High-Speed Data Card
HSSI	High-speed Synchronous Serial Interface
HTML	Hyper Text Markup Language
HTTP	Hyper Text Transfer Protocol
Hz	Hertz (cycles per second)
IAB	Internet Activities Board
IBM	International Business Machines
ICC	International Control Center
ICC	International Communication Conference
ICCP	Inter-control Center Communications Protocol
ICI	Incoming Call Identification
ICM	Intelligent Contact Management
ICMP	Internet Control Message Protocol
ICPT	Intercept Tone
ICR	Intelligent Character Recognition
ICSU	Internal Channel Service Unit
ICW	Interrupt Continuous Wave
IDC	International Data Corporation
IDD	International Direct Dialing
IDF	Intermediate Distribution Frame
IDM	Integrated Diagnostic Modem
IDN	Integrated Digital Network
IDP	Integrated Detector Preamplifier
IDS	Instrusion Detection System
IEC	Interexchange Carrier
IEC	International Engineering Consortium
IED	Intelligent Electronic Device
IEEE	Institute of Electrical and Electronics Engineers
IESG	Internet Engineering Steering Group
IETF	Internet Engineering Task Force
IF	Intermediate Frequency
IFRB	International Frequency Registration Board
IGMP	Interior Gateway Management Protocol
IGP	Interior Gateway Protocol
IIM	Inventory Information Management
IKE	Internet Key Exchange
ILD	Injection Laser Diode
IM	Intensity Modulation
IMP	Interface Message Processor
IMPATT	Impact Avalanche and Transit Time
IMS	Information Management System
IMTS	Improved Mobile Telephone Service
IN	Intelligent Network
INFONET	Network of Computer Sciences Corporation
INTELSAT	International Telecommunications Satellite Organization
I/O	Input/Output
IOD	Identified Outward Dialing
IOM	Input/Output Module
IOP	Input/Output Processor
IP	Internet Protocol
IPv4	Internet Protocol version 4
IPV6	Internet Protocol version 6
IPC	Inter-Process Communications
IPL	Initial Program Load
IPM	Impulses Per Minute
IPN	Instant Private Network
IPN	Integrated Packet Network
IPX	Internetwork Packet Exchange
IR	Infrared
IRC	International Record Carrier
IRLED	Infrared Light Emitting Diode

IRTF Internet Research Task Force
IS International Standard
IS-54 EIA Interim Standard for the U.S. Digital Cellular with Analog Control Channels
IS-95 EIA Interim Standard for the U.S. Digital Code Division Multiple Access
IS-136 EIA Interim Standard 136: U.S. Digital Cellular with Digital Control Channels
ISA Industry Standard Architecture
ISAM Index Sequential Access Method
ISC International Switching Center
ISCC Intelligent System Control Console
ISCF Inter-System Control Facility
ISCSI Internet Small Computer Systems Interface
ISD International Subscriber Dialing
ISDN Integrated Services Digital Network
ISI Intersymbol Interference
ISN Information Systems Network
ISO International Order for Standardization
ISO Intermediate Switching Office
IT Intelligent Terminal
ITB Intermediate Text Block
ITDM Intelligent Time-Division Multiplexer
ITIMS Inservice Transmission Impairment Measurement Set
ITS Invitation To Send
ITU International Telecommunication Union
ITU-R ITU-Radiocommunication sector
ITU-T ITU-Telecommunication sector
IVDT Integrated Voice/Data Terminal
IXC Interchange Carrier Facilities
IXT Interexchange Channel
IXTD International Telex Subscriber Dialing
JCL Job Control Language
JPEG Joint Picture Experts Group
JVM Java Virtual Machine
KAK Key-Auto-Key
KAU Keystation Adapter Unit
KDS Keyboard Display Station
KSO Keyboard Send Only
KSR Keyboard Receive Only
KSU Key Service Unit
KTS Key Telephone System
KTU Key Telephone Unit
LAD Local Area Disk
LAL Leased Access Line
LAMA Local Automatic Message Accounting
LAN Local Area Network
LAP Link Access Protocol
LAPB Link Access Procedure Balanced
LAPM Link Access Procedure Modem
LASER Light Amplifcation by Stimulated Emission of Radiation
LATA Local Access and Transport Area
LCD Line Current Disconnect
LCD Liquid Crystal Display
LCN Logical Channel Number
LCR Least Cost Routing
LCU Line Control Unit
LD Laser Diode
LDAP Lightweight Directory Access Protocol
LDDI Local Distributed Data Interfaces
LDM Limited Distance Modem
LDN Listed Directory Number
LDP Label Distribution Protocal
LEC Local Exchange Company
LED Light-Emitting Diode
LEO Low Earth Orbit
LF Line Feed
LFM Link Framing Module
LHT Long Holding Time
LIFO Last In-First Out
LIM Line Interface Module
LIU Line Interface Unit
LLC Logical Link Control
LMI Local Management Interface
LNA Low Noise Amplifier
LND Local Number Dialing
LO Line Occupancy
LPVS Link Packetized Voice Server
LQA Line Quality Analysis
LRC Longitudinal Redundancy Check
LSA Limited Space-charge Accumulation
LSB Lower Sideband
LSD Line-Sharing Device
LSI Large-Scale Integration
LSR Label Switch Router
LSSU Link Status Signal Unit
LTB Last Trunk Busy
LTE Local Telephone Exchange
LTRS Letter Shift
LU Logical Unit
LUW Logical Units of Work
MAC Medium Access Control
MACU Multidrop Auto Call Unit
MAN Manual
MAN Metropolitan Area Network
MAP Manufacturing Automation Protocol
MAPI Messaging Application Programming Interface
MARECS Maritime Satellite Communication System
MASISAT Maritime Satellite Service
MATR Minimum Average Time Requirement
MAU Media Access Unit
MAU Medium Attachment Unit
MAU Multiaccess Unit
MB Memoryless Behavior
MBZS Maximum Bandwidth Zero Suppression

MCI Microwave Communications Inc.
MCO Multiplexer Control Option
MCS Maintenance Control Subsystem
MD Multiple Dissemination
MDA Mail Delivery Agent
MDF Main Distribution Frame
MDNSC McDonald Douglas Network Systems Company
MDS Multiple Dataset System
MDTS Modem Diagnostic and Test System
ME Mobile Equipment
MERS Most Economic Route Selection
MF Multifrequency Signaling
MFJ Modified Final Judgment
MG Motor Generators
MHD Moving Head Disk
MHP Message Handing
MIC Medium Interface Controller
MICR Magnetic Ink Character Recognition
MILNET Military Network
MIND Modular Interactive Network Designer
MIP Mobile IP
MIPS Million Instructions Per Second
MNCS Multipoint Network Control System
MNDS Multinetwork Design System
MNP Microcom Networking Protocol
MODEM Modulator Demodulator
MOS Metal-Oxide Semiconductor
MOSFET Metal-Oxide Semiconductor Field Effect Transistor
MOU Minutes Of Use
MOV Metal-Oxide Varistor
MP Modem Port
MP3 MPEG 1 Audio Layer 3
MPCC Multiprotocol Communications Controller
MPEG Motion Picture Experts Group
MPL Multischedule Private Line
MPLS Multiprotocol Label Switching
MRO Multiregion Operation
MSD Microwave Semiconductor Devices
MSI Medium Scale Integration
MSM Matrix Switch Module
MSS Modem Substitution Switch
MST Multiple Spanning Tree
MSU Message Signal Units
MSU Modem-Sharing Unit
MT Measured Time
MTA Mail Transfer Agent
MTA Message Transfer Unit
MTAU Metallic Test Access Unit
MTBF Mean Time Between Failure
MTS Mobile Telephone Service
MTR Minimum Time Requirement
MTTR Mean Time to Repair
MTX Mobile Telephone Exchange
MUA Mail User Agent
MUF Maximum Useable Frequency
MUX Multiplexer
MVS Multiple Virtual Storage
NAK Negative Acknowledgement
NAM Network Access Method
NAMS Network Analysis and Management System
NARUC National Association of Regulatory Utilities Commission
NAS Network Attached Storage
NAT Network Address Translation
NATA North American Telecommunications Association
NATD National Association of Telecommunication Dealers
NAU Network Addressable Unit
NBS National Bureau of Standards
NC Network Connect
NCCF Network Communications Control Facility
NCIC Network Control Interface Channel
NCL Network Control Language
NCP NetWare Core Protocol
NCP Network Control Program
NCP Network Control Points
NCR-DNA NCR Corp.–Distributed Network Architecture
NCS National Communications Systems
NCS Network Control Station
NCT Network Control Terminal
NCTE Network Channel Termination Equipment
NDIS Network Driver Interface Specification
NDM Network Database Management
NDS Novell Directory Service
NDT Net Data Throughput
NDTS Network Diagnostic and Test System
NE Network Element
NEC National Electric Code
NEMA National Electrical Manufacturers Association
NEP Noise Equivalent Power
NETBEUI NetBios Extended User Interface
NetBios Network Basic Input/Output System
NETBLT Network Block Transfer
NEWS Network Error Warning System
NEXT Near End Cross Talk
NFS Network File System
NH Non-busy Hour
NID Network Identification/ Interface Device
NIM Network Interface Machine
NIOD Network Inward/Outward Dialing
NIPC National Infrastructure Protection Center
NIS Network Information System

NLDM	Network Logical Data Manager
NMC	Network Management Center
NND	National Number Dialing
NOS	Network Operating System
NPA	Numbering Plan Area
NPDA	Network Problem Determination Application
NPF	Network Partitioning Facility
NRZ	Non-Return-to-Zero
NRZI	Non Return to Zero and Invert on ones
NSP	Network Services Protocol
NSSII	Network Supervisory System II
NSFNET	National Science Foundation Network
NTD	Network Tools for Design
NTP	Network Time Protocol
NTPF	Number of Terminals Per Failure
NTS	Network Tracking System
NTSC	National Television Systems Committee
NTU	Network Terminating Unit
NVRAM	Non-Volatile Random Access Memory
OA	Operator Assistance
OA	Office Automation
OAC	Operational Amplifier Characteristics
OAM&P	Operation, Administration, Maintenance, Provisioning
OAN	Optical Access Network
OC	Optical Carrier
OCA	Open Communication Architecture
OCB	Out-going Calls Barred
OCC	Other Charges or Credits
OCC	Other Common Carriers
OCR	Optical Character Recognition
OCS	Operator Console Services
ODD	Operator Distance Dialing
OEM	Original Equipment Man
OEO	Optical-to-Electrical-to-Optical
OFDM	Orthogonal FDM
OGT	Out-Going Trunk
OHQ	Off-Hook Queue
OIM	Optical Index Modulation
OLE	Optical Logic Etalon
OLT	Optical Line Termination
OLTP	On-Line Transaction Processing
OM	Optical Modulators
ONA	Open Network Architecture
ONI	Operator Number Identification
ONT	Optical Network Termination
ONMS	Open Network Management System
OPEN	Open Protocol Enhanced Network
OPS	Off Premises Station
OPX	Off Premises Extension
OQPSK	Offset Quadrature Phase Shift Keying
ORT	Overload Recovery Time
OS	Operating System
OSI	Open Systems Interconnection
OSS	Operational Support System
OSWS	Operating System Workstation
OTDR	Optical Time Domain Reflectometry
OTQ	Out-going Trunk Queuing
OVD	Optical Video Disk
OWF	Optimum Working Frequency
OWRTS	Open-Wire Radio-Transmission Systems
OWTL	Open-Wire Transmission Line
P2P	Point-to-Point
PABX	Private Automatic Branch Exchange
PAD	Packet Assembler/Disassembler
PAL	Phase Alternate Line
PAM	Pulse Amplitude Modulation
PAT	Port Address Translation
PAX	Private Automatic Exchange
PBX	Private Branch Exchange
PC	Personal Computer
PCEO	Personal Computer Enhancement Operation
PCM	Pulse Code Modulation
PCS	Personal Communication Service
PCS	Plastic Clad Silica
PCU	Packet Control Unit
PDM	Pulse Duration Modulation
PDN	Public Data Network
P/F	Poll/Final Bit
PFEP	Programmable Front End Processor
PFM	Pulse Frequency Modulation
PI	Periphereals Interface
PIA	Peripheral Interface Adapter
PIF	Phase Interface Fading
PIU	Path Information Unit
PKI	Public-Key Infrastructure
PLD	Phase Lock Demodulator
PLIP	Parallel Line Internet Protocol
PLL	Phase Lock Loop
PM	Phase Modulation
PMA	Performance Measurement Analysis
PMBX	Private Manual Branch Exchange
PMC	Passive Microwave Components
PMS	Public Message Service
PMT	Photo Multiplier Tube
PMX	Packet Multiplexer
PND	Present Next Digit
PON	Passive Optical Network
POP	Point of Presence
POS	Point of Sale
POTS	Plain Old Telephone Service
PPL	Plain Position Indicator
PPM	Principal Period Maintenance
PPM	Pulse Position Modulation
PPP	Point-to-Point Protocol
PPV	Pay Per View
PRA	Parabolic Reflector Antenna
PRI	Primary Rate Interface

PRMD	Private Management Domain
PROM	Programmable Read Only Memory
PRTM	Printing Response Time Monitor
PSC	Public Service Commission
PSD	Power Spectral Density
PSE	Power Series Expansion
PSI	Packet Switching Interface
PSK	Phase Shift Keying
PSM	Phase Shift Modulation
PSTN	Public Switched Telephone Network
PTM	Pulse Time Modulation
PTT	Postal, Telegraph, and Telephone Organization
PU	Physical Unit
PUC	Public Utilities Commission
PUC	Peripheral Unit Controllers
PUCP	Physical Unit Control Point
PVC	Permanent Virtual Circuit
PVR	Personal Video Recorder
PWI	Power Indicator
PWM	Pulse Width Modulation
QAM	Quadrature Amplitude Modulation
QOS	Quality of Service
QPCH	Quick Paging Channel
QPSK	Quadrature Phase Shift Keying
QTAM	Queued Telecommunications Access Method
QWEST	Quantum-Well Envelope State Transition
RADAR	Radio Detection and Ranging
RADIUS	Remote Authentication Dial-In User Service
RAM	Random Access Memory
RAID	Redundant Array of Inexpensive Disks
RARC	Regional Administrative Radio Conference
RBHC	Regional Bell Holding Company
RCAC	Remote Computer Access Communications Service
RCD	Receiver-Carrier Detector
RCV	Receiver
RD	Receive Data
RDAU	Remote Data Access Unit
RDC	Remote Data Concentrator
RDS	Remote Data Scope
RDT	Recall Dial Tone
REJ	Reject
RF	Radio Frequency
RFC	Radio Frequency Choke
RFD	Regional Frequency Divider
RFI	Radio Frequency Interference
RFI	Request for Information
RFID	Radio Frequency Identification
RFP	Request for Proposal
RFQ	Request for Quotation
RFS	Remote File System
RIP	Routing Information Protocol
RJE	Remote Job Entry
RLM	Remote Line Module
RLSD	Received Line Signal Detector
RMS	Root Mean Square
RN	Radio Network
RNR	Not Ready to Receive
RO	Receive Only
ROH	Receiver Off-Hook
ROM	Read Only Memory
ROS	Read Only Store
ROTR	Receive Only Typing Reperforation
RPC	Registered Protective Circuitry
RPE	Remote Peripheral Equipment
RPG	Report Program Generator
RPQ	Request for Price Quotation
RR	Ready to Receive
RRT	Reverse Recovery Time
RS	Recommended Standard
RSA	Rivest-Shamir-Adleman
RSTP	Rapid Spanning Tree Protocol
RTMP	Routing Table Maintenance Protocol
RTNR	Ringing Tone No Reply
RTS	Request to Send
RU	Request/Response Unit
RVI	Reverse Interrupt
RX	Receiver
RZ	Return to Zero
SA	Study Administration
SABM	Set Asynchronous Balanced Mode
SABRE	Semiautomatic Business Research Environment
SAGE	Semiautomatic Ground Environment
SAN	Storage Area Network
SAP	Service Advertising Protocol
SAW	Surface Acoustic Wave
SBS	Satellite Business Systems
SBT	System Backup Tapedrive
SCA	Short Code Address
SCA	Subsidiary Communication Authorization
SCC	Specialized Common Carrier
SCFM	Subcarrier Frequency Modulation
SCP	System Control Point
SCPC	Single Channel Per Channel
SCR	Silicon Control Rectifiers
SCS	Silicon Controlled Switches
SCS	Satellite Communications Systems
SCSI	Small Computer System Interface
SCTO	Soft Carrier Turn-Off
SDF	Screen Definition Facility
SDH	Synchronous Digital Hierarchy
SDLC	Synchronous Data Link Control
SDM	Space-Division Multiplex
SDN	Software Defined Networks
SDPO	Sleeve Dial Pulse Originating
SDV	Switched Digital Video

SEED Self-Electro-optic Effect Device
SF Single Frequency
SGMP Simple Gateway Monitoring Protocol
SGND Signal Ground
SH Switch Hook
SHF Superhigh Frequency
SHT Short Holding Time
SI Step Index
SID Switch Interface Device
SID Sudden Ionospheric Disturbance
SIL Semiconductor Injection Laser
SIMP Satellite Information Message Protocol
SINA Static Integrated Network Access
SL Sink Loss
SLC Semiconductor Laser Configurations
SLC Subscriber Line Charge
SLR Service Level Reporter
SLS Sequential Logic Systems
SMAS Switched Maintenance Access System
SMB Server Message Block Protocol
SMDR Station Message Detail Recording
SMRT Signal Message Rate Timing
SNA Systems Network Architecture
SNAP Standard Network Access Protocol
SNMP Simple Network Management Protocol
SNR Signal-to-Noise Ratio (S/N Ratio)
SNS Satellite Navigation System
SOA Safe Operating Area
SOH Start of Header
SONET Synchronous Optical Network
SOP Standard Operating Procedure
SP Space Character
SPADE Single channel per carrier PCM multiple Access Demand assignment Equipment
SPAN System Performance Analyzers
SPC Stored Program Control
SPID Service Provider Identifier
SPMF Servo Play-Mode Function
SPOOL Simultaneous Peripheral Operation On-line
SQD Signal Quality Detector
SQL Structured Query Language
SQUID Super-conducting Quantum Interference Device
SS7 Signaling System 7
SSCP System Service Control Point
SSI Small-Scale Integration
SSI Subsystem Support Interface
SSP Storage Service Provider
SST Single Sideband Transmitter
STDM Statistical Time-Division Multiplexer
STE Signal Terminal Equipment
STI Single Tuned Interstage
STP Spanning Tree Protocol
STP Shielded Twisted Pair
STR Synchronous Transmit Receive
STS Synchronous Transfer Mode
STX Start of Text
SU Signaling Unit
SVC Switched Virtual Circuit
SVD Simultaneous Voice/Data
SWC LEC Serving Wire Center
SXS Step-by-step Switch
SYN Control Character (synchronous idle)
TAAS Trunk Answer from Any Station
TAC Technical Assistance Center
TAC Telenet Access Controller
TACS Total Access Communication System
TADP Tests and Analysis of Data Protocols
TAS Telephone Answering Service
TASI Time Assignment Speech Interpolation
TAT Transatlantic Telephone Cable
TC Terminal Controller
TCAM Telecommunications Access Method
TCM Time Compression Multiplexing
TCP Transmission Central Protocol
TCP/IP Transmission Central Protocol / Internet Protocol
TCU Transmission Control Unit
TD Transmit Data
TDD Temporary Text Delay
TDF Trunk Distribution Frame
TDM Time-Division Multiplex
TDMA Time-Division Multiple Access
TDR Time Domain Reflectometry
TEHO Tail End Hop Off
TELCO Telephone Company
TELSET Telephone Set
TEM Transverse Electromagnetic
TEN Telephone Equipment Network
TGB Trunk Group Busy
TGW Trunk Group Warning
THD Ten High Day
TID Traveling Ionospheric Disturbance
TIFF Tag Image File Format
TIMS Transmission Impairment Measuring Sets
TIP Terminal Interface Package
TLF Trunk Line Frame
TLP Transmission Level Point
TLSPP Transport Layer Sequenced Packet Protocol
TMS Telecommunications Message Switcher
TMS Telephone Management System
TMU Transmission Message Unit
TNS Transaction Network Service
TOP Technical and Office Protocol
ToP Type of Service
TP Teleprocessing
TP Transmission Protocol
TRN Temporary Routing Number

TSO	Terminating Service Office
TSO	Time-Sharing Option
TSPS	Traffic Service Position System
TSS	Time Sharing System
TTD	Temporary Text Delay
TTL	Transistor-Transistor Logic
TTN	Tandem Tie-line Network
TTTM	Tandem Tie Trunk Network
TTY	Teletypewriter
TUCC	Triangle University Computing Center
TUP	Telephone User Part
TV	Television
TVC	Trunk Verification by Customer
TVRO	Television Receive Only
TVS	Trunk Verification by Station
TWX	Teletypewriter Exchange Service
TX	Transmitter
TXK	Telephone Exchange Crossbar
TXS	Telephone Exchange Strowger
TYMNET	Timeshare Inc. Network
UA	Universal Access
UART	Universal Asynchronous Receiver / Transmitter
UCC	Uniform Commercial Codes
UCD	Uniform Call Distribution
UDLC	Universal Data Link Control
UDP	User Datagram Protocol
UDWDM	Ultra Dense Wavelength Division Multiplexing
UHF	Ultrahigh Frequency
UI	User Interface
UJT	Unijunction Field Effect Transistor
UL	User Location
UNI	User Network Interface
UNMA	United Network Management Architecture
UPC	Universal Product Code
UPS	Uninterruptable Power Source
UPSR	Unidirectional Path-Switched Ring
URL	Uniform Resource Locator
USA	Undedicated Switched Access
USACII	United States of America Standard Code for Information Interchange (ASCII)
USB	Upper Sideband
USB	Universal Serial Bus
USITA	United States Independent Telephone Association
USOC	Uniform Service Order Code
UTP	Unshielded Twisted Pair
V	ITU-T Code Designation
V+TU	Voice Plus Teleprinter Unit
VAC	Value-Added Carrier
VAN	Value-Added Network
VDT	Video Display Terminal
VDU	Video Display Unit
VGA	Video Graphics Adaptor
VHF	Very High Frequency
VIP	Visual Information Projection
VLAN	Virtual Local Area Network
VM	Virtual Memory
VM	Voice Messaging
VMS	Voice Message System
VNLF	Via Net Loss Factor
VOD	Video On Demand
VOIP	Voice Over IP
VPN	Virtual Private Network
VRC	Vertical Redundancy Check
VRU	Voice Response Unit
VS	Virtual Storage
VSAM	Virtual Index Sequential Access Method
VSAT	Very Small Aperture Terminal
VSB	Vestigial Sideband
VSF	Voice Store-and-Forward
VSPC	Visual Storage Personal Computing
VSWR	Voltage Standing Wave Radio
VTAM	Virtual Telecommunication Access Method
VTPP	Variable Team Pricing Plan
W3C	World Wide Web Consortium
WACK	Wait for Positive Acknowledgement
WAN	Wide Area Network
WAP	Wireless Application Protocol
WARC	World Administrative Radio Conference
WATS	Wide Area Telephone Service
WCC	World Congress on Computing
W-CDMA	Wideband Code Division Multiple Access
WDM	Wavelengh Division Multiplexing
Wi-Fi	Wireless Fidelity
WIN	Wireless Intelligent Network
WLAN	Wireless Local Area Network
WLL	Wireless Local Loop
WNP	Wireless Number Portability
WDP	Wireless Packet Data
WPM	Words Per Minute
WRU	Who-Are-You? (character)
WSP	Wireless Service Provider
WTP	Wireless Transaction Protocol
WUI	Western Union International
X	ITU-T Recommendation Designation
XMIT	Transmit
XML	Extensible Markup Language
XNS	Xerox Network System
X-OFF	Transmitter Off
X-ON	Transmitter On
XTC	External Transmit Clock
ZF	Zero Forcing

Bibliography

Abbatiello, Judy and Ray Sarch. *Telecommunications & Data Communications Factbook.* New York: McGraw-Hill Information Systems Company, 1987.

Akron, Nicholas. *ABC of the Telephone*, Vol 15. Geneva, IL: ABC Teletraining, 1988.

Ashok, Ambardar. *Analog and Digital Signal Processing*, 2nd ed. Pacific Grove, CA: Brooks/Cole Publishing Company, 1999.

AT&T. *International Telecommunications Guide.* AT&T, 1989.

Badrkhan, Kamiran S. *Video Systems: TV Principles and Servicing.* New York: John Wiley & Sons, 1986.

Basile, C. et al., "The U.S. HDTV Standard: The Grand Alliance," *IEEE Spectrum* (April 1995), 36–45.

Bernard, Josef. "High Definition TV," *Radio Electronics* (August 1987), 48–51.

Black, Uyless. *TCP/IP and Related Protocols: IPv6, Frame Relay, and ATM.* New York: McGraw-Hill, 1997.

Black, Uyless D. *Data Communications and Distributed Networks.* Englewood, NJ: Prentice Hall, 1987.

Blyth, W. John and Mary M. Blyth. *Telecommunications: Concepts, Development, and Management.* Encino, CA: Glenco Publishing Company, 1985.

Bowick, Chris. *RF Circuit Design.* Indianapolis, IN: Howard W. Sams & Co., 1982.

Britz, David. "Free Space Optical Communication." *LIA Today*, Vol. 10(1). (January/February 2002), 1,6.

Datapro. *Datapro Reports on Data Communications, IBM Systems Network Architecture* (SNA). (April 1990), C10-491-101 to 123.

Datapro. *Datapro Reports on Data Communication, ISO Reference Model for Open Systems Interconnection* (OSI) (February 1989), CS93-107-108 thru 111. McGraw-Hill, Datapro Research, Delavan, NJ 08075.

Datapro. "CCITT RECOMMENDATIONS," *Datapro Research*, Appendix 1, March 1989.

Deschler, Kenneth T. *Cable Television Technology.* New York: McGraw-Hill Book Company, 1987.

Douglas, Robert L. *Satellite Communications Technology.* Englewood Cliffs, NJ: Prentice Hall, 1988.

Downes, Kevin et al. *Internetworking Technologies Handbook*, 2nd ed. Indianapolis, IN: Macmillan Technical Publishing, 1998.

Encyclopedia of Associations, 22nd edition, vol. 1. Detroit, MI: Gale Research Co., 1988.

Farkas, Daniel. *Data Communications: Terms, Concepts & Definitions.* Carnegie Press, Madison, NJ: 1983.

Federal Communications Commission Office of Engineering and Technology FCC Online Table of Frequency Allocations. April 19, 2005. http://www.fcc.gov/oet/spectrum/table/fcctable.pdf.

Fike, John L. and George E. Friend. *Understanding Telephone Electronics.* Ft. Worth, TX: Radio Shack, 1983.

Fledman, Len. "High Definition Television," *Radio Electronics* (February 1989), 35–37.

Freeman, Roger L. *Telecommunication Transmission Handbook.* New York: John Wiley and Sons, 1981.

Friend, George E. et al. *Understanding Data Communications.* Ft. Worth, TX: Radio Shack, 1984.

Giancoli, Douglas. *Physics: Principles with Applications*, 5th ed. Upper Saddle River, NJ: Prentice Hall, 2002.

Gowar, John. *Optical Communication Systems.* London: Prentice Hall. 1993.

Grant, William O. *Understanding Lightwave Transmission: Applications of Fiber Optics.* Orlando, FL: Harcourt Brace Jovanovich, 1988.

Greenwood, Greenwood, and Harding. *Business Telecommunications*, Dubuque, IA: Wm. C. Brown Publishers, 1988.

Grob, Bernard. *Basic Television and Video Systems*, 5th ed. New York: McGraw-Hill Book Company, 1984.

Guide to Networking Essentials, 3rd ed. Boston: Course Technology, Thomson Learning, 2003.

Gurrie, Michael and Patrick J. O'Conner. *Voice/Data Telecommunications Systems: An Introduction to Technology*. Englewood Cliffs, NJ: Prentice Hall, 1986.

Halsall, Fred. *Data Communications, Computer Networks and OSI*, 2nd ed. Wokingham, England: Addison-Wesley Publishing Company, 1988.

Hamsher. *Communication System Engineering Handbook*. New York: McGraw-Hill Book Company, 1982.

Hawkes, J.F.B. and J. Wilson. *Optoelectronics: An Introduction*. London: Prentice-Hall International, 1983.

Hecht, Eugene. *Schaum's Outline of Theory and Problems of Optics*. New York: McGraw-Hill, 1975.

Hentschel, Christian *Fiber Optics Handbook*. Germany: Hewlett Packard Gmbh, 1988.

Hioki, Warren. *Telecommunications*, 4th ed. Upper Saddle River, NJ: Prentice Hall, 2001.

Hjorth, Linda, Barbara Eichler, Ahmed S. Khan, and John Morello. *Technology and Society: A Bridge to the 21st Century*. Upper Saddle River, NJ: Prentice Hall, 2003.

Intelsat Report 1988–89, Washington, DC: INTELSAT, 1989.

ISDN Handbook and Buyers Guide, 1st ed. Vol. I. Boston: The Information Gatekeepers Group, 1988.

On-Line Education, Tutorials. International Engineering Consortium. February 20, 2005. http://www.iec.org/online/tutorials/.

Internet World Stats, Usage and Population Statistics. February 20, 2005. http://www.internetworldstats.com/stats.htm.

Jurgen, Ronald K. "High Definition Television Update," *IEEE Spectrum* (April 1988),56–62.

Kellejian, Robert. *Applied Electronic Communication: Circuits, Systems, Transmission*. Chicago: SRA, 1980.

Kelso, T. "Basics of the Geosynchronous Orbit." *Satellite Times*, May 1998.

Kennedy, George. *Electronic Communication Systems*, 3rd ed. New York: McGraw-Hill Book Company. 1985.

Khan, Ahmed S. *Fiber Optic Communication: An Applied Approach*. Boston: Pearson Custom Publishing, 2002.

Killen, Harold. *Modern Electronic Communication Techniques*. New York: Macmillan Publishing Company, 1985.

Lacy, Edward A. *Fiber Optics*. Englewood Cliffs, NJ: Prentice Hall, 1982.

LaGuardia, Cheryl, ed. *Magazines for Libraries*, 13th ed. New Providence, N.J: R.R. Bowker, 2004.

Lane, Malcolm G. *Data Communication Software Design*. Boston: Boyd and Fraser Publishing Company, 1985.

Langley, Graham. *Telephony's Dictionary*. Chicago: Telephony Publishing Corporation, 1982.

Lenk, John D. *Handbook of Data Communications*. Englewood Cliffs, NJ: Prentice Hall, 1984.

Liff, Alvin A. *Color and Black & White Television Theory and Servicing*, 2nd ed. Englewood Cliffs, NJ: Prentice Hall, 1985.

Lio, Samuel Y. *Microwave Solid-State Devices*. Englewood Cliffs, NJ: Prentice-Hall, 1985.

Mackey, D. *Web Security for Network and System Administrators*. Boston: Thomson Course Technology, 2003.

Maggiora, P. and J. Doherty. *Cisco Networking Simplified*. Indianapolis, IN: Cisco Press, 2003.

Mano, Morris M. *Digital Logic and Computer Design*. Englewood Cliffs, NJ: Prentice Hall, 1979.

Marshall, S. V. and G. G. Skitek. *Electromagnetic Concepts and Applications*. Englewood Cliffs, NJ: Prentice Hall, 1982.

Martin, James and Joe Leben. *Data Communications Technology*. Englewood Cliffs, NJ: Prentice Hall, 1988.

Maynard, Jeff. *Computer and Telecommunications Handbook*. New York: Granada Publishing, 1984.

Miller, Gary. *Modern Electronic Communication*, 2nd ed. Englewood Cliffs, NJ: Prentice Hall, 1988.

Mynbaev, Djafar and Lowell Scheiner. *Fiber-Optic Communications Technology*. Upper Saddle River, NJ: Prentice Hall, 2001.

Newton. *The Teleconnect Dictionary*. Chelsea, MI: Bookcrafters, 1987.

Palais, Joseph. *Fiber Optic Communications*. Upper Saddle River, NJ: Prentice Hall, 2001.

Putman, Byron W. *RS-232 Simplified*. Englewood Cliffs, NJ: Prentice Hall, 1987.

Rappaport, Theodore. *Wireless Communications—Principles and Practice*, 2nd ed. Upper Saddle River, NJ: Prentice Hall, 2002.

Reynolds, George W. and Donald Riecks. *Introduction to Business Telecommunications*, 2nd ed. Columbus, OH: Merrill Publishing Company, 1988.

Rolzen, Joseph. "Dubrovnik Impasse Puts High Definition TV on Hold," *IEEE Spectrum* (September 1986), 32–37.

Roody, Dennis and John Coolen. *Electronic Communications*, 3rd ed. Reston, VA: Reston Publishing Company, 1984.

Schoenbeck, Robert J. *Electronic Communications—Modulation and Transmission*. Columbus, OH: Merrill Publishing Company, 1988.

Seyer, Martin D. *RS-232 Made Easy*. Englewood Cliffs, NJ: Prentice Hall, 1984.

Senior, John. *Optical Fiber Communications*, 2nd ed. London: Prentice Hall, 1992.

Sherman, Ken. *Data Communications—A User's Guide*, 3rd ed. Englewood Cliffs, NJ: Prentice Hall, 1990.

Shrader, Robert L. *Amateur Raio—Theory and Practice*. New York: McGraw-Hill, 1982.

Sinnema, William. *Digital, Analog and Data Communication*. Reston, VA: Reston Publishing Company, 1982.

Sinnema, William *Electronic Transmission Technology: Lines, Waves and Antennas*, 2nd ed. Upper Saddle River, NJ: Prentice Hall, 1988.

Soloman, M. and M. Chapple. *Information Security Illuminated*. Sudbury, MA: Jones and Bartlett Publishers, 2005.

Stallings, William. *Business Data Communications*. New York: Macmillan Publishing Company, 1990.

Stallings, William. *Wireless Communications and Networks*, 2nd ed. Upper Saddle River, NJ: Prentice Hall, 2005.

Stamper, David A. *Business Data Communications*, 2nd ed. Redwood City, CA: The Benjamin/Cummings Publishing Company, 1989.

Standard Periodical Directory, 14th ed. New York: Oxbridge Communications, 1991.

Tanenbaum, Andrew. *Computer Networks*, 4th ed. Upper Saddle River, NJ: Prentice Hall: 1996.

Technology and Architecture Trends in Optical Networking, 2004 IEEE Distinguished Lecturer Talk, Telcordia Technologies, 2004. April 19, 2005. http://www.eie.polyu.edu.hk/eEvents/IEEE_Talk_2004_General1.pdf.

Tillman and Yen. "SNA and OSI: Three Strategies for Interconnection." *Communications of ACM* (33) (February 1990), 214–224.

Tomasi, Wayne. *Electronic Communications Systems: Fundamental Through Advanced*. Upper Saddle River, NJ: Prentice Hall, 1988.

Tomasi, Wayne and Vincent F. Alisouskas. *Digital and Data Communications*. Englewood Cliffs, NJ: Prentice Hall, 1985.

Trends Datasets, *Signposts 2004*, CD-ROM, Washington, D.C: WorldWatch Institute, 2004.

Tugal, Dogan A. and Osman Tugal. *Data Transmission: Analysis, Design, Applications*. New York: McGraw-Hill Book Company, 1982.

Uffenbeck, John. *Microcomputers and Microprocessors: The 8080, 8085, and Z-80 Programming, Interfacing, and Troubleshooting*, 3rd ed. Upper Saddle River, NJ: Prentice Hall, 1999.

Witte, R. *Spectrum and Network Measurements*. Upper Saddle River, NJ: Prentice Hall, 1993.

Index

A
Acronyms, telecommunications, 357–368
Agencies, telecommunication, 314–317
American National Standard, 221
Americas-1, 224–225
Amplitude modulation formulas, 253–256
Antennas, formulas, 259–260
ASCII, 299t
  comparison of information/data codes and, 294t
Asynchronous transfer mode (ATM), 272
AT&T, 235, 343
Availability and reliability of components/systems formulas, 266–267

B
Baudot code, 296t
  comparison of information/data codes and, 294t
BCD. *See* Binary coded decimal (BCD)
Bell Laboratories, 235, 343
Binary coded decimal (BCD), 295t
  comparison of information/data codes and, 294t
Binary number system, 243, 243t–244t
Boolean algebra, 243
  basic laws of, 244t
Boolean logic. *See* Logic gates
Brillouin amplifiers, 234
Broadband integrated services digital network (B-ISDN), 272

C
Carrierless amplitude phase (CAP), 276
CATV channel designations and carrier frequencies, 329t–330t
Cellular technology, 343
  cellular phone subscribers worldwide (1980–2003), 356f
Center National deludes des Telecommunications (CENT), 235–236
Clarke, Arthur C., 331
Classless Interdomain Routing (CIDR), 280
Codes
  ASCII, 299t
  Baudot code, 296t
  binary coded decimal, 295t
  comparison of information/data, 294t
  EBCDIC, 297t–298t
  for information/data transmission, 294t–299t
  international telephone dialing, 305t–313t
  Morse Code, 296t
Communications satellites
  characteristics of GEO satellites, 333t
  communication and other modes of VHF/UHF propagation, 334f
  comparison of LEO, MEO, and GEO, 332f, 332t
  evolution of, 331
  evolution of INTELSAT, 338t–340t
  fiber optics technology versus satellite technology, 337t
  INTELSAT, as example of service provider, 341
  interference prevention, 336f
  modes of transmission, 334f–335f
  provision of global coverage by GEO, 336f
  types of, 331
  useful Websites for, 341t–342t
Companding formulas, 263
Computer networks, 346–352
  classification, 346f
  components of a LAN, 347t
  IEEE 802.11 specifications, 351t–352t
  inter-networking devices and backbone technologies, 350t–351t
  LAN access methods, 350t
  topologies, 348t–349t
  types of, 347t
Constants, physical, 248t

D
Digital circuitry, 243
Digital communications formulas, 262
Digital compression, 302
Digital logic, 243
Digital subscriber line (DSL), 276
  characteristics of xDSL configurations, 276t–277t
Digital television (wDTV), 302
Discrete multitone (DMT), 276
Doped-fiber amplifiers (DFAs), 230–231
  advantages of rare-earth, 231
Dotted decimal notation, 280

E
EBCDIC, 297t–298t
  comparison of information/data codes and, 294t
Electrical amplification, 220, 227
Electromagnetic spectrum, 238f, 241
  frequency bands, 322t
Electromagnetic waves, 238–242
  direction of propagation, 238f
  electromagnetic spectrum, 241
  energy in, 240–241
  generation of, 238
  modes of propagation, 241, 242t
  velocity of propagation, 239–240
Electronic devices, symbols for, 250–252
Electronics Industry Alliance (EIA), 287
Electron-to-photon conversion, 220, 227
Erbium doped fiber amplifiers (EDFAs), 220, 231–232
  advantages of, 232
  block diagram of, 232f
  determination of characteristics of, 233
  EDFA-PDFFA comparison, 234t
  energy levels of trivalent erbiums, 233f
  increased transmission capacities of, 227
  limitations of, 232
  long-haul transmission system, 232f
Explorer I, 331

F
Fabry-Perot amplifiers, 229–230, 230f
FCC. *See* Federal Communications Commission (FCC)
Federal Communications Commission (FCC), 303
  CATV channel designations and carrier frequencies, 329t–330t
  electromagnetic spectrum: frequency bands, 322t
  frequency allocations, 322t–330t
  frequency allocations for AM, FM, TV, radar, citizen, amateur radio, and personal communications services (PCS), 325t
  frequency prefixes, 322t
  high frequency, 323t
  medium frequency, 323t
  radar bands, 325t
  TV channel frequencies, 326t–328t
  ultrahigh frequency, 324t–325t
  very high frequency, 324t
Fiber CATV, 221
Fiber Link Around the Globe (FLAG), 225
Fiber optics communications, 205–207
  advantages of, 207, 208t
  applications, 221
  comparison of, with copper wires, 208t
  condition for viability of, 220t
  construction, 209, 209t
  design and conditions for system viability, 218
  development history, 205t–206t
  flowchart of design process, 219f
  formulas, 263–266
  LEDs versus laser diodes, 217t
  oceanic, 221–222. *See also* Oceanic fiber optics systems
  optical amplifiers/repeaters, 220
  optical detectors, 218
  optical sources for, 217
  system block diagram, 207t
  system components, 206, 207t
  technology versus satellite technology, 337t
  useful Web resources for, 226

Fiber Raman amplifiers (FRAs), 220
Fiber to the curb (FTTC), 221
Fiber to the home (FTTH), 221
Fourier series formula, 268
Fourier transform formula, 268
Frequency allocations, FCC, 322t–330t

G
Geostationary orbit, 331
Geosynchronous earth orbit (GEO) satellites, 331
  characteristics of, 333t
  compared with LEO and MEO satellites, 332f, 332t
  providing global coverage, 336f
Grand Alliance
  compression techniques proposed by, 303
  HDTV standard defined by, 302
Greek alphabet, 247t

H
Hasegawa, Akira, 235
Hertz, Heinrich, 239
Hicks, W., 234
High-bit-rate digital subscriber line (HDSL), 276
High-definition television (HDTV), 302–303
  formats, 303t
  system characteristics, 304t
  systems and makers, 303t
Hybrid modes (HE and EH), 241

I
IBM, 275
IEEE 802.11 specifications, 351t–352t
Integrated services digital network (ISDN), 269–272
  CCITT's reference model for, 271t
  channel characteristics, 271t
  defined by ITU-TS (CCITT), 269
  evolution of, 269t
  integrated access versus non-integrated access, 270t
  ITU interface standards for, 270t
  potential applications for, 272t
INTELSAT
  satellites, evolution of, 338t–340t
  service provided by, 341
International Organization for Standardization (ISO), 273
International Telecommunication Union (ITU), 281
  sectors of, 281
International telephone dialing codes, 305t–313t
International television systems
  and HDTV, 302, 303t–304t
  NTSC system and users, 302t
  PAL system and users, 300t–301t
  SECAM system and users, 301t
Internet
  usage, worldwide, 354f
  users, top ten nations by number of, 355f
Internet Engineering Task Force, 280
Inverse Fourier transform formula, 268
IP addresses, 278
  characteristics of, 279t
  class A-class E, 279f
  classes of, 278
  dotted decimal notation, 280t
  IPv4 addressing scheme, 278f
  new scheme for, 280
  version 6 (IPv6), 280
IPng, 280. *See also* IPv6
IPv4
  addressing scheme, 278f
  interoperability of IPv6 and, 280
IPv6, 280
  interoperability of IPv4 and, 280. *See also* IPv4
ISDN. *See* Integrated services digital network (ISDN)
ITU-TS (CCITT), 281
  interface standards for ISDN, 270t
  ISDN defined by, 269
  recommendation examples, 282t–286t
  recommendations, 281t–282t
  reference model for ISDN, 271t

J
Journals, telecommunication, 318–321

K
Kerr non-linearity, 235

L
Laser diode amplifiers, 229
  Fabry-Perot amplifiers, 229, 230f
  traveling wave amplifiers, 229, 230f
Laser diodes, 217
Light-emitting diodes (LEDs), 217
Local area networks (LANs), 221
  access methods, 350t
  components of, 347t
Logic gates, 243, 245t
  types of, symbols, and truth tables, 245t
Logic states, 243, 243t
Low earth orbit (LEO) satellites, 331
  compared with MEO and GEO satellites, 332f, 332t

M
Magazines, telecommunications, 318–321
Marconi, Guglielmo, 239, 343
Mathematical symbols, 247t
Maxwell, James Clerk, 239
Medium earth orbit (MEO) satellites, 331
  compared with LEO and GEO satellites, 332f, 332t
Modulation Scheme, 345
  BPSK, 345t
  DQPSK, 345t
  FM, 345t
  GMSK, 345t
Mollenauer, Linn, 235
Monomode step-index fiber, 209, 211f
Morse Code, 296t
  comparison of information/data codes and, 294t
MPEG-2 compression, 303
Multimode graded-index fiber, 209
Multimode step-index fiber, 209, 211f
Multiple transmission format, 302

N
Narrowband ISDN (N-ISDN), 269. *See* integrated services digital network (ISDN)
National Television System Committee (NTSC). *See* NTSC
Network architecture, 273–275
  OSI reference model, 273–274
  SNA, 275
Next Generation Internet Protocol, 280.
  See also IPv6
Noordung, Herman. *See* Oberth, Herman
NTSC, 302t
  system, users of, 302t
NTT, 235

O
Oberth, Herman, 331
Oceanic fiber optics systems
  Americas-1, 224–225
  Fiber Link Around the Globe (FLAG), 225
  major, 223t–224t
  SEA-ME-WE-2, 225
  TAT-12/13, 225
  transatlantic telephone cable-8 (TAT-8), 222, 222t–223t
Open Systems Interconnect (OSI) reference model. *See* OSI reference model
Optical amplifiers, 227–228. *See also specific amplifiers*
  advances in, 236
  amplifier-regenerator comparison, 228t
  application of, 228f
  applications of, 227
  development history, 228t–229t
Optical fibers
  attenuation and dispersion in, 212f
  attenuation mechanisms in, 213t–215t
  characteristics, types, and applications of, 210t
  dispersion mechanisms in, 216t
  light transmission principles in, 211t
  monomode step-index fiber, 209, 211f
  multimode graded-index fiber, 209
  multimode step-index fiber, 209, 211f
  principles of light transmission, attenuation, and dispersion mechanisms in, 210, 213, 216–217
  regeneration requirements, 220, 227
  silica, attenuation versus wavelength relationship for, 212f
  single-mode step-index fiber, 209
  types of, 209
Organizations, telecommunication, 314-317
OSI reference model, 273
  layers, 274t
  protocols and interfaces, 273f

P
Packetization, 302
PAL, 300t
  system, users of, 300t–301t
Phase alternate line (PAL). *See* PAL
Photon-to-electron conversion, 220, 227
Physical constants, 248t
Power of ten, prefixes for, 246t
Praseodymium doped fluoride-fiber amplifiers (PDFFAs), 233
  EDFA-PDFFA comparison, 234t
Prefixes for power of ten, 246t
Pulse-shaping, 220, 227

R
Radar bands, 325t
Radio-communication (ITU-R), 281
Radio wave propagation formulas, 260–261
Raman amplifiers, 234, 235f
Re-timing, 220, 227

S

SECAM, 301t
  system, users of, 301t
Semiconductor laser amplifiers (SC), 220, 229–230
Sequential color and memory (SECAM). *See* SECAM
Single-mode step-index fiber, 209
SI units, 248t
SNA, 275
  functions of layers, 275t
  software layers, 275f
Solitons, 235–236
SONET
  hierarchy and line rates, 221t
Southeast Asia-Middle East-Western Europe-2 (SEA-ME-WE-2), 225
Speed/velocity, frequency, and wavelength formulas, 257
Sputnik I, 331
Stimulated Brillouin scattering (SBS), 234
Stimulated Raman scattering (SRS), 234
Symbols
  for electronic devices, 250–252
  mathematical, 247t
Synchronous optical network (SONET), 221. *See also* SONET
System effectiveness formula, 267
Systeme International d'Unites (SI). *See* SI units
Systems Network Architecture (SNA). *See* SNA

T

TAT-12/13, 225
TCP/IP protocol, 278
Technical Study (SGs), 281
Telecommunication Development (ITU-D), 281
Telecommunications
  acronyms, 357–368
  agencies and organizations, 314–317
  journals and magazines, 318–321
Telecommunications formulas
  amplitude modulation, 253–256
  antennas, 259–260
  availability and reliability of components/systems, 266–267
  companding, 263
  digital communications, 262
  fiber optics communications, 263–266
  Fourier series, 268
  Fourier transform, 268
  inverse Fourier transform, 268
  radio wave propagation, 260–261
  speed/velocity, frequency, and wavelength, 257
  system effectiveness, 267
  traffic measurement, 267–268
  transmission lines, 257–259
Telecommunications Industry Association (TIA), 287
Telecommunication Standardization (ITU-T), 281
  recommendation examples, 282t–286t
Telecommunications technology
  cellular phone subscribers worldwide (1980–2003), 356f
  evolution of number of transistors on CPUs, 353f
  number of host computers worldwide (1986–2003), 354f
  semiconductor sales (1954–2002), 356f
  telephone lines worldwide, 355f
  top ten nations by number of Internet users, 355f
  trends and statistics, 353–356
  worldwide Internet usage, 354f
Telephones
  international dialing codes, 305t–313t
  worldwide lines, 355f
Television channel frequencies, 326t–328t
TIA/EIA standards, 287–289
  232-D specifications, 292t
  basic DTE/DCE interface using RS-232-D, 289f
  commonly used, for DTE/DCE interfaces, 288t
  data transmission, classification of, 287f
  DTE/DCE transfer specifications, 288f
  mechanical specifications of TIA/EIA-232 and TIA/EIA-449, 290t
  RS-232-D interface as null modem, 289f
  signal comparison RS-232-D, RS-449, and CCITT standards, 291t
Traffic measurement formulas, 267–268
Transatlantic telephone cable 8 (TAT-8), 222, 222t–223t
Transmission lines, formulas, 257–259
Transverse electric (TE), 241
Transverse electromagnetic mode (TEM), 241
Transverse magnetic (TM), 241
Traveling wave amplifiers, 229, 230f
Truth tables, 243
Tsiolkovsky, Konstantin, 331

U

Unicode, 294t
United Nations (UN), 281
Units
  conversion factors, 249t
  SI, 248t

W

Wavelength-division multiplexing (WDM), 229
  Raman amplifiers with, 235
Websites
  for additional information about TIA/EIA standards, 288
  resources for fiber optics communications, 226
  for satellite communications, 341t–342t
  for wireless communications, 345t
Wireless technologies (CMDA, FDMA, TDMA)
  comparison of major wireless standards, 345t
  comparison of wireless access (air interface) technologies, 344t
  comparison of wireless generations, 343t
  and standards (AMPS, N-AMPS, CDMA, GSM) 343–345
  useful Websites for wireless communications, 345t

* f refers to figure; t refers to table